MICROBIAL ASPECTS OF
POLLUTION

THE SOCIETY FOR APPLIED BACTERIOLOGY
SYMPOSIUM SERIES NO. 1

MICROBIAL ASPECTS OF POLLUTION

Edited by

G. SYKES

AND

F. A. SKINNER

1971

ACADEMIC PRESS · LONDON · NEW YORK

ACADEMIC PRESS INC. (LONDON) LTD
24-28 OVAL ROAD
LONDON N.W.1

U.S. Edition published by
ACADEMIC PRESS INC.
111 FIFTH AVENUE,
NEW YORK, NEW YORK 10003

Copyright © 1971 By The Society for Applied Bacteriology

Third printing 1976

Library of Congress Catalog Card Number: LCCCN 73-180799
ISBN: 0-12-648050-8

Printed in Great Britain by
The Whitefriars Press Ltd., London and Tonbridge, England

Contributors

J. ANTHEUNISSE, *Laboratorium voor Microbiologie der Landbouwhogeschool, Wageningen, The Netherlands*

R. E. CRIPPS, *Shell Research Limited, Borden Microbiological Laboratory, Sittingbourne, Kent, England*

W. J. H. CROMBACH, *Laboratorium voor Microbiologie der Landbouwhogeschool, Wageningen, The Netherlands*

C. R. CURDS, *Department of the Environment, Water Pollution Research Laboratory, Stevenage, Herts, England*

H. M. DARLOW, *Microbiological Research Establishment, Porton Down, Salisbury, England*

J. G. DAVIS, *J. G. Davis & Partners, 9 Gerrard Street, London, W.1, England*

A. L. DOWNING, *Department of the Environment, Water Pollution Research Laboratory, Stevenage, Herts, England*

H. O. W. EGGINS, *Biodeterioration Information Centre, University of Aston in Birmingham, England*

R. T. HAUG, *Department of Environmental Engineering, Loyola University, Los Angeles, California, U.S.A.*

H. A. HAWKES, *Applied Hydrobiology Section, University of Aston in Birmingham, England*

R. A. HERBERT*, *Department of Microbiology, University of Edinburgh, Edinburgh, Scotland*

P. N. HOBSON, *The Rowett Research Institute, Bucksburn, Aberdeen, Scotland*

A. J. HOLDING, *Department of Microbiology, University of Edinburgh, Edinburgh, Scotland*

A. HOLT, *Department of Chemistry, University of Aston in Birmingham, England*

D. R. KEENEY, *Department of Soil Science, University of Wisconsin, Madison, Wisconsin, U.S.A.*

P. L. McCARTY, *Department of Environmental Engineering, Stanford University, Stanford, California, U.S.A.*

J. H. McCOY, *Public Health Laboratory, Hull Royal Infirmary, Kingston-upon-Hull, England*

J. MILLS, *Biodeterioration Information Centre, Department of Biological Sciences, University of Aston in Birmingham, England*

B. MOORE, *Public Health Laboratory Service, Heavitree, Exeter, England*

E. G. MULDER, *Laboratorium voor Microbiologie der Landbouwhogeschool, Wageningen, The Netherlands*

E. B. PIKE, *Department of the Environment, Water Pollution Research Laboratory, Stevenage, Herts, England*

K. ROBINSON, *Bacteriology Division, School of Agriculture, Aberdeen, Scotland*

* Present address: Department of Bacteriology, The University, Dundee, Scotland.

G. SCOTT, *Department of Chemistry, University of Aston in Birmingham,
England*

B. G. SHAW, *The Rowett Research Institute, Bucksburn, Aberdeen, Scotland*

J. A. STEEL, *Metropolitan Water Board, Water Examination Department,
London, England*

S. J. L. WRIGHT, *School of Bacteriological Sciences, University of Bath,
Somerset, England*

Preface

EACH year the Society for Applied Bacteriology devotes the major part of its Summer Conference to a symposium on a microbiological subject of general topical interest. The choice of subject generally alternates between those dealing with selected groups of organisms and those in which microbial activities are of primary importance.

Hitherto these symposia have been published in the *Journal of Applied Bacteriology* but it has now been decided to publish them separately in book form. This volume constitutes the first of the new series.

The subject of the 1971 symposium "Microbial Aspects of Pollution" is particularly topical. Pollution is an environmental problem and almost invariably arises from the activities of man. As is common, micro-organisms have their part to play, both advantageously and disadvantageously, and the 16 contributions, written by recognized experts in the field, range widely over the subject. They include considerations of the health hazards of pollution, embracing the consequences of sewage pollution of our water supplies and, a most important topic to the laboratory worker, the safe disposal of infected material. A series of papers deal with the problems of water purfication and of the disposal of sewage and other wastes and their effects on the waters of rivers and lakes. Special attention is given in this context to the disposal of industrial wastes. Other contributions deal with the disposal of the newer industrial products of the organic chemist, namely, pesticides, herbicides, fungicides and plastic materials.

G. SYKES
3 Grosvenor Court
Egerton Road
Weybridge
Surrey, England

F. A. SKINNER
Rothamsted Experimental Station
Harpenden
Herts, England

December, 1971

Contents

Microbial Aspects of Pollution
Some General Observations

J. G. DAVIS

*J. G. Davis and Partners, 9 Gerrard Street,
London W1, England*

CONTENTS

1. Introduction

IN DISCUSSING any problem it is advisable first to define the terms involved. According to the dictionary "to pollute" means "to defile, to soil, or to make unclean". This may be suitable for general use, but today pollution is a technical problem and a more technical definition is desirable. The following is more appropriate: "Pollution is the introduction of materials or effects at a harmful level".

In this definition the word "introduction" covers both production (as in manufacture of chemicals) and transfer (as in the treatment and disposal of sewage). It is necessary to include both "materials" and "effects" to cover the aspects of pollution by radiation, noise and other nonmaterial pollutants. "Harmful level" asserts that a condition becomes a pollution when the intensity reaches a point at which harmful effects are experienced. An analogy in the food field is with poisons and other undesirable additives.

There is intrinsically no line of demarcation between a poison and a safe substance. It is entirely a matter of concentration. Thus arsenic and lead are innocuous at a concentration of 10^{-9} g (we have them in our bodies at this level), but they reach the threshold of toxicity at $c.$ 10^{-6} g and are strongly lethal at 10^{-3} g. Such common substances as salt and oxygen can be toxic if the concentration is high enough, and this fundamental reasoning applies to all food additives.

A further point of interest is whether Nature itself can cause pollution. The emission of ash and sulphur fumes by a factory would undoubtedly be regarded as pollution, as would also the killing of human beings and other animals and leaving their bodies to rot in the streets. But what is our attitude if these are caused by a volcano or an epidemic disease? Should we not restrict our definition to the activities of man? If so, the definition should be amended to: "Pollution is the introduction by man of materials or effects at a harmful level".

A simple, but common, example is the effect of phosphates in industrial effluents. These are frequently discharged into streams where they can so enhance the growth of algae that the stream becomes clogged, and so deprived of its oxygen supply, thus supressing the ecological balance in the stream. Moreover all effluents are ultimately discharged to a lake or the sea and the factors controlling the consequent types of microflora and their numbers are complex (see Gameson, Bufton & Gould, 1970; Gameson, Munro & Pike, 1970).

2. Historical Development of Pollution Studies

Only recently has pollution become a problem of major significance but it has been with us to a less extent for centuries and it is interesting to trace the various phases of its evolution from earliest times.

(a) *Population*

Pollution is highly correlated with density of population and technical development. The world population is likely to increase from the present 3500×10^6 to 7000×10^6 by 2000 A.D. The earth can absorb or nullify a certain amount of polluting materials, irrespective of their source or nature, thus acting as a protecting and buffering agent, and this mechanism was adequate to take care of the problem until recently. But the now rapidly increasing population and its concomitant advancing technologies are giving rise to increasing amounts of pollution, much of which has a residual or persistent effect.

The total volume of air and water and the total area of land remain constant but the extent of pollution is increasing rapidly. The land aspect is of particular interest. Every year we lose 60,000 acres of land in the U.K. for the construction of new roads, buildings, reservoirs, etc., with the result that the area of cultivated and naturally wild land is steadily diminishing. It is this "biological land" which hitherto has been able to deal with pollution by the natural processes of absorption, "scrubbing", filtering and metabolizing noxious substances. This loss of biological land is a further contributory factor in the pollution problem; it only adds to that created by the increasing population. Considering only the human and not the technological aspect of pollution it might be thought that doubling the population in the next 30 years should only double the extent of pollution, but this is a fallacy. If we accept the "absorbance" or "buffer" theory, the degree of pollution can be equated, not with the total population, but with the population in excess of a certain number. For example, if we assume that a population of 3000×10^6 causes no pollution, it is the present excess of 500×10^6 which is responsible for the existing pollution. From this it follows that a further increase in world population to 4000×10^6 will actually double the extent of this type of pollution.

The increase may be even worse than indicated because any increase in one kind of pollution gives rise to increases in other kinds and so affects, directly or indirectly, the absorbing power of the natural resources. For example, toxic chemicals in an effluent may affect the biological activity of a river, pesticides may upset the ecology or balance of different forms of life in a particular environment; similarly, factory chimney deposits may affect the fertility of a field and even poison animal life.

(b) *Organization*

The date when man began to organize himself as a social animal is not known, but it probably began *c*. 5000 B.C. in caves. Man is unique as a living organism in that he has learned how to adapt his environment to his needs. If plants cannot flourish in one environment they wither and die; if animals cannot get food they move away to find it, but man has learned to cultivate, build and in countless ways adapt conditions to his requirement. This first took the form of small organized communities or hamlets and the cultivation of the ground and the rearing of animals for meat, milk and clothing. All these activities led to a concentration of human beings and of their activities into small areas. Man ceased to be a roving animal, getting his food where he could find it, and became a settler.

(c) *Latitude*

One finds little or no pollution at the North Pole or in the Antarctic, partly because the human population is negligible and partly because temperature is a factor affecting pollution. Pollution is greatest in the temperate and subtropical regions where man is most numerous and most active technologically.

(d) *Legislation*

The law can, and frequently has to, help in dealing with pollution problems. Today we are in the throes of devising such legislation but, in fact, it has been embodied in religious teachings for some 3000 years. The Mosaic laws, for example, deal with hygiene and propound rules for the taking of food, and purification rites are a feature of many religions.

(e) *Urbanization*

The congregation of human beings in large towns is the ultimate development of the first organization of man into communities. Alexandria and Rome are examples at the beginning of the Christian era, and today cities of $> 10^6$ inhabitants are found in many regions, not only in developed, but also in developing, countries. Where the technical provisions for hygiene are inadequate the chances of pollution are considerable, both bacteriologically and chemically. Even in the newer advanced countries there is the tendency for people to seek habitation in the towns rather than in the country. For example, Australia is one of the largest countries in the world with ample space for a large population, yet of the 13×10^6 inhabitants no fewer than 5×10^6 live in 3 cities. Such a state of affairs, whether we consider it as upsetting the balance of nature or not, inevitably increases the probability of pollution. A dense population not only creates public health problems but also increases chemical pollution, pollution of the air and water, and it intensifies noise.

(f) *Technology*

Some of the earliest papers read to the Royal Society, from the time of its foundation in 1660, dealt with applied science, particularly agriculture. Some at least of these communications hinted at what we would today call technology or the application of scientific knowledge to industry and these developments, in turn, have contributed significantly to the problems of pollution. The technological aspects of pollution can be considered under 2 headings, industrial and public health.

(i) *Industrial*

The real impetus came with the industrial revolution, dating from *c.* 1750. Primitive engineering, the building of factories, the extensive use of coal and the production of iron and steel steadily increased the chemical pollution of land, water and air. These beginnings culminated in the wonderful achievements of the synthetic organic chemist and of other scientists from *c.* 1900 onwards, and this has been probably the biggest factor in increasing pollution. The amounts of recognized pollutants in the environment is staggering. It has been estimated that the oceans contain 100×10^6 tons of mercury, and that 5000 tons are released into the air every year by the burning of coal and oil. Of the 9000 tons of mercury produced annually some 3000 tons are lost to the environment and so pollute the land, the water and the air.

(ii) *Public health*

Diseases in general and epidemics have been common in the history of mankind from the earliest times. The contributions of public health and hygiene to the health and longevity of man have been outstanding and far surpass those of any other branch of medicine. The expectation of life at birth in the U.K. has almost doubled since 1800, and this is almost entirely due to developments in the public health sciences, supported by advanced engineering processes and the law.

The beginning of this achievement can be traced to the "pure food" Act of 1860, followed by successive Public Health and Food and Drugs Acts, all designed to protect the public against disease and impure food. Medical men and engineers, and later bacteriologists and chemists, have all made considerable contributions to this pioneer work. The essential achievements were in the disposal of sewage and the provision of pure water supplies.

There is, however, a disadvantage in this successful solution of a public health problem. Before the modern system of sewage treatment and disposal came into being, all human excreta, like that of animals, were returned to the soil so that all the nitrogen and minerals taken in food were returned automatically to the earth, thus maintaining a continued cycle of soil fertility. On the assumption that the average individual takes in 70 g of protein, 1 g of calcium and 20 mg of iron daily it can easily be calculated that our modern sewage disposal systems result in a loss from the soil of nitrogen equivalent to 10^6 tons of protein as well as 20,000 tons of calcium and 400 tons of iron annually. These quantities are not lost permanently as they find their way into rivers, lakes and seas. From here they pass into marine plants and fishes, but they are lost as far as human food grown on the land is concerned. There may not be a shortage of calcium and iron in the world as a whole but these nutrients are becoming decreasingly available for land cultivation, which is where they are needed. This is a public health achievement which must be balanced against a nutritional loss.

(g) *Information*

It is most desirable that all thinking persons should have up to date and accurate information on matters affecting the community, and this applies equally to pollution, to safety in foods and to aspects of nutrition. Unfortunately it often requires an expert in communications to impart scientific information to the nonscientific public, and the public is surprisingly gullible on many matters. This state of affairs has enabled a few enterprising people to achieve considerable publicity and even material benefits by acting as crusaders for a cause, and one of these is the prevention of pollution or "poisoning by chemicals in food". The evidence is usually presented in a biased manner, and it is behoven on all scientists to give the facts as objectively as possible to all interested persons. The present symposium is seeking to do this so that others may give it their thought and so help to alleviate the situation. Opinion is free, but facts should always be regarded as sacred.

(h) *Opinion*
The political history of democracies with true parliamentary government shows quite clearly that overwhelming public opinion can force the authorities, governmental and local, to take action in a cause which is to the public interest. The fight against pollution is undoubtedly such a cause.

(j) *New polluting products*
The most recent serious type of pollution, and one which has increased considerably in the last decade, is the rapidly growing production of new products which do not disintegrate or cannot be broken down by natural reactions. The most prominent examples are certain detergents, plastics, oil residues, chlorinated hydrocarbons, trace toxic metals and radioactive wastes. These are amongst the persistent chemicals which are produced as byproducts in chemical manufacture.

3. The Microbiologist's Role
The microbiologist has a role to play in several aspects of waste disposal and of pollution control, and frequently he works in collaboration with chemists, physicists and engineers.

(a) *The disposal of radioactive wastes and toxic heavy metals*
There is no evidence so far to suggest any microbiological solution to the problems of disposing of these 2 groups of materials. They are essentially physical and chemical problems. However, it is of interest to note that plants may become adapted to grow in the presence of normally toxic concentrations of heavy metals (Bradshaw, 1970).

(b) *Disposal by chemical and microbiological actions*
Where pollution is caused by the accumulation of substances which are not biodegradable, it may be possible to modify them by chemical means to substances which can be broken down by micro-organisms. One example is the

change in formulation of synthetic detergents to give straight chain alkyl compounds which can be attacked in this way.

(c) *Microbial adaptions*

There are many examples of micro-organisms adapting themselves to carry out unsuspected chemical reactions and thereby being able to thrive in hitherto unfavourable conditions. Ability to attack polysaccharides and nitrogenous compounds not previously broken down are amongst these achievements, and this fundamental property is of course the basis of many effluent disposal schemes. Such systems may come to a halt if a germicide enters the effluent, but in time organisms may even become adapted to degrade such agents. An example is found in the problem of an effluent containing phenols. At first a low concentration c. 100 p/m inhibited micro-organisms, but in time a microflora became established which was capable of dealing with several hundred p/m of the phenols.

One of the gravest new problems is that of plastics. Already some countries are banning the use of cups, etc., made of plastics because of the rapid accumulation of these indestructible items. They cannot be burnt easily as can paper cups.

An interesting example is the fast developing use of blow-moulded milk bottles. Every year some 11×10^9 bottles of milk are delivered to households in the U.K. If these were all in plastic bottles, each weighing 15 g, it would mean an annual accumulation of 165,000 tons of polythene in bulky items. Even plastic sachets weighing 5 g each would produce 55,000 tons. The food industry as a whole would probably use 10^6 tons of plastics for packaging every year if all foods were wrapped or contained in plastics.

Biodegradable plastics would be a solution to the problem, but so far the unique properties of these polymers have defied micro-organisms. Doubtless scientists will solve this problem in time. Possible solutions are the development of light- or oxygen-degradable plastics or even built in disintegrators. Another possible method of attack is that suggested by Booth, Cooper & Robb (1968) and Booth & Robb (1968) who showed that *Pseudomonas* and *Brevibacterium* spp. can degrade plasticized pvc, although the polymer itself is not attacked.

(d) *Disposal by utilization*

By far the most satisfactory solution to material pollution problems would be the development of methods which utilize the waste materials to produce energy, food or other useful materials. An enterprising example is the suggested hydrolysis of paper to sugars and then their fermentation to alcohol by yeast (Porteous, 1971). Another equally enterprising suggestion is to grow yeasts on petroleum wastes and then extract protein from the yeast for animal and possibly human food, the so called "single cell protein". Both schemes have been used successfully in pilot scale production, but much yet remains to be done.

(e) *Outstanding problems*

It is reasonable to conclude that in time it may be possible to dispose of all or most forms of organic material pollution by microbiological methods. There is therefore considerable scope for the microbiologist in combating the present pollution problems. The ecology of micro-organisms, their relationship to their environment and to other forms of life, is not only one of the most interesting aspects of microbiology but is also at the heart of microbial methods for the prevention of pollution.

It is well established that when the decomposition of any type of food or other material is allowed to take its natural course, a particular type of flora becomes established. Thus carbohydrate-containing animal foods such as milk become dominated by souring lactic acid bacteria, fruits allow the growth of yeasts to produce an alcoholic fermentation, proteinaceous animal foods such as meat are colonized by anaerobic bacteria and watery systems are invaded by pseudomonads. These characteristic end conditions are the results of millions of years of evolution and, assuming favourable physical conditions and adequate basic nutrients, the deciding factor is probably not the general thermodynamic *efficiency* of the organisms' metabolic processes but the speed or *rate of release* of energy. Thus the lactic and alcoholic fermentations are far less efficient thermodynamically than the oxidative systems of the higher animals but permits a speedy release of energy. This may be regarded as a general law controlling the establishment of a dominant microflora (Davis, 1932).

Not only do microbial activities act favourably in relation to pollution but also they can act unfavourably by changing one type of pollutant to another. Amongst the unfavourable activities may be mentioned those of the sulphate-reducing bacteria which, by producing H_2S, can blacken heavy metals and their solutions, reduce the E_h value of water to kill fish and poison the respiratory systems of higher forms of life.

Micro-organisms may sometimes intensify a type of pollution by making the pollutant more dangerous. An example of this is the ability to convert mercury compounds to the highly toxic dimethyl derivative. Another example is the oxidation of nonacidic substances to acids, which are not only toxic to most forms of life but also corrosive to metal, calcareous and other types of material. The ideal is for some type of micro-organism, or a succession of different types of organism, to break down material to harmless end products such as CO_2 and water. The degradations of animal excreta, paper, textiles, detergents, insecticides, herbicides and numerous other products of the synthetic organic chemist are examples of what we would like micro-organisms to do for us. We cannot expect, however, these organisms to "learn" or become adapted to effect these changes in a few generations, bearing in mind that the highly specialized catabolic processes involved, such as the lactic acid fermentation, have evolved over many generations. However, the skilful microbiologist can shorten the period for such desirable types to evolve in a practicable and controllable form, and this line of research will play an increasingly important role in the pollution microbiology of the future. We shall use micro-organisms not only to synthesize foods, antibiotics and organic acids, but also to break down those organic substances which are not wanted.

4. References

Booth, G. H., Cooper, A. W. & Robb, J. A. (1968). Bacterial degradation of plasticized pvc. *J. appl. Bact.* **31**, 305.

Booth, G. H. & Robb, J. A. (1968). Bacterial degradation of plasticized pvc—effect in some physical properties. *J. appl. Chem.* **18**, 194.

Bradshaw, A. (1970). Pollution and plant evolution. *New Scientist* p. 497.

Davis, J. G. (1932). Studies in anaerobic metabolism Ph.D. thesis, London University.

Gameson, A. L. H., Bufton, A. W. J. & Gould, D. J. (1970). Studies of the coastal distribution of coliform bacteria in the vicinity of a sea outfall. *Wat. Pollut. Control.* **66**, 501.

Gameson, A. L. H., Munro, D. & Pike, E. B. (1970). Effect of certain parameters on bacterial pollution at a coastal site. *Wat. Pollut. Control* **00**, 000.

Porteous, A. (1971). Sweet solution to domestic refuse. *New Scientist* and *Science J.* p. 736.

The Health Hazards of Pollution

B. MOORE

Public Health Laboratory Service, Church Lane, Heavitree, Exeter, England

CONTENTS

1. Introduction

THE PUBLIC health administrator who sets out to reduce the incidence of communicable disease will first identify the main prevalent diseases in the community for which he is responsible and then apply the most effective measures available for their prevention. Limitations of manpower and resources usually force him to allocate priorities, and these tend to be determined by which diseases have the highest morbidity and mortality, and by whether effective methods of prevention are in fact known.

All this seems simple and straightforward in principle. When, however, health hazards of the environment come under discussion, a difficulty arises. Public opinion begins to allocate priorities, and these may have little to do with prevalences of disease. Contrast for instance the relative apathy about deaths caused each year by the motor car with public concern about the postulated hazards of bathing in sewage-contaminated sea water. Several reasons may account for this. Some are aesthetic or psychological. Not surprisingly, anxious parents whose child has been taken ill after exposure to a polluted environment tend to blame this. A more subtle temptation, to which medical and applied bacteriologists are not immune, is to assume that the time trends that have given rise to concern with some environmental pollutants must necessarily apply to all. With the growth of cities and of technology, we have seen more smog, more

11

insecticides, more chemical pollutants in general. It is sometimes taken for granted that the risk of microbial infection from the environment must also be increasing. This encourages the fallacy, so characteristic of current literature on water pollution, that the isolation by sensitive techniques of a pathogen from a sewage effluent presages a disastrous epidemic of the corresponding disease in the community before long. One has only to reflect for a moment on what has happened to waterborne diseases in Britain during the past 150 years or to poliomyelitis during the past 15 years to realize that unqualified pessimism is not called for here.

Two questions are posed in the following discussion. (i) How should one assess whether a polluted environment is causing or is likely to cause one or other of the communicable diseases? (ii) By the relevant criteria, what are the present hazards of microbially polluted environments and how should they be tackled?

2. Preliminary Orientations

The London fog of December, 1952, a notorious example of an abnormal environment causing disastrous ill effects in an exposed population, serves to identify at the outset some of the parameters of environmental health hazard assessments. The main features of the episode are summarized in Fig. 1 from

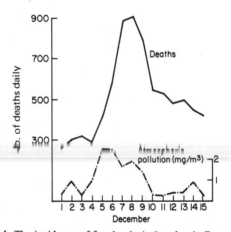

Fig. 1. The incidence of fog deaths in London in December, 1952.

data given in the paper by Logan (1953). The top graph shows the daily London mortality figures during the first fortnight of December 1952. Those who died were mainly the elderly with chronic heart or chest complaints. Between 4 and 7 December, the number of daily deaths rose rapidly from 300 to 900. Peak mortality occurred on 8 December and the number of deaths then fell, at first rapidly and then more gradually, till the 15th. The lower graph plots for the same period the atmospheric pollution in mg/m³ of air.

As the Figure shows, high mortality and a severe fog both occurred in London during the first fortnight of December, 1952. Did the fog cause these deaths, or was the association between them coincidental? Certain obvious pointers to a cause-and-effect relationship are (i) the strength of the association between the change in environment and the high mortality; (ii) the time relationship, an increase in atmospheric pollution from 3 December onwards being followed after a 2 days' lag period by the rapid rise in mortality; (iii) the suggestion of a dose-response curve, with increase in pollution above a certain baseline causing a related rise in mortality; (iv) the fact that the excess of deaths was accounted for largely by respiratory and cardiac complications in the elderly, as one would indeed expect of a respiratory irritant such as fog, rather than by, say, diarrhoea and vomiting. On these arguments, the case for a causal association between fog and the high mortality might seem selfevident. Some doubts were in fact expressed, e.g. it was pointed out at the time that an unprecedented increase in applications to the London Emergency Bed Service towards the end of 1952 started a whole month before the fog, suggesting that it might simply have aggravated the ill effects of some harmful influence already operating in the community, such as an epidemic respiratory infection. Later analysis has also suggested that the most important trigger of the high mortality may have been the sudden fall in ambient temperature rather then the atmospheric impurities as such.

Figure 1 also provides two useful pointers to the relationship between environmental monitoring and the assessment of health hazards. Inspection of the two graphs suggests that if the level of atmospheric pollution remained < 1 mg/m³ a disastrous episode of this kind might not recur. The important point here is not whether this is true, but that such a prediction depends on a quantitative relationship between pollution and the prevalence of disease having already been established. Secondly, the adoption of 1 mg/m³ as a permitted upper limit of atmospheric pollution in other cities would imply that fog was constant in composition, at least in respect of its toxic components.

The assessment of microbial health hazards of the environment follows a similar sequence of argument. The first step is an observed association between a clinical disease and a physical feature of the environment. This was fog in the episode just cited. For Snow (1855) and Budd (1856), it was an association between the occurrence of cholera or typhoid fever and polluted water supplies. In Budd's experience, the strength of the association was often conclusive. A classic example of this was the typhoid outbreak at Richmond Terrace, Clifton, Bristol, described by Budd (1873). Of the 34 houses in this terrace 13, spread over the whole length of the terrace, used water from a pump at one end. During one September, the pump water became polluted with sewage and cases of typhoid fever occurred in every household using the polluted supply, while the contiguous households all escaped. Episodes like this convinced Budd that typhoid fever was often waterborne, and this conclusion preceded by 30 years the laboratory identification of the causal agent.

The interval between exposure to a polluted environment and the onset of disease takes on a special importance in the assessment of microbial health hazards. In particular, when the interval corresponds to the known incubation period of a given communicable disease, a postulated cause-and-effect sequence becomes the more convincing.

The problem of establishing a quantitative relationship between the occurrence of microbial disease and exposure to a polluted environment is discussed in some detail below. The foregoing discussion will already have made it clear, however, that technical bacteriology is less important here than might have been supposed. Budd and Snow showed that typhoid fever and cholera were waterborne some 30 years before the causal organism of either disease had been identified in the laboratory. Even at the present time, water is almost invariably incriminated as the cause of an outbreak of typhoid fever without the causal organism being recovered from the supply at the relevant time. Again, the monitoring parallel has its microbiological counterpart. The isolation of a pathogen from a polluted site cannot be deemed a health risk unless a quantitative relationship between exposure to the environment that contains it and the incidence of the corresponding disease has already been established.

3. The Assessment of Microbial Hazards of Pollution

We return now to the first question posed in the introduction to this paper. How should one assess whether a polluted environment is causing or likely to cause one of other of the communicable diseases? Establishing a framework for such assessments must take account of the general epidemiological criteria, some of them mentioned above, for postulating a cause-and-effect relationship rather than a coincidental association between a polluted environment and disease. Some special complicating features of microbial disease must also be borne in mind. These are discussed in turn below.

(a) *Epidemiological criteria*

The assessment starts from a suspicion that an excess of a particular communicable disease above normal expectation is occurring in association with exposure to a polluted environment. This association may be undramatic in the sense that not everyone exposed to the environment contracts the disease. When this is so, highly significant associations in the statistical sense can still be established by appropriate prospective or retrospective surveys. Significant associations are not necessarily cause-and-effect sequences, however, and must be further scrutinized before they are accepted as such.

Bradford Hill (1965) has listed 9 criteria for judging whether an environmental pollutant is causing disease, and these have been applied elsewhere by the writer to the special context of sea bathing (Moore, 1970a, b). Briefly, an association between a polluted environment and a communicable disease is more likely to be deemed one of cause-and-effect; (i) the stronger the association, as in the Richmond Terrace outbreak cited; (ii) the more it has been previously reported by others; (iii) when the disease is not associated with other environments or the particular kind of polluted environment with other communicable diseases; (iv) when the disease follows and does not precede exposure to the polluted environment, and in the present context particularly when the interval between exposure and the onset of disease corresponds to a known incubation period; (v) when a dose-response curve showing an increasing incidence of the disease with increasing exposure to the environment can be demonstrated; (vi) when the postulated association is compatible with; or (vii) at least not in conflict with other knowledge of the natural history and biology of the disease in question; (viii) when the cause-and-effect relationship has been supported by experiment, e.g. when removing the source of pollution has an immediate impact on the incidence of the disease; and (ix) when there is a fair analogy with some parallel context.

(b) *The special complications of microbial hazard assessments*

The foregoing criteria apply in principle to any environmental hazard, phsyical, chemical or microbial. The assessment of communicable disease hazards is complicated by two additional factors.

(i) Microbial disease is a highly complex host-parasite reaction; there is no simple correspondence between a pathogenic organism and the associated disease in man. The problem of extrapolating from the laboratory culture of a pathogen to the corresponding microbial disease is not one that suddenly obtrudes itself after the pathogen has found its way into drinking water, on to a bathing beach or into the air above a sewage treatment plant. The medical bacteriologist faces it every day. One of the major fallacies of uncritical clinical bacteriology is that a pathogen isolated from a swab or from pus is necessarily the cause of the patient's current illness. Such a conclusion is rarely justified without additional evidence, and often the interpretation remains uncertain. No general statement covers all the combinations of relevant variables, but clearly this biological complexity must be reckoned with even more when the importance of a pathogen as an environmental pollutant is being assessed.

A few examples will illustrate the difficulties of assessing the importance of different pathogens in a polluted environment. Perry, Siegel & Rammelkamp (1957) showed that haemolytic streptococci liberated into the environment rapidly lost their virulence and became irrelevant in the epidemiology of streptococcal infection. *Staphylococcus aureus*, although seriously cited by Brisou (1968) as a possible seawater hazard, is a ubiquitous organism carried in the nose of almost half the normal population; the risk of staphylococcal disease depends therefore on personal susceptibility and certainly not on exposure to the organism in the environment. *Clostridium welchii* again is a ubiquitous organism, and clostridial disease does not occur unless devitalized tissues are available in the host to provide the anaerobic conditions necessary for clostridial multiplication. Poliovirus, to cite one more example, can be isolated from polluted environments only when the disease is present in the population, and when this is so the chance of infection by personal contact is vastly greater than that of acquiring the disease from the polluted environment.

Only one of the difficulties of microbial hazard assessments has so far been stressed, i.e. that of extrapolating from an organism in the polluted environment to the associated disease. The host-parasite reaction has many other complexities. Cruickshank (1963) has pointed out that in the study of communicable disease with a view to its control, the environmental reservoir in which the pathogen lurks is one aspect but there are three others, *viz.* the source of the organism, the precise route by which the patient becomes infected, i.e. respiratory, gastrointestinal, percutaneous or otherwise, and the nature of the susceptible human population. The upshot of all this is that each postulated health hazard must be discussed separately in terms of all the characteristics of the disease in question.

(ii) The second complication of microbial hazard assessments is that recovery of a pathogenic organism from a polluted environment indicates the associated disease to be already circulating in the adjacent community. The disease may tend therefore to be ascribed to an obviously polluted environment when it has been contracted in fact from an undetected personal contact or from some other less obvious infecting source. A good example of this confusion was the paratyphoid B outbreak described by Moore (1948) and popularly ascribed to sea bathing although in fact it was transmitted by ice cream. After the outbreak in question, the paratyphoid carrier responsible was identified by tracing the organism back from a polluted beach to the house from which it was entering the sewerage system. Ironically, the use of bacteriological monitoring in this way to find an infecting source or to monitor the presence of a particular disease in a population has been misconstrued by the lay public as evidence that the polluted environment itself was a cause of disease, a fallacy similar to that of supposing that seismographic equipment was dangerous in its own right rather than an indicator that earthquakes were taking place elsewhere.

4. Pollution: Past and Present

World concern about chemical pollution has grown in parallel with the technological explosion of the past quarter of a century and the increasing use of synthetic chemicals in every aspect of community life. To keep a sense of proportion, it is important to bear in mind that the present microbial hazards of pollution are evidently trivial in countries like Britain by comparison with a century ago. The central theme of Chadwick's famous report (Chadwick, 1842) was the association of disease and death with the filthy conditions that prevailed in the poorer quarters of cities and towns, and the conviction that what was needed was "drainage, proper cleansing, better ventilation and other means of diminishing atmospheric impurity". For Chadwick, an adequate water supply was principally an effective means for washing the labourer and cleaning his environment. It was soon found, however, that better water supplies and sewerage not only helped to improve domestic and personal hygiene. They led also to a sharp fall in the incidence of waterborne diseases in Britain long before the causal agents had been identified in the laboratory.

The only serious drawback to communal water supplies has been that occasional failure to maintain their purity implicates many more people, as the great waterborne outbreaks of the past century testify. By comparison, the removal of excreta from the home environment, given the integrity of the associated drinking water supply, has had few associated hazards. These can be discussed under two broad headings, those associated with the recreational use of polluted water and those due to direct or indirect contamination of food. Diseases due to polluted drinking water are discussed briefly in the next section and those due to pollution of recreational waters and of food later.

5. Diseases Due to Pollution of Drinking Water Supplies

Waterborne disease is far too wide a subject for comprehensive discussion in a short paper. The following aspects are touched on briefly below: (i) the diseases that have been associated with polluted drinking water; (ii) the cause-and-effect association between drinking water and disease in the light of Bradford Hill's criteria; and (iii) the monitoring of water supplies in relation to health hazards.

(a) *Diseases associated with polluted drinking water*

Enteric fever, bacillary dysentery and cholera form the classic triad of waterborne disease. Cholera is at least for the moment not a problem in developed countries and is not discussed here, although it is a sobering reflection that exactly a century after a Royal Commission listed "the supply of wholesome and sufficient water for drinking and washing" as the first of 10 minimal requirements for civilized social life, it is not yet available to a large slice of the world's population. In recent years infectious hepatitis has also become well established as a waterborne disease. Mosley (1966) accepted as satisfactory the evidence cited for at least 30 of the 50 reported outbreaks. In all of these, massive and sustained contamination of the water supply with sewage had occurred. This was particularly the case for the Delhi outbreak of 1955, in which some 28,000 cases of clinical hepatitis with jaundice were reported (Viswanathan, 1957). On the day of maximal pollution of the Delhi water supply, virtually the entire water intake from the relevant channel of the Jamna River was drawn from a canal which was virtually an open sewer receiving the excreta of a large population of temporarily housed refugees already in the throes of an infectious hepatitis outbreak.

Weibel *et al.* (1964) in their report on waterborne diseases in the U.S.A. during 1946-1966 cited 142 outbreaks characterized by gastroenteritis or diarrhoea and not due to any known pathogen, 39 typhoid, 23 infectious hepatitis and 11 dysentery outbreaks, in a grand total of 228 notified outbreaks.

For various reasons these figures give only the broadest indications of the extent of the waterborne hazard in the United States. They are probably most reliable for typhoid fever. Some of the questions of specificity posed by the symptomatic diagnosis of "gastroenteritis or diarrhoea" are discussed below.

Diseases that have rarely been ascribed to waterborne infection include poliomyelitis, paratyphoid B and *Salmonella typhimurium* infections. The contrast between the rarity of waterborne infection ascribed to these two salmonellae and the long history of waterborne typhoid fever is discussed later. Poliomyelitis was at one time considered by certain Swedish workers to be primarily a waterborne disease but this has long since been discounted. Two possible waterborne outbreaks of poliomyelitis are still cited in the literature, that reported from Lincoln, Nebraska, by Bancroft, Engelhard & Evans (1957) and the outbreak at Edmonton, Alberta, reported by Little (1954). Even for these two outbreaks the evidence is inadequate in various respects, e.g. waterborne transmission as the cause of the Nebraska outbreak postulates very different incidence rates of clinical disease on opposite sides of one particular water main.

(b) *Causation of disease by polluted drinking water*

Various aspects of the association of polluted drinking water with disease are now looked at in sharper focus.

(i) Linking a communicable disease with polluted water supplies is in some ways easier and in some more difficult than establishing a cause-and-effect association between the same disease and an activity such as bathing. The known topography of water distribution systems can help in several ways. First, the unidirectional flow of water from a point of supply to the consumer may provide an obvious link in a cause-and-effect sequence, as in the Richmond Terrace outbreak already cited. Second, outbreaks due to polluted water are usually explosive because a large number of consumers in the area of supply are exposed more or less simultaneously. Third, the affected population may show a selective prevalence of the disease in different parts of the affected town which can be related to a particular branch of the distribution system or to the use of one contaminated source of supply while other sources remain uncontaminated. Fourth, when neighbouring communities which draw on the same milk and food supplies but have a different water supply escape, this differential location of affected patients may again point to polluted water, an argument used among others to incriminate the Montrose water supply in the outbreak described by Green *et al.* (1968).

(ii) As shown by the figures of Weibel *et al.* already cited, and also in an earlier report on waterborne diseases in England and Wales (*Report*, 1939), some $\frac{2}{3}$ of the waterborne outbreaks described in the literature have been characterized by gastroenteritis not due to any known pathogen. The ingestion of large numbers of viable sewage bacteria has been held responsible for these symptoms, although the time interval between exposure and the onset of symptoms has often been more suggestive of the incubation period of a microbial disease.

During the past 25 years, growing evidence has been found of the existence of filtrable agents that cause gastroenteritis but have not yet been reliably propagated in the laboratory. Gordon, Ingraham & Korns (1947) reported an outbreak at Marcy State Hospital in New York State characterized by the sudden onset of nausea, vomiting or diarrhoea, with profuse watery diarrhoea as the dominant symptom. Infection was apparently spread from person to person and most recovered within 2–3 days. No bacterial pathogens were isolated and no viruses could be demonstrated by the techniques available at the time. When faecal filtrates were fed to human volunteers, however, they developed watery diarrhoea after an average interval of 60 h, and the infective agent was eventually transmitted through 8 serial generations in volunteers. A temporary immunity followed the experimental disease. Later, Jordan, Gordon & Dorrance (1953) reported similar oral transmission of an agent that differed from the Marcy agent and was obtained from an individual in the Cleveland family study. This FS agent was also passaged through human volunteers by feeding them with filtrates. The experimental disease differed in several respects from that caused by the Marcy agent. The average period before symptoms developed was 27 h, and fever and constitutional symptoms were prominent. Independently of the Americans, Japanese workers (Kojima *et al.*, 1948) described an agent with properties similar to those of the Marcy agent, and Gordon & Whitney (1956) reported that crossimmunity tests showed the two agents to be closely similar, or identical.

A recent waterborne outbreak of gastroenteritis described by Lobel, Bisno,

MAP–2

Goldfield & Prier (1969) has greatly reinforced the argument for a viral aetiology of this condition. This large outbreak occurred in visitors to a Pennsylvania State Park during the first week of June 1966. Several thousand illnesses probably followed, but the writers traced 454 affected persons suffering from vomiting or diarrhoea. The average interval between exposure and the onset of symptoms was 29 h, and the attack rate in those exposed was 59%. Epidemiological and laboratory findings incriminated the park water supply. More interesting in pointing to an infective agent was that secondary cases of gastroenteritis occurred in household contacts who had not visited the park. The average incubation period in the secondary cases was 63 h, and the secondary attack rate in affected households was 44%. The longer time interval in the secondary cases would be compatible with a smaller infecting dose or with transmission by a different route from the primary cases to members of their families.

The search for a virus agent of gastroenteritis has been taken up again more recently by Dolin et al. (1971). These workers collected material from typical outbreaks of the condition variously known as 'winter vomiting disease', epidemic nausea and vomiting, gastric 'flu or acute nonbacterial gastroenteritis. Acute gastroenteritis followed in 2 of 3 volunteers after oral administration of bacteria-free stool filtrates from a naturally occurring case of the disease. Subsequently, a stool filtrate from one of the sick volunteers reproduced the disease in 7 of 9 more volunteers. The original rectal swab from the naturally occurring case of the disease was passaged 3 times in human foetal intestinal organ culture. Material from the third passage caused symptoms in 1 of 4 volunteers. The authors are careful to point out that the dilution of the original material through 3 passages was no greater than $10^{-4.9}$. Propagation of the agent therefore has not yet been conclusively achieved; the symptoms in the single volunteer might have been due to persistence of an agent originally present in high titre and not yet eliminated by serial dilution from the successive organ cultures. Further results of this promising study will be awaited with interest.

One consequence of the present uncertainty as to the cause of waterborne gastroenteritis must be stressed here. The isolation of various enteroviruses from a certain proportion of affected patients does not justify the conclusion that these viruses cause gastroenteritis. Sewage always contains whatever agents are circulating in the population. When it is swallowed, the recovery of these agents from the stools of patients does not fulfil Koch's postulates as to their causal role in the symptoms of those concerned.

(iii) As already pointed out, the recovery of a pathogen from a polluted site is an indication that it is already circulating in the adjacent population. The risk of its causing disease through environmental contamination must therefore be measured against the likelihood of the same disease being acquired by other routes. In sanitarily advanced countries, in fact, a diminishing proportion of diseases like typhoid fever, dysentery or infectious hepatitis is attributable to infection from polluted drinking water. Weibel *et al.* (1964), for instance, estimated that in the U.S.A. between 1946 and 1960 only 1·4% of the total typhoid morbidity could be ascribed to waterborne infection, as compared to an estimated 40% cited by Whipple in 1908. Mosley (1966) thought the proportion of infectious hepatitis cases in the U.S.A. transmitted by water was <1% and the same probably holds for shigellosis. The low proportion of waterborne infections is a tribute to the effectiveness of water purification methods. This was well shown in figures cited by Frost (1941). Eleven cities on the Ohio watershed were still in 1914 drawing water without purification from the same sources as in 1906. Their average death rate from typhoid fever, 76·8/100,000 in 1906, was still 74·5/100,000 in 1914. Sixteen other cities had since 1906 started to treat their water supplies without any change in the source of supply. In these cities the death rate from typhoid had fallen during the same span of years from 90·5 to 15·3/100,000.

(iv) A quantitative measure of exposure to a postulated hazard is required before a dose-response curve can be plotted relating exposure to the incidence of disease. For cigarette smoking this was the number of cigarettes smoked/day, to which mortality from lung cancer was found to bear a linear relationship. No such quantitative measure is available for the analysis of waterborne infection, even for waterborne typhoid fever. If typhoid bacilli were uniformly distributed throughout a distribution system at the relevant time—a highly unlikely event for a particulate pollutant excreted by a very small proportion of the population—the volume of water drunk by every patient who contracted the disease might give some indication of the number of organisms ingested. This would presuppose, however, that typhoid had been contracted directly by swallowing polluted water and not indirectly through food contamination with the organism in the kitchen of the affected household. Virtually no information is available to guide us here, partly no doubt because of the long incubation period of the waterborne disease.

The long history of waterborne typhoid fever and the rarity of waterborne disease due to *Salmonella paratyphi B* has usually been explained in terms of fewer typhoid bacilli being required to cause clinical infection. Paratyphoid fever is much more a foodborne than a waterborne disease, and this is compatible with a larger infecting inoculum achieved by prior multiplication in the contaminated food. All this may well be so, but recent experimental findings and a curious waterborne outbreak of *Salm. typhimurium* infection invite some re-appraisal. The experimental findings were those of Hornick & Woodward (1967), who fed typhoid bacilli in milk to volunteers, aborting the disease when it developed by prompt chemotherapy with chloramphenicol. In their hands, a dose of 10^3 organisms failed to cause symptoms in any of 20 volunteers. A dose of 10^7 organisms led to clinical disease in 16 of 32 volunteers, and as many as 10^9 organisms were needed for a 95% clinical attack rate in a group of 42 volunteers.

These findings were closely similar to those of an earlier study in chimpanzees by Edsall *et al.* (1960), who had concluded from the high doses required to induce typhoid fever in chimpanzees that the disease in that species must be essentially different from the corresponding disease in man.

By contrast, a unique outbreak of waterborne *Salm. typhimurium* infection that occurred in Riverside, California in 1965 (Ross, Campbell & Ongerth, 1966. Greenberg & Ongerth, 1966; *Report*, 1971) has been cited as indicating that many fewer salmonellae than hitherto believed necessary may be required to cause gastroenteritis in man. In this outbreak, which was brought to an end by urgent chlorination of a previously untreated supply, >16,000 people were affected from a population of about 133,000. Nine days before the main outbreak occurred, the principal water source for the city was cut off for 42 h to allow new connections to be made for a booster pump station. The main transmission line was entered at 3 points. Contamination of this supply was excluded however and it became clear from a study of the areas of high incidence that the pollution must have been associated with one or other of 4 wells within the city boundaries. During the period when the structural work was in progress, water was being drawn in unprecedented quantities from these wells, and the causal organism was in fact isolated by membrane filtration and enrichment procedures from a number of samples of water collected from the part of the town which these wells supplied. Intensive tests failed, however, to demonstrate whether or how the wells had been polluted. One well and its booster station were near pasture land where young calves were being grazed, and the run off from this land entered a stormwater drain near the booster station. It was known, moreover, that some of the calves had diarrhoea at the relevant time, but all attempts to demonstrate pollution of the supply from this source were unsuccessful. In one of their earlier papers, the authors suggested that having entered the distribution system by some undetected route, the causal agent had multiplied in nutrient-rich dead ends on the system before invading the mains that supplied the affected population. In the more recent collaborative report by the epidemiologists concerned, the possibility that organisms ingested in water pass through the stomach quickly and may thus cause disease from a very small inoculum is put forward as a possible explanation of this episode. They also point out, however, that the samples from which *Salm. typhimurium* was isolated were collected 8 days after the outbreak started and had been kept at room temperature for some days before being cultured. The original level of contamination of the affected supply was therefore speculative. They do not discuss the possibility of indirect foodborne infection from the kitchens of affected households, and the data given cannot be assessed from this standpoint.

The experimental findings of Hornick & Woodward could be reconciled with earlier views on the inoculum of typhoid bacilli required to cause disease by the physiological explanation of the Californian investigators that an organism ingested in water bypasses the acid barrier of the stomach and the likelihood of successful invasion from a small inoculum is thus enhanced. Alternatively, it could be argued that a larger inoculum of typhoid bacilli than hitherto believed necessary is required for waterborne disease to occur and that the typhoid carrier plays the special role here by polluting the environment, carriers of the other salmonellae being few and far between. In this event, contracting waterborne typhoid fever would be either a question of ingesting a particle of heavily infected faecal matter or of some multiplication occurring in the kitchen of the household concerned. If the experimental findings of Hornick & Woodward apply to the natural disease in man, one would expect relatively simple measures to bring typhoid fever under control in a polluted environment, and this was certainly the experience of Budd. Over many years he found that even when the disease was well established in a closed community, simple hand washing and the disinfection of excreta and of bed and body linen rapidly brought the outbreak under control. A celebrated example of this was the Bristol reformatory outbreak in 1863. Budd's advice was sought when 30 of 126 girls in this reformatory had gone down with typhoid following the return of one of their number with typhoid fever. Within a further 48 h another 20 went down with the disease. Some 70 girls were still well but could not be moved under the rules of this penal institution. With the adoption of Budd's simple measures, the outbreak rapidly subsided, with only 3 additional cases of typhoid among the girls in contact with the disease.

(c) *The monitoring of water supplies in relation to health hazards*

Routine bacteriological tests on water supplies provide a useful control on adequacy of treatment and a sensitive method for checking whether the supply has been recently contaminated with sewage. Coliform and faecal coli counts on water samples do not bear any simple relationship to the health hazard of waterborne infection. However high such counts may be, typhoid fever cannot result from ingestion of the polluted supply unless a typhoid carrier happens to be polluting the sewage at the time, and this is equally true of the other waterborne diseases, apart perhaps from nonspecific gastroenteritis, if a part or all of this is due to swallowing large numbers of nonpathogenic organisms rather than to a specific agent not yet identified. Again, many waterborne outbreaks affect only one or other part of the distribution system, and a negative coliform test on a main that has not been contaminated will be irrelevant to the hazards incurred in the area supplied by the contaminated water. Hence the curious finding that during the period of the Riverside outbreak mentioned above, coliform tests were quite satisfactory at the time when the outbreak was taking place. This latter finding prompted the comment by Greenberg & Ongerth (1966) that, while no better test than the coliform was available at present for routine control of water supplies, "no bacteriological test can substitute for sound engineering practice to insure maximum source and distribution system safety". Perhaps the main lesson to be derived from the long tally of waterborne disease is a catalogue of the various failures of water undertakings in the choice of source, the maintenance of the structural integrity of distribution systems and their methods of water treatment.

The incrimination of infectious hepatitis as a waterborne disease does however raise the question whether the faecal coli test necessarily reflects the likelihood of viruses from sewage being present in a water supply. Little progress can be expected here until the virus of infectious hepatitis has been reliably propagated in the laboratory. Other viruses are relevant to the extent that no water undertaking could tolerate on general grounds contamination of the supply for which it was responsible with any pathogenic organism irrespective of whether it was a recognized cause of waterborne disease or not. The point becomes even more relevant with the necessity of continual re-use of water from sewage effluents for drinking water supplies, given the enormous demands on quantity as well as quality of supply.

The possible limitations of the coliform test as an indicator of adequate treatment and chlorination have also been widely discussed. Clearly, further work on the susceptibility of all sewage pathogens to standard methods of treatment is essential.

6. Health Hazards of the Recreational Use of Water

The possible hazards of bathing in sewage contaminated sea water have been the subject of active discussion in Britain for many years. A Public Health Laboratory Service (P.H.L.S.) working party was set up in 1953 to study various aspects of bathing beach contamination. In that year the total number of polimyelitis notifications in England and Wales was 4538, with 2970 paralytic cases and 320 deaths. Not surprisingly, an uncritical emphasis on monitoring studies for poliovirus in sewage effluents led to public concern about the possibility that poliomyelitis could be contracted from bathing. The working party, in an investigation discussed below that would not have been practicable a few years later when poliovaccine reduced the incidence to an insignificant level, failed to find any significant association between poliomyelitis and bathing. The only evidence they could find of a health hazard from sea bathing between 1953 and the publication of their final report (Public Health Laboratory Service, 1959) amounted to 4 cases of paratyphoid fever, all associated with grossly fouled beaches where direct contact with faecal matter could occur. The working party had also had reported to it an additional 4 cases of paratyphoid fever and 2 of typhoid fever ascribed to sea bathing and notified between 1946 and 1952.

Perhaps a score or two in all of cases of enteric fever ascribed to sea bathing have been reported in the world literature, and this has been reviewed by the writer (Moore, 1954, 1970a). There is now general agreement that on beaches that meet the lowest aesthetic criteria of acceptability the risk of contracting enteric fever is minimal.

Here then is a field where there is virtually no hard evidence of a significant health hazard, and where pessimistic assessments are largely speculative and based on the monitoring fallacy already discussed earlier. The few aspects discussed below have been selected for their general interest and as illustrations of a number of the points already made on the establishment of cause-and-effect relationships between environmental pollution and disease.

(i) An investigator who decided to look into the health hazards of sea bathing could make his assessment in various ways. In an authoritarian society he might take a population of adequate size, divide it into comparable test and control groups, debar the control group from swimming and allocate various subgroups of his test group to a number of swimming regimes involving a varying degree of exposure to polluted sea water. Given time, his results would be meaningful. A less rigorous prospective survey was carried out twenty years ago by the Environmental Health Centre at Cincinnati and described in summary by Stevenson (1953). Here the approach was to pick the first instance 3 communities each with a population 5000–10,000: one was on a great lake, one on an inland river and one on a tidal water. Each community had one satisfactory and one polluted bathing beach. Records were kept during the bathing season by a member of each participating household, and the findings were later analysed in relation to the illness experienced by different bathing groups and by people who had not bathed during the relevant periods. An appreciably higher incidence of illness was found in the swimming groups, whatever the quality of the bathing water. More than half the illnesses were eye, ear, nose and throat ailments, but, as in other reports of illnesses of this kind in relation to bathing, no evidence was obtained to suggest that such infections were not due to the patients' own nasopharyngeal flora rather than to polluting organisms. In two of the studies, somewhat tenuous evidence emerged that water with an average coliform count $>2500/100$ ml was associated with more illness than was purer water. A recent proposal to base bathing water "standards" in the U.S.A. on these findings was criticized by the writer elsewhere (Moore, 1970b).

The P.H.L.S. working party organized a retrospective study of bathing histories in poliomyelitis patients and matched control healthy children to seek presumptive evidence of any association between sea bathing and the onset of poliomyelitis. This was done with the help of medical officers of health of coastal districts. They were asked to obtain a record of the bathing history, for the three weeks before the onset of illness, of locally resident children under 15 who contracted poliomyelitis. At the same time, they were to choose from the birth or school registers, as a matched control for each patient, the name of a healthy child. The child must not live in the same household as the patient, but should be of the same sex, as near as possible in age to the ill child, and should preferably live in the same street, but certainly in the same administrative area. A similar bathing history would be taken from this control child and both histories recorded on one card. In 1957, 98 paired records were collected, and in 1958 an additional 52, making in all 150 patients and their matched controls. The two groups were found on analysis to be well matched for relevant characteristics. Bathing histories were analysed in two ways. First, the bathing histories were tabulated according to whether there was a history of bathing at any time within 3 weeks of the onset of symptoms in the patients. This showed that the bathing histories of both groups were very similar. Thus, 45 of the 150 poliomyelitis patients had bathed during the 3 weeks before the onset of symptoms. Doubtless, some of the parents of these 45 children might have wondered whether the bathing had caused the illness. But 44 of the corresponding healthy control children had also bathed during these 3 weeks. No special significance therefore could be attached to a previous event that was just as likely to be reported by a child who was perfectly well. The bathing histories were also analysed in terms of the intervals between dates of bathing and the dates when symptoms first appeared in the ill child. Again, the comparison showed that there was no material difference between the two groups.

(ii) When the P.H.L.S. working party's report was published, information was not yet available on a small Australian outbreak of typhoid fever in 1958, later to be reported briefly in the annual report for 1958 of the Commissioner of Public Health for Western Australia. This outbreak, ascribed to sea bathing, occurred at Perth during the early months of 1958 (Kovacs, 1959; Snow, 1959). The outbreak occurred against a background of rapidly declining incidence of typhoid fever in the area. From 1950 onwards, the average annual number of typhoid cases in Western Australia was 9. Early in January 1958, at the height of the Australian summer, the first case in the outbreak was notified in a boy of 13 who was a keen swimmer and had been swimming at City Beach near Perth, perhaps the most popular beach in the state. Between then and the end of March, 15 cases of typhoid fever in all occurred in the area, 10 of them in persons who had swum at City Beach. The strains of typhoid bacillus isolated from the 10 patients fell into 5 different phage types. This was interpreted as indicating that the patients had been infected by a sewage-contaminated source, since the great majority of individual typhoid carriers harbour only one phage type of *Salm. typhi*.

The beach in question was at the time subject to sewage pollution from an old sewer outfall 1 mile to the south. The outfall pipe was 300 ft long and discharged 4 ft below low water mark. The sewer was overloaded as a result of rapid growth of population during the previous decade, and the outfall pipe was also found to be fractured. The beach was closed to bathers on 12 March 1959. Heavy chlorination of the effluent was started, and a new 3000 ft long outfall came into service in October 1960. Routine epidemiological inquiries failed to uncover alternative sources of infection. The significance of these negative findings is difficult to assess, as Snow comments in his report that in previous years the sources of infection of typhoid fever cases had rarely been detected. The bacteriological tests reported do not permit evaluation of the intensity of pollution, as sufficiently high dilutions of sea water were not tested, but the visual evidence cited certainly indicates that pollution of the beach must have been heavy. Typhoid bacilli were not isolated from the sewage or from the sea water.

On the evidence cited, the association of this episode with sea bathing remains tantalizingly uncertain in a number of respects. First, in an area where summer temperatures are high and a large number of people frequent the beaches, particularly the favourite in question, some attempt at assessing the significance of a history of bathing per se would have been valuable. Second, the multiplicity of phage types in those affected would certainly point to sewageborne infection provided there was independent proof that all the cases constituted one outbreak, but this was precisely the point at issue. In fact, 5 of the 15 cases in the outbreak had no connection with City Beach. Third, the lack of any supporting bacteriological evidence of environmental contamination with typhoid bacilli was unfortunate, although as indicated earlier this is not a strong counterargument. The association of bathing with disease is the essential epidemiological requirement. Positive bacteriological evidence would have supported the proposed aetiology, but its lack does not refute it.

So much for the weak points in the evidence. Many of the details of individual case histories, on the other hand, certainly suggest cause-and-effect sequences when judged by Bradford Hill's criteria. Thus, 5 of the first 8 cases had not only bathed at City Beach but had done so more frequently and for longer periods than the average beach visitor. Again, the first boy to develop typhoid fever was always accompanied by his two brothers, but they by contrast swam very little. Here is a clear suggestion of a dose-response curve, with typhoid fever occurring particularly in persons exposed for a long time to the polluted environment. This makes the point that "bathing" in the climate of Western Australia may be essentially a different activity from its counterpart in Britain and other colder countries, and that corresponding caution is required in extrapolating epidemiological findings from one to the other.

(iii) As mentioned earlier, the test of experiment is available in the assessment of environmental health hazards, when a source of pollution can be eliminated and the resulting impact on disease incidence observed. A claim that the laying of a new sewer outfall at the Black Sea resort of Yalta had significantly lowered the incidence of gastrointestinal disease during the bathing season was therefore of interest. Gorodetskij & Raskin (1966) reported that until June 1956 crude sewage from Yalta was discharged from an outfall on the shore. A new sewer then came into use, which now carries the town sewage some 200 m out to sea and discharges it at a depth of 10 m below the surface. The consequent improvement in the appearance of the Yalta beaches prompted a study of morbidity figures for gastrointestinal disease before and after the new sewer was installed. The index used was the over-all morbidity/10,000 inhabitants from gastrointestinal disease, obtained by combining notifications for dysentery, gastroenteritis and enteric fever. In addition, the authors compared the morbidity figures for Yalta during the years 1953-9 with corresponding figures for another Crimean resort, Eupatoria, a town of roughly similar population. The towns were also alike in various other respects. At both, the summer population was inflated 2½-fold. Both towns had good water supplies and regular refuse collection. In Yalta, however, 96% of the population was connected to the main sewerage system, whereas Eupatoria had not yet completed a municipal sewerage scheme and because of this has no beach pollution problem. The authors' claim rests entirely on graphs similar to those shown in Figs 2 and 3.

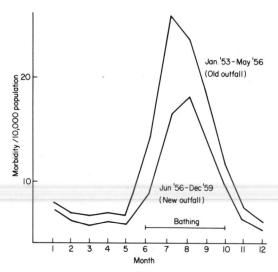

Fig. 2. The indicence of gastroenteritis in Yalta before and after the installation of a new sewage outfall.

The first of these purports to show a significant fall in monthly incidence of gastroenteritis when data for the 4 summers after the installation of the new sewer outfall were averaged and compared with those obtained for the 3 summers beforehand. Even ignoring the difficulties of achieving adequate notification of gastroenteritis in seaside resorts (Moore, 1962), it is clearly impossible to distinguish in the diagram between a falling time-trend in gastroenteritis and any change referable to the new outfall. The argument in Fig. 3 is even more nebulous. The authors suggest that, as the graph shows a greater

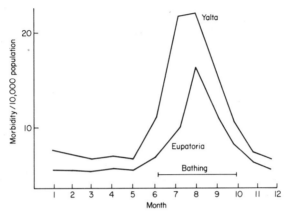

Fig. 3. The incidence of gastroenteritis in Yalta and Eupatoria.

relative increase in gastroenteritis at Yalta during the summer than in Eupatoria, therefore the greater increase at Yalta must be due to the factor known to be different there, i.e. the presence of a sewer. The naiveté of this argument calls for no further comment. It must be concluded that the claim of Gorodetskij & Raskin quite fails to stand up to scrutiny.

7. Environmental Pollution and Food

Typhoid fever from shellfish is the classic example of food contamination from a polluted environment. It was recognized as a major hazard by the Royal Commission on Sewage Disposal at the turn of the century, but the incidence in Britain has been very low in recent years, although enteric fever from shellfish is still very common in Mediterranean countries. The severity of the disease is the result of concentration of the pathogen by filter-feeding molluscs.

The contamination of various rivers and streams by the residuum of enteric carriers in the population has from time to time given rise to infections. Martin (1947) reported a series of paratyphoid B infections in boys and girls aged 10-13 who had bathed in the river Waveney, and Lendon & MacKenzie (1951) described 5 cases of typhoid fever occurring between 1944 and 1948 in boys who had drunk water from the Wallington River. They isolated typhoid bacilli from the river water with the use of Moore swabs, and traced the organism 2 miles upstream and a further ½ mile along a sewer to the house drain of a typhoid carrier.

Moore, Perry & Chard (1952) speculated on the possible relationship of the use of polluted river water on riverside allotments near the River Sid to a long history of endemic typhoid and paratyphoid fever in Sidmouth. Certainly, the substitution of the town supply was followed by a virtual freedom from enteric fever in the ensuing 20 years, but whether this was a cause-and-effect sequence can only be a matter of opinion.

To what extent other salmonella infections in the community, particularly the large number of unexplained sporadic cases, are due to the circulation of salmonellae in rivers and streams and the contamination by polluted water of raw vegetables, remains uncertain. Much of the day-to-day work of health departments and public health laboratories is concerned, however, with the gradual detection and sealing off of a host of routes along which the barrier between the human population and its sewage can be breached. A recent report by the Ministry of Housing and Local Government's working party of sewage disposal (*Report*, 1970) lays stress on the importance of improving the quality of sewage effluents in the context of the increasing re-use of our limited water supplies. The combination of good food hygiene with environmental decontamination is required for a continuing improvement of the public health.

8. References

Bancroft, P. M., Engelhard, W. E. & Evans, C. A. (1957). Poliomyelitis in Huskerville (Lincoln), Nebraska. *J. Am. med. Ass.* **164**, 836.

Bradford Hill, A. (1965). The environment and disease: association or causation? *Proc. R. Soc. Med.*, **58**, 295.

Brisou, J. (1968). La pollution microbienne, viralle et parasitaire des eaux littorales et ses consequences pour la sante publique. *Bull. Wld Hlth Org.* **38**, 79.

Budd, W. (1856). On the fever at the clergy orphan asylum. *Lancet*, ii, 617.

Budd, W. (1873). *Typhoid Fever: Its Nature, Mode of Spreading and Prevention.* London: Longmans Green.

Chadwick, E. (1842). General Report on the Sanitary Condition of the Labouring Population of Great Britain. London: H.M.S.O.

Cruickshank, R. (1963). *Infection in Hospitals: Epidemiology and Control.* Oxford: Blackwell.

Dolin, R., Blacklow, N. R., DuPont, H., Formal, S., Buscho, R. F., Kasel, J. A., Chames, R. P., Hornick, R. & Chanock, R. M. (1971). Transmission of acute infectious nonbacterial gastroenteritis to volunteers by oral administration of stool filtrates. *J. infect. Dis.*, **123**, 307.

Edsall, G., Gaines, S., Landy, M., Tigertt, W. D., Sprinz, H., Trapani, R. I., Mandel, A. D. & Benenson, A. S. (1960). Studies on infection and immunity in experimental typhoid fever. I. Typhoid fever in chimpanzees orally infected with *Salmonella typhosa*. *J. exp. Med.* **112**, 143.

Frost, W. H., (1941). Papers of W. H. Frost, M.D. *A Contribution to Epidemiological Methods.* Ed. K. F. Maxcy. New York: Commonwealth Fund.

Gordon, I., Ingraham, H. S. & Korns, R. F. (1947). Transmission of epidemic gastroenteritis to human volunteers by oral administration of faecal filtrates. *J. exp. Med.* **86**, 409.

Gordon, I. & Whitney, E. (1956). Virus diarrhoeas of adults and their possible relationships to infantile diarrhoea. *Ann. N.Y. Acad. Sci.* **66**, 220.

Gorodetskij, A. S. & Raskin, B. M. (1966). Gigiena pribrezhnykh morskikh vod. (Leningrad).

Green, D. M., Scott, S. S., Mowat, D. A. E., Shearer, E. J. M. & Thomson, J.M. (1968). Waterborne outbreak of viral gastroenteritis and sonne dysentery. *J. Hyg., Camb.* **66**, 383.

Greenberg, A. E. & Ongerth, H. J. (1966). Salmonellosis in Riverside, California. *J. Am. Wat. Wks Ass.* **58**, 1145.

Hornick, R. B. & Woodward, T. E. (1967). Appraisal of typhoid vaccine in experimentally infected human subjects. *Trans. Am. clin. & clim. Ass.* **78**, 70.

Jordan, W. S., Gordon, I. & Dorrance, W. R. (1953). A study of illness in a group of Cleveland families. VII. Transmission of acute nonbacterial gasroenteritis to volunteers: evidence for two different etiologic agents. *J. exp. Med.* **98**, 461.

Kojima, S., Fukumi, H., Kusama, H., Yamamoto, S., Suzuki, S., Uchida, T., Ishimaru, T., Oka, T., Kuretani, S., Ohmura, K., Nishikama, F., Fujimoto, S., Fujita, K., Nakano, A. & Sunakawa, S. (1948). Studies on the causative agent of the infectious diarrhoea. Records of the experiments on human volunteers. *Jap. med. J.* **1**, 467.

Kovacs, N. (1959). Enteric fever in connection with pollution of sea water. Western Australia. *Report of the Commissioner of Public Health for the Year 1958.*

Lendon, N. C. & Mackenzie, R. D. (1951). Tracing a typhoid carrier by sewage examination. *Mon. Bull. Ministr Hlth* **10**, 23.

Little, G. M. (1954). Poliomyelitis and water supply. *Can. J. Publ. Hlth* **45**, 100.

Lobel, H. O., Bisno, A. L., Goldfield, M. & Prier, J. E. (1969). A waterborne epidemic of gastroenteritis with secondary person-to-person spread. *Am. J. Epidemiol.* **89**, 384.

Logan, W. P. D. (1953). Mortality in the London fog incident, 1952. *Lancet, i*, 336.

Martin, P. H. (1947). Field investigation of paratyphoid fever with typing of *Salmonella paratyphi B* by means of Vi. bacteriophage. *Mon. Bull. Ministr Hlth* **6**, 148.

Moore, B. (1948). The detection of paratyphoid carriers in towns by means of sewage examination. *Mon. Bull. Ministr Hlth* **7**, 241.

Moore, B. (1954). Sewage contamination of coastal bathing waters. *Bull. Hyg., Lond.* **29**, 689.

Moore, B. (1962). Acute nonbacterial gastroenteritis at seaside resorts. *Publ. Hlth, Lond.* **77**, 28.

Moore, B. (1970a). The present status of diseases connected with marine pollution. *Rev. intern. Oceanogr. med.* **18-19**, 193.

Moore, B. (1970b). Public health aspects. In *Water Pollution Control in Coastal Areas.* London; Institute of Water Pollution Control.

Moore, B., Perry, E. L. & Chard, S. T. (1952). A survey by the sewage swab method of latent enteric infection in an urban area. *J. Hyg., Camb.* **50**, 137.

Mosley, J. W. (1967). Transmission of viral disease by drinking water. In *Transmission of Viruses by the Water Route.* Ed. Berg. London: Interscience Publishers.

Perry, W. D., Siegel, A. C. & Rammelkamp, C. H. (1957). Transmission of group A streptococci II. The role of contaminated dust. *Am. J. Hyg.* **66**, 96.

Public Health Laboratory Service (1959). Sewage contamination of coastal bathing waters in England & Wales. *J. Hyg., Camb.* **57**, 435.

Report (1971). A waterborne epidemic of salmonellosis in Riverside, California. 1965. *Am. J. Epidemiol.* **93**, 33.

Report (1939). Memorandum and circular on the safeguards to be adopted in day-to-day administration of water undertakings. London: Ministry of Health.

Report (1970). Taken for granted. Report of the working party on sewage disposal. London: H.M.S.O.

Ross, E. C., Campbell, K. W. & Ongerth, H. J. (1966). *Salmonella typhimurium* contamination of Riverside, California, water supply. *J. Am. Wat. Wks Ass.* **58**, 165.

32 B. MOORE

Snow, D. J. R. (1959). Typhoid and City Beach. Western Australia. *Report of the Commissioner of Public Health for the Year 1958*, p.52.
Snow, J. (1855). *On the Mode of Communication of Cholera,* 2nd Ed. London: J. & A. Churchill.
Stevenson, A. H. (1953). Studies of bathing water quality and health. *Am. J. Publ. Hlth* **43,** 529.
Viswanathan, R. (1957). Infectious hepatitis in Delhi (1955-56): A critical study. Epidemiology. *Int. J. med. Res.* **45,** Suppl. 1.
Weibel, S. R., Dixon, F. R., Weidner, R. B. & McCabe, L. J. (1964). Waterborne disease outbreaks 1946-1960. *J. Am. Wat. Wks Ass.* **56,** 947.

Sewage Pollution of Natural Waters

J. H. McCoy

*Public Health Laboratory, Hull Royal Infirmary,
Kingston-upon-Hull, England*

CONTENTS

CRUDE SEWAGE consists essentially of human excreta suspended in the waste waters of the community. The hazards arising from the disposal of crude sewage to natural waters are the creation of nuisances, the destruction of animal, plant and fish life. and the transmission of infectious disease.

1. Nuisances and the Destruction of Aquatic Life

Natural waters always contain oxygen dissolved from the atmosphere. In natural waters animal and vegetable organic matter discharges are converted by aerobic bacterial action to simple organic salts which act as fertilizers for plant and animal growth. During this process oxygen in the water is converted into CO_2 which under the influence of light is absorbed by vegetation. The carbon required for plant growth is fixed and oxygen is returned to the water. The cycle of selfpurification thus requires in the first place a sufficiency of dissolved oxygen for utilization by the bacteria, which bring about the ultimate oxidation of organic matter to CO_2 and mineral salts, and this must be concomitant with a vigorous growth of vegetation to return oxygen to the water. The cycle is broken by a deficiency of oxygen in the water, by turbidity of the water which limits the penetration of light and by the absence of vegetation. Excessive organic matter in discharges depletes the oxygen: excessive suspended matter limits the penetration of light and when deposited, blankets and kills vegetation.

When the oxygen cycle is broken, the reduction of organic matter is affected by anaerobic bacteria. Such digestion is characterized by the production of foul smelling gases and compounds, and by failure to oxidize organic matter completely. The self purifying capacity of a natural water is thus a delicate balance between the quantity of oxygen absorbed from the atmosphere, the quantity used in the oxidation of added organic matter and the quantity returned by vegetation.

The balance is easily upset by the discharge to the natural water of organic matter in quantity sufficient to deplete the oxygen more rapidly that it can be restored. A large amount of organic matter discharged continuously into a small stream may produce permanent anaerobic conditions for many miles downstream: the same quantity discharged into a large river may be dealt with in a short distance without effect on the river as a whole. The continuous discharge of crude sewage to a watercourse in excess of the stream's capacity for self purification is rapidly followed by the establishment of anaerobic conditions, with consequent destruction of animal, plant and fish life, through the absence of oxygen, and with silting of the watercourse, through the deposition of partially digested organic matter. The natural water thus becomes a nuisance, turbid and foul smelling, and devoid of natural life.

Such nuisance may be prevented by limiting the quantity of organic matter discharged to well below the capacity for self purification of the water receiving the discharge.

The standards are simple. The Royal Commission on Sewage Disposal (1912) recommended a general standard for effluents limiting the suspended matter to ⊁30 p/m, the Biochemical Oxygen Demand (BOD) to ⊁20 p/m and it related these standards to both the volume and BOD of the water receiving the discharge. If the dilution of the effluent after discharge was low, a specially stringent standard might be prescribed: if very great the general standard might be relaxed or suspended altogether. Where the dilution exceeded 500 volumes it was suggested that all tests might be dispensed with and that crude sewage be discharged without treatment other than simple screening and the removal of grit.

The function of sewage treatment is therefore to remove organic matter and to produce an effluent satisfying a physical and chemical standard. It is essential that a sharp distinction be made between the chemical and the bacteriological aspects of sewage pollution. The distinction is necessary as many industrial waste regarded chemically as highly polluting are free from bacteria, whilst many crude sewages and sewage effluents satisfying the recommended chemical standards, and so regarded as nonpolluting, contain many millions of intestinal bacteria/ml. The distinction is also necessary because the primary purpose of sewage treatment is to remove as much as possible of the putrescible organic matter present. The greater part, however, of the organic matter in human excreta is present in the form of the bacteria which inhabit the human intestine.

The average weight of moist faeces excreted daily by the average individual is 100 g, some 20-30% of which is composed of undigested food residues, the remainder consisting of water and bacteria. In the healthy individual, these bacteria consist of the normal inhabitants of the intestine of which *Escherichia coli* I is regarded as the most characteristic. The number of these organisms is of the order of 10^9/g of faeces. In actute intestinal disease and in the carrier state, these normal inhabitants may be replaced by pathogenic organisms, but, as the sick, convalescent and chronic carriers form only a small portion of the community, pathogenic organisms in crude sewage are always outnumbered by the normal inhabitants of the intestine. In sewage treatment the removal of organic matter is accompanied by the removal of organisms present, but a proportion of the organisms originally present in the crude sewage remains and can easily be demonstrated both in the sludge removed from crude sewage and in the effluent discharged after treatment.

The removal of bacteria is not selective, so that sewage effluent resulting from the treatment of crude sewage contains the same proportions of all the varieties of organisms, normal and pathogenic, present before treatment but in diminished total numbers.

There is no bacterial standard for sewage effluents discharged after treatment; instead stringent standards are prescribed for potable waters. To ensure absence of pathogenic organisms the general standard for potable waters is the absence of *E.coli* from 100 ml. of water. As the numbers of *E.coli* in wet faeces may approach 10^9/g, this standard implies a dilution of crude sewage of $1 : > 10^9$ parts of water. Such dilution is rarely obtained, or indeed necessary, as the numbers of intestinal organisms in effluent are reduced after discharge to natural waters by sedimentation, by exposure to light and low temperatures, by predation by water animals and by simple starvation through lack of nutriment. In other words, intestinal organisms survive in water for a period, but do not multiply. This is to be expected, as intestinal organisms are highly specialized parasites, living in a high concentration of nutriment in darkness and at a constant temperature of $37°$.

In fresh water the time of survival may be measured in weeks, depending largely on the initial numbers discharged: in sea water the period is measured in hours. As the normal inhabitants of the intestine are present in greater numbers than pathogenic organisms, a standard for potable water based on the absence of the most characteristic organism normally present in faeces ensures the absence of pathogenic organisms.

2. Pathogens in Sewage

Crude sewage contains all the agents causing infectious disease in man —

bacteria, viruses, protozoa, intestinal parasites excreted through the intestinal tract. Of these, organisms of the *Salmonella* group, which are widely distributed in man and animals, are by far the most common in developed countries. Organisms of this group produce in man and animals a variety of infections ranging from acute septicaemic fevers through acute intestinal infections to subclinical or symptomless infections in which the individual presents few signs or symptoms of infection but excretes the infecting organism.

In man by far the most important infections produced by salmonellae are the enteric infections, typhoid and paratyphoid fever, although food poisoning is in England and Wales by far the most common manifestation of salmonellosis.

Salmonellae of human origin are derived from acute and convalescent cases, from symptomless excretors and from chronic carriers among the population. Salmonellae of animal origin in towns are derived from trades and industries processing animals and their products for human or animal use. These sources include abattoirs, butchers' shops, the meat products industry, poultry processing plants, egg breaking plants, bakeries, tanneries, knackers' premises, animal fat extraction plants, meatmeal and bonemeal plants, animal feeding-stuffs plants and fertilizer plants. In general, the pathogenic organisms derived from human sources are of greater importance than those from animal sources, because man is the sole reservoir of typhoid and paratyphoid bacilli, and of the organisms of dysentery and cholera. These organisms are incapable of infecting animals, and animal infection plays no part in the propagation of these purely human diseases.

3. Bacterial Content of Crude Sewage and Sewage Effluent

The bacterial content is directly related to the strength of the sewage, that is the amount of excreta contained in a constant volume. The strength shows hourly, daily and seasonal variations. Generally speaking, hourly variations are mainly the result of the pattern of defaecation in a community; daily variation the result of variation in domestic and industrial water usage, seasonal variation the result of variation in rainfall.

(a) *Hourly variation*

The pattern of defaecation in a community is more or less constant, being the result of activity and food intake reinforced by habit training from early childhood. In general in a large community 50–60% of the daily load is present in the crude sewage between 6 a.m. and noon, and 15–20% in the period noon to 6 p.m. In a small community, most of the load may be concentrated in the crude sewage before noon.

(b) *Daily and seasonal variations*

As the average individual excretes daily some 100 g of faeces, the faecal excretion of a population is constant within narrow limits and it averages *c.* 1 ton/10,000 population. In a constant population therefore daily and seasonal variations result from variations in the volume of water in which excreta are carried. Domestic and industrial water usage generally conform to predictable patterns. The pattern of rainfall is predictable, however, only within very wide limits with a large margin of error. This is of some importance as the design of sewage treatment works is based on the dry weather flow of sewage and arrangement is made to accommodate only part, usually 3–6 times that of the dry weather flow, of storm sewage. Storm sewage in excess of these volumes is discharged directly. In combined sewage systems treating both crude sewage and run off from rainfall, the faecal content of storm water may be high.

Table 1 presents the *E.coli* content of sewage effluent after treatment at the point of discharge to a river in a wet season and in a dry season when the plant was operating at its design capacity. In the wet season, the numbers of *E.coli* in hourly samples are seen to vary >100-fold from 0·18 to >10^6/100 ml. In the dry season the variation in hourly samples was within much narrower limits, > 0·18–0·40 x 10^6/100 ml. Both sets of samples were collected at the same plant, treating the crude sewage of a constant population of 7000 persons. In the examples given, the faecal content of the sewage remained constant; only the water content varying.

To determine the relationship between numbers of *E. coli* and of salmonellae in the same effluent, hourly samples during a 24-h period were collected once weekly from March 1968 to February 1969. Samples were bulked in 4 lots of 6 successive samples and the *E. coli* and salmonella content of the composite samples determined by a dilution method. In all, 42 days' samples were available for analysis. Salmonellae were isolated from at least one of each day's composite samples. During the period only one acute human salmonella infection is known to have occurred in the population. The salmonellae present belonged to many different serotypes and can be regarded as the "background" salmonellae present in any sewage effluent. These were derived principally from an abattoir, 2 poultry processing plants, an animal feed plant, a plant manufacturing bakers' and confectioners' sundries.

No constant relationship appeared to exist between the numbers of *E. coli* and the numbers of salmonellae present in the samples.

A detailed study of the relationship between numbers of *E. coli* and salmonellae in the sampling days during which *Salm. brandenburg* was present in the effluent from an acute human infection, and the preceding 5 sampling days (Table 3) illustrates the sudden unpredictable increase in the numbers of salmonellae following the appearance of acute infection in the community and the duration of excretion. It will be appreciated that the occurrence of successive

Table 1

The effect of variation in volume on the E. coli content of sewage effluent

Date		Effluent volume (10^6 gall/day)		E. coli content ($\times 10^6$/ml) of effluent at (time of day)			
				10.00–15.00	16.00–21.00	22.00–03.00	04.00–09.00
15/16	January 1970	2·1	Median	5·5	2·9	0·3	0·6
			Range	1·8–>18·00	0·4–6·9	<0·18–1·10	0·18–1·10
3/4	September 1970	0·6	Median	0·18	0·2	0·3	0·3
			Range	<0·18–0·18	<0·18–0·40	<0·18–0·40	<0·18–0·40

Table 2

Most probable numbers of salmonellae & E. coli in sewage effluent at point of discharge to river. March 1968–February 1969

Salmonellae (MPN/100 ml)	No. of samples in groups	No. of samples with MPN/$\times 10^6$/100 ml of *E. coli* of										
		<0·18	0·18	0·40	0·69	1·10	1·80	4·05	6·93	11·00	18·00	>18·00
<1	20	7	6	2	2	2						1
1	27	2	3	9	3	2	3	3	1			1
2	21	1	4	1	5	3	3		2			2
4	22	3	2	5	6	3	1		2			
6	27	4	5	4	6	3	5					
10	21		3	4	1	5	6	1				1
20	12		6			1	3	1	1			
40	5		1			1	1	1		1		
60	6				2		1		2	1		
100	5				1		1			2	1	
>100	2					2						
Total	**168**	17	30	25	26	22	24	6	8	4	1	5

Table 3

Acute human infection—MPN of salmonellae in sewage effluent during a period of excretion in September 1968.

Sampling day	Organism	MPN at period of day				Effluent volume (×10⁶ gall/day)
		10.00–15.00 h	16.00–21.00 h	22.00–03.00 h	04.00–09.00 h	
August 1/2	Salmonellae	6	4	<1	<1	0·60
	E. coli	1·10	0·18	0·18	0·18	
8/9	Salmonellae	1	1	<1	<1	0·63
	E. coli	0·69	1·80	>18·00	0·40	
15/16	Salmonellae	2	23	10	10	1·02
	E. coli	0·18	0·18	>18·00	0·18	
22/23	Salmonellae	10	10	6	1	0·68
	E. coli	1·80	1·10	0·69	0·40	
29/30	Salmonellae	<1	1	<1	<1	0·68
	E. coli	1	Not estimated			
			acute human infection of Salm. brandenburg			
September 5/6	Salmonellae	>100	100	100	60	0·62
	E. coli	>100	>100	60	23	↑
12/13	Salmonellae	>100	6·93	1·80	6·93	0·61
	E. coli	18·00	60	10	60	
19/20	Salmonellae	40	11·00	0·40	6·93	1·56
	E. coli	1·80	2	2	1	
26/27	Salmonellae	1	1·10	0·69	0·69	0·56
	E. coli	1·10	1	1	<1	
October 3/4	Salmonellae	<1	0·40	0·40	0·18	0·60
	E. coli	>0·18	0·40	0·40	0·18	

Table 4

Natural purification of a river during 1954/65

Criterion	Observations on sewage effluent at point of discharge (Undiluted)	Observations on river water at miles below discharge			
		2 (diluted × 40)	4 (diluted × 40)	6 (diluted × 300)	7 (diluted × 300)
E. coli I/100 ml (median)	$2 \cdot 5 \times 10^6$	$9 \cdot 2 \times 10^3$	$1 \cdot 1 \times 10^3$	$2 \cdot 7 \times 10^2$	Not examined
Salmonellae in 50 ml Samples (total) No. (and %) positive	570 316(55)	—	—	—	—
Salmonellae in 2500 ml Samples (total) No. (and %) positive		466 43(9)	584 21(3·5)	582 18(3)	416 2(0·5)

sporadic human salmonella infections at intervals of 3—4 weeks is sufficient to maintain the salmonella content of sewage and sewage effluent at a high level.

(c) Natural purification of a river

After discharge to a natural water, sewage is first diluted and then transported. The distance from the outfall at which particulate matter is deposited depends on the size of particles, their specific gravity and the velocity of the stream into which the sewage is discharged. The larger the particle and the smaller the velocity of the stream, the sooner does sedimentation occur: the smaller the particle and the greater the velocity of the stream, the later does sedimentation occur. Thus particulate matter is carried downstream in suspension for greater distances during November–March, when rainfall and stream velocity are greatest, than during April–October, when rainfall and stream velocity are least. During transport the bacteria in sewage are subjected to other influences, exposure to light and low temperature, lack of nutriment, and eventually they die.

Table 4 shows the natural purification in a river receiving such a sewage effluent. The fall in the numbers of *E. coli* and in the proportion of samples containing salmonellae with distance cannot be ascribed to dilution alone. The intake of a water works is situated at the sampling point 7 miles below the entry of effluent. The water abstracted is subjected to full treatment, storage, softening, filtration and chlorination before delivery to the consumer. Salmonellae were isolated from 2 samples at this point during the wet months when stream velocity was greatest.

4. Salmonellae in an Estuary

In rivers the direction of flow is constant; downstream. In an estuary the direction of flow reverses 6-hourly with the tide. In a river, stream volume and stream velocity change slowly at well defined intervals. In an estuary the stream velocity is least at high and low water, and greatest around mid tide. In the 3 h after low water, the water velocity changes from nil to maximum, whilst the volume of water doubles. In the succeeding 3 h, the velocity falls from maximum to nil, whilst volume again doubles. It is thus apparent that the leisurely processes of self purification, other than by dilution and precipitation, seen in rivers and static fresh water can play little part in the natural purification of estuaries. The main processes involved would appear to be transport of particulate material during flow, with fallout at slack water. Transport of suspended matter seaward is effected by the prolongation of the outgoing tide which occurs in shallow waters. Survival of intestinal organisms in estuaries is of less importance than in rivers or streams as a given point below or above an outfall receives freshly polluted water every alternate 6-hourly period.

The estuary sampled receives the sewage of 350,000 persons, discharged continuously, without treatment other than removal of inorganic solids, through 2 outfalls 3 miles apart. The population served by the Western outfall is 250,000; by the Eastern it is 100,000. Water usage is 27×10^6 gall/day more or less equally divided between the Eastern and Western parts of the city: the lower domestic usage in the Eastern part being more than balanced by higher industrial usage. Salmonellae in the sewage discharged are derived, in addition to human sources, from abattoirs, tanneries, broiler processing plants, egg breaking plants and from plants processing human and animal food.

Samples were collected at the same time on the same days thrice weekly from the same sampling points. The distribution during 1970 of salmonellae in the crude sewage at the point of discharge from the Western outfall and at the sampling point 1 mile below the discharge shows 6% of the samples at the outfall and, 22% of the samples one mile below the outfall to contain no salmonellae/100 ml (Table 5). Thirtynine % of the samples at the outfall, and 4% 1 mile below the outfall contained salmonellae in each 4 ml of sample. The degree of dilution of the crude sewage, as far as salmonellae are concerned 1 mile below the sampling point, appears small, somewhere in the region of 10-fold. The actual dilution of the crude sewage is certainly many times greater, but bacteriologically this dilution is probably masked by the breakdown of particulate matter after discharge, releasing the contained bacteria.

Table 5

Distribution of salmonellae in crude sewage and in estuary in 1970

Salmonellae content	Result from			
	Crude sewage at point of discharge		Estuary 1 mile below discharge	
	No. samples	% of total	No. samples	% of total
Absent from 2·5 1	5	3	22	14
Present in 2·5 1				
Absent from 100 ml	4	3	12	8
MPN 1–5	24	16	84	54
6–10	10	7	13	8
Present in 100 ml				
MPN 11–23	48	32	19	12
>23	59	39	6	4
Total no. of samples	150		156	
Mean salmonellae content 100 ml	38·0		4·5	

Human salmonella infection is seasonal and is closely related to ambient temperatures (Table 6). Infections reappear in April, when ambient temperatures

Table 6

Seasonal distribution of known human salmonella infections in the population during 1960–70 from which untreated crude sewage was discharged continuously to estuary.

Year	Number of known excretors in												Total
	January	February	March	April	May	June	July	August	September	October	November	December	
1960	5	–	1	3	1	8	15	3	4	10	1	1	52
1961	2	–	–	17	3	30	7	2	5	1	1	–	68
1962	41	12	4	5	5	9	7	2	6	5	–	1	97
1963	–	–	5	5	13	5	18	4	6	3	1	7	67
1964	2	2	3	–	3	11	2	4	1	8	46	7	89
1965	2	5	5	3	7	1	2	6	2	–	3	2	38
1966	2	6	–	2	6	4	4	8	9	1	9	3	54
1967	–	1	7	5	1	1	9	14	2	7	4	4	55
1968	1	–	1	4	7	15	17	5	18	6	2	2	78
1969	23	4	1	1	13	4	15	99	21	3	–	4	188
1970	–	2	–	5	–	5	25	14	7	2	3	15	78

begin to rise after the winter; they reach a peak in June-July-August, when a more or less constant high ambient temperature is maintained; they then decline slowly in September and October as temperatures begin to fall and are at minimum in November–February or March. The pattern shows variation in that winter as outbreaks occur, as in January and February, 1962, when they were associated with the distribution of dried milk to schools for cooking, and as in November, 1964, and in January, 1969, when there were institutional incidents associated with faulty food preparation. In general, however, the incidence of sporadic cases and family infections conform closely to the seasonal pattern described.

This seasonal pattern is clearly seen in the monthly numbers of salmonellae present in the estuary (Table 7). The transition from low to high numbers is more smooth than the numbers of human infections occurring monthly would suggest, an effect resulting from the seasonal variation in the fresh water discharged to the estuary. The greater volume discharged in the winter months tends relatively to reduce the numbers of salmonellae present: the smaller volume discharged in the warmer months tends relatively to exaggerate the numbers. It will be appreciated that the warmer months, when the numbers of pathogens present in sewage-polluted natural waters are greatest, are associated also with bathing in sea and river water. During the months May–August, however, oysters are traditionally not eaten. This tradition at least is soundly based.

The general increase in the numbers of salmonellae present in the estuary waters in 1970 can be attributed to the establishment of plants processing meat and poultry products for human and animal food and discharging their effluents either directly into the estuary or indirectly into the general sewerage system.

Not all samples collected from the estuary contained salmonellae (Table 8). In the years 1960–70, salmonellae were not isolated from 38% of 1719 x 2·51 samples: salmonellae were present in numbers $> 10/1$ in $c.$ 25% of the samples.

The numbers of salmonellae in individual samples are small. Superimposed, however, on the seasonal variation in the numbers of salmonellae (Table 7) is the hourly variation according to the state of the tide (Table 9). In general the numbers tend to be greatest at or about half tide, when transport of suspended solids is greatest, and at low water, when dilution and transport are least. Conversely numbers tend to be least at and around high water when dilution is greatest and transport least.

The salmonellae recorded in Table 9 were isolated from 2 outfalls, at a sampling point roughly equidistant from East and West outfalls. Salmonellae isolated between high water and 5–7 h later at low water, are derived from the Western outfall and from the Eastern outfall between low water and $\geqslant 11$–12 h after high water. The numbers derived from the Eastern outfall were smaller, but show the general pattern of variation in numbers according to the time of sampling in the tide cycle. The smaller numbers isolated in samples from the

Table 7

Seasonal distribution of salmonellae in an estuary during 1960–70

Salmonellae (MPN/1) in

Year	January	February	March	April	May	June	July	August	September	October	November	December
1960	5	23	7	6	2	4	31	24	10	6	29	7
1961	13	8	11	38	30	54	20	25	2	13	8	3
1962	12	6	6	8	36	19	10	10	2	5	7	2
1963	3	1	4	2	15	33	35	14	23	19	5	6
1964	4	4	11	8	13	31	47	5	13	40	8	12
1965	7	5	5	10	3	14	7	7	5	11	26	9
1966	5	7	3	8	11	21	8	11	5	3	5	4
1967	6	6	15	15	3	2	6	7	9	4	3	10
1968	8	5	1	7	24	34	9	27	41	36	16	6
1969	7	7	7	27	34	25	48	119	105	61	60	17
1970	35	60	34	82	19	74	91	35	31	46	33	52

Table 8

Distribution of salmonellae in samples

Year	No. of samples with salmonellae (MPN/l) of												No. of samples in group	Mean MPN/l
	0(1)	5(2)	10	22	36	51	69	92	120	161	230	>230		
1960	65	44	26	8	5	1	3	3	—	1	—	2	158	13
1961	65	27	27	16	7	3	3	1	1	1	2	3	156	18
1962	78	31	30	6	5	2	1	—	—	—	3	—	156	10
1963	65	31	29	12	6	4	3	4	2	1	—	1	157	13
1964	49	47	22	19	5	5	4	3	1	—	1	1	157	16
1965	77	29	23	14	3	5	2	2	—	—	—	—	155	9
1966	72	37	25	19	2	1	—	1	—	1	—	—	157	7
1967	81	27	32	6	5	3	1	1	—	—	—	—	156	7
1968	54	28	28	24	6	5	4	3	2	1	2	—	157	18
1969	21	26	26	18	15	17	6	9	2	3	6	5	154	43
1970	21	14	34	23	18	8	5	7	11	4	5	6	156	48
Total no. samples	649	339	302	165	78	54	33	33	19	11	19	17	1719	
% all samples	39	20	17·5	9·5	4·5	3·0	2·0	2·0	1·0	0·5	1·0	1·0		

(1) Salmonellae not isolated from 2·5 1 sample: MPN/l, <0·4.

(2) Salmonellae isolated from 2·5 1 sample, but not isolated from 100 ml: MPN/l, 0·4–9·0

Table 9

Distribution of salmonella in the tide cycle in an estuary

Salmonellae (MPN/l) in samples taken (h) after high water

Year	0–1	1–2	2–3	3–4	4–5	5–6	6–7	7–8	8–9	9–10	10–11	11–12	>12
1960	5	5	16	10	23	25	2	25	5	2	7	5	5
1961	5	16	31	18	22	43	45	9	5	5	8	10	6
1962	4	16	3	22	6	11	6	33	10	3	10	10	—
1963	5	15	28	15	21	22	17	12	9	3	2	4	6
1964	2	2	6	16	20	24	29	49	9	41	7	9	6
1965	2	4	10	12	21	8	21	3	3	5	5	7	6
1966	4	4	5	6	4	11	15	4	5	3	7	7	7
1967	5	3	4	9	13	16	12	9	5	2	7	2	4
1968	4	10	7	17	19	39	60	6	16	16	12	9	3
1969	12	22	30	71	85	75	64	46	23	49	6	28	11
1970	11	47	45	111	85	80	41	31	15	42	13	29	19

Eastern outfall are attributed directly to the smaller population sewered as described above. As the estuary dries out to as much as 2 miles offshore at low water, sampling points are limited to easily accessible piers and quays fronting directly on to the deep water channel. It did not prove possible to sample in succession at increasing distances below the discharge of crude sewage as in the river described above. The estuary differs from the river in that the volume of water downstream of the discharge is not increased by tributary streams and that after initial dilution of the sewage on discharge, no further dilution occurs.

5. Sewage Pollution of Sea Water

In its turn, the investigations of the estuary serve as a model for assessing the degree of pollution of bathing beaches to whose waters crude sewage is continuously discharged by outfalls discharging at or below low water mark.

As no further dilution of crude sewage occurs after initial discharge, the only factor apart from the strength and volume of the sewage discharged determining the degree of pollution of sea water is the rate of the tidal currents. These vary in regular and predictable fashion within seas and between seas.

6. Risks to Human Health

The risk to human health arises directly from the ingestion of pathogenic organisms in sewage-polluted water, or from the consumption of foods contaminated by sewage or sewage-polluted water during growth or preparation. The risks are well documented and have been recognized, if not from time immemorial, at least from well before the advent of bacteriology. They include enteric fever, cholera and dysentery from the drinking of sewage-polluted water; enteric fevers transmitted by shellfish grown or stored in sewage-polluted water; fevers transmitted by vegetables fertilized with crude sewage, by raw milk contaminated directly or indirectly, by sewage-polluted water and by canned meats cooled after sterilization in sewage-polluted water. The methods of transmission are limited but recurrent.

One important point emerges from a study of these outbreaks: the majority have been traced to a single infected individual whose excreta have been discharged, not to the sewage system of a community, but to an individual cess pit, septic tank or stream.

In the sewage from a community the excreta of any individual are suspended in a very large volume of water: in the sewage from an individual cess pit, or septic tank, they are in a very small volume. In the latter circumstances, the numbers of pathogenic bacteria in the effluent may number many millions/ml.

In general the risk is least where sewage effluent after treatment, conforming to accepted standards, is discharged to natural waters providing sufficient dilution to allow the natural processes of purification to take place: it is greatest where crude sewage is discharged to natural waters of limited volume.

7. Further Reading

This paper is based mainly on earlier publications by the author (McCoy, 1962*a*, 1963, 1965). Reports of surveys in general and in particular can also be found in *Report* (1959) and *Report* (1970). The methods used for isolating salmonellae from crude sewage, sewage effluent and polluted natural waters are those described by McCoy (1962*b*) and McCoy & Spain (1969).

Basic information on tides in the seas and in estuaries around the British Isles is contained in the *Admiralty Manual of Tides* (1952) and in other Admiralty publications and charts such as those for the Channel Pilot, North Sea Pilot, West Coast of England Pilot and the Irish Coast Pilot, all of which are available from agents authorized to sell Admiralty charts.

8. References

Admiralty Manual of Tides (1952). London: H.M.S.O.

McCoy, J. H. (1962*a*). Salmonellae in crude sewage, sewage effluent, and sewage polluted natural waters. *Int. J. Air Wat. Pollut.* **7**, 597.

McCoy, J. H. (1962*b*). The isolation of salmonellae. *J. appl. Bact.* **25**, 213.

McCoy, J. H. (1963). River pollution. *J. Roy. Soc. Hlth* **83**, 154.

McCoy, J. H. (1965). Sewage pollution of rivers, estuaries and beaches. *Sanitarian* **74**, 79.

McCoy, J. H. & Spain, G. E. (1969). Bismuth sulphite media in the isolation of salmonellae. In *Isolation Methods for Microbiologists*, Soc. Appl. Bact. Tech. Series No. 3. Eds D. A. Shapton & G. W. Gould. London: Academic Press.

Report (1959). Sewage contamination of coastal bathing waters in England & Wales. *J. Hyg., Camb.* **57**, 435.

Report (1970). Taken for Granted. Working Party on Sewage Disposal. London: H.M.S.O.

The Scope of the Water Pollution Problem *

A. L. DOWNING

*Water Pollution Research Laboratory,
Stevenage, Herts, England*

CONTENTS

1. Introduction

IT WOULD be hard to overemphasize the importance of microbiology in relation to the problem of pollution. To make the point it would surely be necessary simply to mention that, on the one hand, among the most serious consequences of failure to control pollution, can be outbreaks of waterborne disease of microbial origin, such as those that afflicted the citizens of England in the 19th century, and yet on the other hand that the activities of micro-organisms play a major role in preventing the accumulation of pollutants in the environment to levels that would otherwise become intolerable.

In this paper attention is concentrated on water pollution because of the predominance in the Symposium of other papers devoted to that aspect, though brief reference is made to aspects of the problems of pollution of other media, where these are closely linked with problems of water pollution.

The subject could be dealt with in many ways but it seems easiest to establish a coherent theme by examining the influence of microbial activity in relation to the way in which water resources are exploited in the U.K. according to the pattern indicated in Fig. 1. Micro-organisms exert an influence at nearly every stage of the cycle, the events thought to be of particular importance being briefly itemized in Table 1. Broadly, these influences can be distinguished in terms of (a) those situations in which the release of micro-organisms to the

* Crown copyright: Reproduced by permission of H.M.S.O.

MAP–3   51

environment causes or aggravates pollution problems; (b) those in which the
release of other pollutants causes proliferation of natural microbial populations
which generate such problems and (c) those in which the activities of
micro-organisms reduce the problems of pollution. However, the distinction
between these situations at each stage is to some extent more a matter of degree
of importance than of absolute difference.

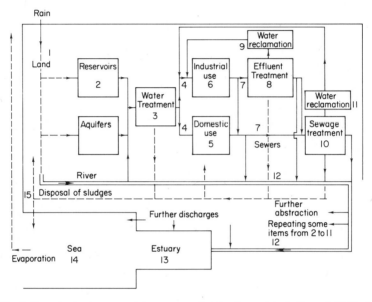

Fig. 1. Some basic elements of the water cycle. Numbers relate to items in Table 1.

Although the picture offered is a simplification of the actual situation it still
retains far more facets than could be explored in detail in the space available.
Some selection of topics has been made, therefore, which reflects both the
author's personal responsibilities and his subjective judgement of the problems
which most intrigue those directly concerned with pollution control at the
present time. It may also be useful to mention that it is assumed that an
adequate definition of pollution is to regard it as having occurred when human
activity, directly or indirectly, has caused natural waters to be less suitable for
any purpose required of them than they were in their natural state; also, by the
same token, that some degree of contamination of natural waters can be
accepted without interference with their use, and that the object of pollution
control is to try to find the most effective compromise between their utilization
as a means of safe disposal of the waste products of human activity and their use
for other purposes.

Table 1

Some of the principal microbial processes in various phases of the water cycle (numbers in first column refer to Fig. 1)

	Phase	Microbial processes occurring or activity and influencing microbial population
(1)	Passage of rain water from land to	Nitrogen fixation, biodegradation of nitrogenous organics, nitrification and denitrification. Bacterial reduction of sulphate and bacterial oxidation of sulphide and ferrous iron. Aerobic and anaerobic degradation of organic matter.
(2)	Storage in reservoirs	Removal of pathogens (predation, lysis). Bio-oxidation of organic and inorganic impurities many of which can create nuisance. Growth of algae. Anaerobic decomposition of organic matter in bottom strata; dissolution of iron and manganese.
(3)	Water treatment	Biological oxidation of impurities especially in slow sand filters. Elimination of pathogens by disinfection.
(4)	Distribution of treated water	Growth of organisms (especially bacteria) within water mains on adventitious impurity.
(5)	Domestic use of water	Intestinal bacteria and other organisms introduced, including pathogens in large numbers together with high concentrations of nutrient materials and some inhibitors.
(6)	Industrial use of water	Certain processes, for example industrial fermentations, introduce micro-organisms in considerable numbers; many introduce nutrients and many introduce growth inhibitors; the same situation also applies to the special case of the agricultural industry. Problems of transmission of pathogens are sometimes involved, especially in agriculture. In some cases where relatively polluted sources are used growth of microbial slimes on surfaces causes problems.
(7)	Transport of sewage and industrial effluents in drainage systems	Depletion of dissolved oxygen by bio-oxidation produces anaerobic conditions in slimes leading to release of sulphide. This is oxidized by bacteria when transported to aerobic regions with the production of sulphuric acid. (Injection of inhibitors to control sulphide formation has been tried but without much success.)

Table 1 (cont.)

(8) Treatment of industrial effluents	Aerobic processes involving mixed microbial populations are extensively used for a wide range of effluents including many in which the principal impurities have inhibitory properties. Anaerobic processes are used for treatment of certain strong organic wastes.
(9) Re-use of water in industry	Problems of slime formation sometimes occur, e.g. on the surface of cooling towers and heat exchangers. Inhibitors are used to control such growth. Disinfection is required in certain industries such as food manufacturing.
(10) Sewage treatment	Aerobic processes (as in 8) are used whenever substantial purification is required. Nitrification is often essential. Factors affecting the ability of micro-organisms to form flocculent aggregates are of profound importance in the activated sludge process and those affecting adherent films in biological filtration. Anaerobic digestion is a common process of sludge treatment and aerobic digestion is also practised at small works. Processes for removal of nitrogen by biological denitrification are under active development. In some countries this is accomplished to a degree by treatment in lagoons in which algae flourish.
(11) Reclamation of water from sewage	There is interest in several processes in addition to those mentioned under 10, in which microbial action can play a minor part including sand filtration and carbon filtration. Successful application of reverse osmosis depends on the prevention of slime formation and attack on membranes by cellulolytic bacteria. Disinfection is generally important. One promising process involves the use of ozone for oxidation of impurities and disinfection.
(12) Self purification in rivers	The events which occur include those listed under 2. Microbial processes in benthal deposits can exert a considerable effect on oxygen resources and may be important in determining the fate of heavy metals and other toxins. Formation of unsightly slimes often dominated by *Sphaerotilus* ('sewage fungus') is a problem in some rivers.
(13) Self purification in estuaries	The problems are similar to those in rivers except that sewage fungus is rarely much in evidence, the effluents discharged are often less well-purified than those released to rivers, and the activities of both marine and terrestrial organisms are important.

(14) Self purification in the sea	The situation is similar to that for estuaries except that effluents discharged customarily receive little pretreatment; mortality of sewage organisms is a feature of particular interest as also is the ability of marine organisms to decompose substances resistant to biodegradation such as chlorinated pesticides and PCBs, and slowly degradable material such as oil.
(15) Ultimate disposal of sludges and concentrated water liquor	Where sludges are used for agriculture, transmission of human and animal pathogens must be avoided. The influence of micro-organisms on the fate of potentially toxic materials in sludges dumped at sea requires further elucidation.

2. The Supply of Water

Detailed consideration of the problems of water supply is outside the scope of this paper. Broadly this aspect can be considered in relation to 4 main phases of the water cycle, namely those during passage of the water: (i) from the point at which it reaches the ground as rain to a reservoir, stream or aquifer; (ii) through these natural or manmade features; (iii) through water treatment plants; (iv) through distribution systems.

In phase (i) a variety of natural processes and human activities can be involved in which micro-organisms play a part. Of the natural processes which occur, among the most important are those governing the cycling of nitrogen (N) within the soil such as N fixation, bacterial decomposition of nitrogenous organic matter, nitrification and denitrification. These have an important effect on the content of inorganic N in waters percolating through the ground and, in catchments where effluent or solid waste disposal does not take place, result principally in an increase in the nitrate content of the water. In some types of ground microbial reactions result in the solution of iron and manganese and reduction of sulphate to sulphide; the converse process of oxidation of sulphide to sulphate also occurs in some situations.

Where effluents or solid wastes are disposed of on land, or where materials such as fertilizers and pesticides are applied to land by farmers, water percolating through, or running off, the ground can carry contaminating material with it. In some cases microbial activity may have adverse effects, as for example where contaminants were brought into solution that would otherwise have remained locked up in dumps. On the whole, however, micro-organisms have a beneficial effect in that they result in many organic impurities being degraded to innocuous products and in the elimination of some inorganic pollutants. A case particularly worthy of mention in which microbial activity tends to be detrimental is the problem of production of acid mine waters. This arises in disused mines when water infiltrating through them becomes contaminated with acid ferrous sulphate released from rock containing iron sulphide which has been exposed by the mining operations; the mechanism of the reactions is complex but both chemical and bacterial processes are involved.

In phase (ii), micro-organisms are involved in a range of processes of self purification in which many of such pollutants as may have entered the system are degraded to innocuous products. In upland catchments in areas of low population density the degree of pollution is normally minimal; in the lower reaches of rivers, however, it is not uncommon for a high proportion of the flow to consist of effluent, especially in dry weather. The self purification of natural waters is a topic of major importance, and this is dealt with later. One aspect of the role of micro-organisms that has captured much interest in the last few years concerns the growth of algae in surface waters, especially relatively static bodies such as lakes and impoundments.· Such growths can give rise to a variety of problems, including unsightly masses of rotting debris or an excessively accentuated green appearance of the water, damage to fisheries as a result of release of algal toxins or of deoxygenation, interference with processes of water treatment, and tainting of waters passing through to supply. The problems tend, however, to be sporadic, and the precise conditions leading to excessive algal growths are not well understood. There seems little doubt that in certain circumstances they can be stimulated by enrichment (eutrophication) of water by nutrient materials such as salts of inorganic N and phosphorus (P), which can be derived from many sources, including effluents and natural microbial processes occurring in the soil such as those to which reference has already been made. Much attention is being given determining the way in which excessive growths can best be avoided including the possibilities for biological control by the use of algal viruses. There is clearly considerable scope for microbiological investigations in this field.

With regard to phase (iii), it is perhaps sufficient to note that the performance of slow sand filters, much used for treatment of waters from lowland rivers, is greatly dependent on microbial processes, particularly those that occur in the Schmutzdecke at the surface of the filter; some biological activity also commonly occurs in rapid gravity sand filters.

With increasing necessity to exploit rivers subject to pollution as sources of supply, interest is growing in the opportunities for removing ammonia by encouraging the growth of nitrifying bacteria in sludge-blanket sedimentation tanks and in the possibilities of removing sulphate by bacterial reduction to sulphide followed by stripping with air.

Disinfection is, of course, a vital process in treatment of water for public supply and there is a wealth of information on the subject. Chlorination is by far the most common processes used in the U.K. but currently there is considerable interest in the scope for the wider use of alternative disinfectants, especially in situations where chlorine can cause problems; this occurs, for example, in the presence of phenolic contaminants which may be converted to chlorophenols, of which some can give rise to unpleasant tastes in low concentrations, or for significant concentrations of material with a high chlorine demand, such as ammonia.

Problems involving microbial processes also arise in distribution systems [phase (iv)]. One that has been troublesome in the past being the growth of bacteria on the greases applied to the flexible rings used for sealing pipe joints, though this is perhaps hardly a problem of pollution in the normal sense. The difficulty has been solved by the introduction of greases which do not support bacterial growth (Hutchinson, 1971).

3. Treatment and Disposal of Waste Waters and Sludges

Processes based on microbial activity form one of the chief means of control of the pollution resulting from the release to the environment of the waste waters arising from domestic and industrial (including agricultural) use of water. A large proportion of the 70×10^6 m^3/day (1500×10^6 gall/day) used in the U.K. by the domestic population, together with around 55×10^6 m^3/day (1200×10^6 gall/day) released to the public sewers from industry, after removal of coarse solids in preliminary stages including sedimentation, receives treatment at sewage works by one of the 2 main aerobic processes, biological filtration and the activated sludge process, or by a combination of the 2. A substantial proportion of the industrial effluent released to the sewerage system and a further 15×10^6 m^3/day discharged directly to natural water courses also receives treatment by these processes before release; there are in fact relatively few industries, from which the waste waters require purification before disposal, where biological processes have not been found useful as a method of treatment. Indeed it can be said that these processes provide the largest scale applications of controlled continuous culture in most developed countries. Essentially they depend on the utilization of pollutants in the waste water as nutrients for the growth of micro-organisms, though purely physical processes of adsorption and entrainment are also involved; the combination of these processes leads to the conversion of impurities into suspended matter which can be separated as a sludge. In the U.K. probably c. 3000 tonnes/day of organic and other biodegradable impurity, including perhaps 100 tonnes/day of NH_4^--N, are removed in such processes by microbial action with the production of c. 2000 tonnes/day of what is termed secondary sludge (to distinguish it from primary sludge which is obtained by settling the raw waste water before biological treatment); the dry mass of organisms maintained continually in the biological treatment plant is probably similar to the daily mass of organic matter removed.

Such processes are relatively simple to engineer and to operate and considering what is accomplished are reasonably cheap, at least as far as can be seen by comparison with the alternatives that would have to be used in their absence. They have undergone gradual evolution in the last 60 years during which many ways have been found of improving their efficiency and extending their range of application.

Most plants discharging to rivers in the U.K. are required to produce an effluent containing ⊁30 mg/1 of suspended matter and ⊁20 mg/1 BOD (5 days biochemical oxygen demand, a measure of the tendency of the effluent to take up dissolved oxygen from the water into which it is discharged). In the case of treatment of settled sewage of the strength normally found in the U.K. this usually means achieving removals of c. 90%. It is also customary for pollution control authorities to place specific restrictions on the discharge of pollutants with toxic properties, such as phenols, cyanides and heavy metals. Where several plants discharge to rivers used as sources for potable supply the concentration of NH_3 (as N) released must not exceed 10 mg/1 which in practice means maintaining an active nitrifying flora capable of oxidizing at least 75% of the incoming NH_4-N. Such plants are usually also required to limit the concentration of suspended solids and BOD in the effluent to $\not> 10$ mg/1, though this is usually achieved by the treatment of a 30:20 standard effluent by a "polishing" process depending more on physical removal of suspended matter than on biological action.

It is noteworthy that during sewage treatment removal of 90–99% of the coliform bacteria is achieved, as well as of those pathogens whose fate has been examined. Thus, although the treatment processes provide a useful barrier against release of pathogens it is far from complete.

A practical problem encountered at sewage works is that constituents of industrial effluents released to the sewerage system are sometimes highly toxic to the microbial populations on whose activity the treatment processes depend, and it is often necessary to restrict the release of such substances to tolerable levels. The ability of the mixed cultures which develop in plants to deal with such substances is, however, remarkable, especially when they are given opportunity to adapt to the presence of such materials.

This ability is often of vital importance also in the pretreatment of industrial effluents at factories and there are many examples of successful biological purification of waste waters containing materials which would normally be thought of as highly inhibitory to micro-organisms.

The detailed kinetics of the conventional processes are inevitably very complex, or apparently so, and much of the development that has taken place has been achieved empirically. The complexities arise partly because waste waters commonly contain many constituents whose identity and concentration are difficult to specify without much more elaborate analysis than would be possible on a routine basis, and partly because the flows involved are too large to consider providing the conditions of asepsis that would be necessary if one wished to use pure cultures of micro-organisms. Thus the populations which develop inevitably tend to be intricate mixtures of the wide range of organisms that usually gain access to the system, including bacteria, protozoa and to a less extent fungi, the detailed composition depending on the conditions.

It seems reasonable to think that if the events which occur were better understood it would be possible to improve process efficiency though it might be of course that the knowledge would merely confirm that the best practicable means of treatment have already been chosen as a result of empirical experimentation. Some encouragement for the view that improvements could be expected has been afforded by the practical advantages that followed from an investigation in the early 1960s which led to a better understanding of the factors of importance in determining the efficiency of removal of NH_4-N by nitrifying bacteria and the development of a rationale for the design of nitrifying plants. More recent work suggests that it may be feasible to extend this type of approach to elucidate the kinetics of removal of carbonaceous substrates, and indeed a hypothetical theory has be ,n partially successful in accounting for at least some observations of the removal of anionic surface active agents of the type used in household detergents. Interestingly enough it appears that it may be possible to improve the success of this approach, not by introducing additional complications, but by adopting a rather simpler hypothesis than the one originally proposed.

Still more recently, as is described by Pike & Curds (1971), in a later paper in this symposium, models have been developed for simulating the behaviour of hypothetical populations thought to have a basic similarity to those which develop in actual plants in that they include flocculating and nonflocculating bacteria and predatory protozoa both of a type assumed to remain attached to flocs and of a free-swimming type. Promising qualitative agreement has been observed between the predicted variations in the structure of these populations with operating conditions and in those observed in full scale plants.

(a) *Some problems*

In reviewing the topics to which it would seem profitable to direct attention in the future it needs to be borne in mind that the performance of the process will be judged not only by the standard of effluent achieved but also by the quantities of sludge produced and the ease with which it can be dewatered, because all of these factors have a bearing on overall costs. Moreover, in the activated sludge process the ease with which the sludge can be separated by gravity settling and recycled is of vital importance.

Essentially what is most needed to assess the potential of progress for further development is a better identification of the rate-limiting steps. The number of potentially useful lines of enquiry toward this end is considerable. Perhaps it is sufficient to note here that among them one can envisage investigation of the factors which determine the way in which micro-organisms bind together in flocs or slimes, the extent to which organization of organisms in such agglomerations affects their rate of metabolization of pollutants and the cell yield (which influences sludge production), the possibilities of modifying these rates in an economically feasible manner by altering the chemical or physical environment, the possibilities of finding organisms which could be maintained in the process in such a way as to achieve the treatment required more effectively than a "natural" population (which in turn requires development of methods of identifying and enumerating those that normally arise), and the possibilities of finding ways in which in the activated sludge type of process organisms can be separated from water more quickly and in more concentrated form than can readily be obtained at present.

Other lines of enquiry may need to be pursued to meet new requirements arising from changes in the pattern of pollution or in the exploitation of natural waters. In general it can be expected that the volume of waste waters will increase at roughly the same rate as the increase in water usage, currently 2–2·5%/annum, and this alone involves application of more restrictive standards in some areas if the present condition of natural waters is to be maintained. Undoubtedly, there will be changes in the nature of waste waters, especially perhaps in those from the chemical industry, the forecast growth of which is exceptionally large, and this may well present new problems, for example of devising methods of dealing with entirely new compounds; there may well also be requirements for removal from effluents of substances normally present now but not the subject of standards.

In connection with this last point it is already evident that there may be a future requirement for the removal of inorganic N and P from effluents in order, for example, to prevent excessive growth of algae in receiving waters and possible methods for achieving such removals during sewage treatment are already under study. There appears to be some promise in a biological process for removing inorganic N which depends on conversion of the N to oxidized forms followed by denitrification to N gas in an anaerobic reactor, to which it may be necessary to add degradable organic matter to accelerate the reaction sufficiently to permit a plant of economical size to be used. Finally one can envisage that, as effluent standards become more restrictive and new processes have to be used to attain them, it may be possible to optimize design in such a way as to impose entirely new conditions of operation on biological stages of treatment.

(b) *Sludge treatment and disposal*

Something of the order of 10^6 tons dry wt of sludge are produced annually in the U.K. at the works of local authorities alone. One of the most common methods used for treatment of such sludges is the process of thermophilic anerobic digestion which is carried out in heated digesters at $c.$ $35°$. The process, which serves $c.$ 40% of the population, renders sludges much less offensive and usually results in conversion of about half the organic matter present to CO_2 and methane, the latter being commonly used as fuel for generation of power in the larger installations. (The process is also used for the treatment of strong industrial effluents though the quality of effluent is usually too poor to permit discharge to inland waters without further treatment.) "Cold" digestion, i.e. at ambient temperatures, is also practised at small works.

Much has been learned empirically about the factors which determine process efficiency. A good deal is known of the broad characteristics of the fermentation process which takes place in 2 stages, the first involving conversion of organic matter to organic acids and the second the conversion of the acids to methane (CH_4) and CO_2. Just as in the aerobic processes, however, a full description of the events that occur has not yet been obtained, though some useful progress has been made in the formulation of approximate models. In one respect elucidation of the kinetics might be thought to be easier than in the case of aerobic systems because a narrower range of organisms might be expected to be involved in the anaerobic process; this potential advantage, however, may be more than offset by the greater difficulty in working with anaerobes. Possibly because of the dependence of the process on a narrow range of organisms with rather restricted abilities it appears more susceptible to calamitous failure than aerobic processes, especially as a result of inhibition of the activity of the CH_4-forming bacteria. Many types of substances produce such inhibition, among those having the most potent effect/unit of concentration being chlorinated hydrocarbons such as chloroform, methylene chloride and trichlorethylene commonly used as industrial solvents. The surface active agents used as proprietary household detergents are also inhibitory, but fortunately the concentration which has to be present in sewage to produce a sludge which will suppress gas formation completely c.20 mg/l, is only rarely exceeded in sewage in the U.K. Furthermore it has been found that by addition of a chemical known commercially as stearine amine (consisting principally of n-octadecylamine) in equimolecular proportion to the surface active agents present, with which it forms a stable complex, digestion can be fully restored in plants previously completely inhibited by the action of the surface active materials. Happily the development of this technique, and of the expertise necessary to deal with industrial inhibitors, and the general growth of experience of the susceptibility of the process to inhibition is helping to keep the problem in check.

About $\frac{1}{3}$ of the sludge produced in the U.K. is discharged by pipeline or dumped at sea. Of the remainder about one half is used as an agricultural fertilizer and the remainder is tipped on land or used as land fill. Use of raw sludge on land to be used for production of foodstuffs, especially crops to be eaten without cooking, is generally regarded as undesirable because of risk to public health. Treatment of sludge by digestion largely eliminates pathogens, however, and use of digested sludge on agricultural land is regarded as reasonably safe, though it is usual to seek veterinary advice when the land is to be employed for grazing livestock. Composting of sewage sludge with municipal refuse is practised by a number of local authorities and this process which involves development within the fermenting mixture at temperatures of 60-70° (140–160°F) also results in the elimination of many pathogens.

(c) *Water reclamation*

In many industries it has long been found worthwhile to re-use waste waters for factory processes, usually some degree of purification being required to facilitate the operation. No special comment on this practice is needed in the present context, other than to mention that disinfection is often required in the case of its use in food manufacturing industries. In some industries water reclaimed from sewage effluent is used industrially, and it is a common practice in such cases to disinfect the renovated effluent if there is any possibility of it coming into contact with the product, chlorine usually being relied on for such a purpose. It is of interest in this connection that one process that has been demonstrated in a large scale pilot plant results in the recovery of a sparklingly clear, "sterile" water, suitable for many industrial purposes, from highly purified sewage effluent, using a combination of microstraining, ozonation, sand filtration and chlorination. The action of ozone alone appears sufficient to secure disinfection but chlorine was included because this might well be necessary to eliminate contamination through a distribution system.

An aspect of particularly topical interest concerns the possibility of augmenting supplies of potable water from conventional sources by the renovation and re-use of sewage effluent. This is already being done at places, such as in South Africa, where acute shortage of fresh water has been hindering development (Vuuren et al., 1970). Naturally much consideration has been given to the possible hazards of transmission of disease and the fate of many pathogens, including viruses, during chlorination is being studied intensively. Some pathogens are undoubtedly more resistant to chlorination than are the coliform organisms on whose presence or absence the suitability of treated water for potable purposes is normally judged. Possibly, however, in the U.K. the case for or against re-use of sewage effluent for potable purposes might turn more on considerations other than risks from the presence of pathogens, including appreciation of the potential hazards from the presence of chemical impurities in the reclaimed water, especially where this is derived from a river draining an industrial area, of aesthetic factors, and of the relative costs of supplying water by other means.

4. Self Purification in Natural Waters

It has long been known that many pollutants when released to natural waters become subject to processes of self purification in most of which micro-organisms are at some stage involved. One of the most important of these processes is the biological oxidation of pollutants brought about by aerobic micro-organisms, much as in biological treatment plants but, because of the much smaller population densities involved, at far lower rates. Unfortunately, while these serve the beneficial purpose of removing impurities they can at the same time cause problems because the organisms involved utilize dissolved oxygen to meet their respiratory needs. This can result in the concentration of dissolved oxygen falling to levels inadequate for the survival of fish, or in the extreme case, when the water becomes completely deoxygenated, to the release of the poisonous and foul-smelling H_2S to the atmosphere by the action of anaerobic bacteria on sulphate and other sulphur-containing compounds.

Other aspects of importance concern the fate of micro-organisms, including pathogens, that might be released to natural waters and the stimulatory effects of pollutants on the growth of microbial populations whose presence can cause problems such as certain types of algae and organisms which accumulate on surfaces forming unsightly slimes.

The extent of such problems varies according to the type of natural water and it will be convenient to deal separately with the freshwater and marine environments.

In order to make optimum use of the resources of these waters it is important to know the extent of the self purifying capacity so that the amount of pollution that can be discharged without producing adverse effects can be determined, and much research has been and is being devoted to developing methods by which this can be done.

(a) *Fresh water*

Although the activities of micro-organisms play a determining role in self purification, the structure of the populations present in effluents and river waters has seemed so complex that with certain reservations it has been thought quite impracticable to attempt to forecast rates of removal of pollutants and consumption of dissolved oxygen by identifying and enumerating the individual species and coupling this with information about their characteristic behaviour. For routine work especially it has been necessary to rely on empirical methods, such as the familiar BOD test, which measures activity of the population present in gross terms. An assumption commonly made in making use of such data is that the variation in demand with time takes place according to a simple exponential curve which can be calculated from the 5 day BOD and the rate constant for the "reaction". In fact, experiments in respirometers show that, as one would expect, the actual course of the reaction is much more complex and depends a good deal on circumstances. It appears to coincide most closely with the simple exponential trend when the mixture incubated contains little nutrient and the population consists mainly of micro-organisms in the endogenous phase. This situation is most commonly encountered when well purified effluent from a biological treatment plant, or dilutions of such effluent, are tested.

One case in which it might be worthwhile in practice to attempt determination of the concentration of the bacterial species involved, concerns that of nitrification of NH_4-N in rivers. This is because only 2 types of organism, *Nitrosomonas* and *Nitrobacter*, are believed to be involved and because observations of the course of nitrification in incubated samples of water can be very much better described in terms of kinetics of growth of these organisms than of the simpler exponential type of relation commonly adopted in the past on the assumption that the rate of oxidation of NH_4-N was merely proportional to the concentration present.

As can no doubt be readily imagined predictions about the fate of ammonia and the effect of nitrification on the oxygen resources of rivers can be very different for the 2 types of kinetic model. If first order kinetics apply then the rate of removal of NH_4-N and dissolved oxygen would be expected to decrease directly as the concentration falls. On the other hand if bacterial growth kinetics apply, a fall in concentration of NH_4-N that would be produced as a result of development of a nitrifying population in the treatment plant might actually be associated with an increased rate of removal in the river as a result of a release to it of nitrifying organisms in the suspended matter remaining in the effluent (provided of course that the ammonia were not removed completely). Techniques have been devised for estimating the concentration of nitrifying bacteria in effluents from treatment plants, or at least their equivalent nitrifying ability, so that in principle it is possible to predict the course of nitrification in rivers assuming bacterial growth kinetics. Unfortunately insufficient work has been done in rivers themselves to determine whether the assumed model is correct. One requirement for this to be so would be that the organisms would remain in uniform suspension and would travel through the river channel at the same speed as the river itself. In practice deposition may occur or organisms may conceivably adhere to surfaces, producing local variations in rates of nitrification differing from those predicted. Further information on these aspects is now being urgently sought.

(i) Formation of slimes

Another effect of pollution which also seemed to offer scope for detailed microbiological examination concerns the formation in rivers of unsightly slimes, commonly known as sewage fungus. Such infestations can spoil amenity and for management purposes it is important to be able to forecast the extent of growth and the effect on the quality of water in the river that will result from any given degree- of pollution. Fortunately in the U.K. such outbreaks appear to be of limited extent. Investigations in collaboration with river authorities revealed the presence in 1968–9 of 178 outbreaks in rivers in England, Scotland and Wales, some 3/4 of these being confined to stretches < 800 m long (Curtis & Harrington, 1971). In about half *Sphaerotilus natans* was the dominant or co-dominant organism, the other most frequently found type of organism being zoogloeal bacteria of various species.

Study of the nutritional requirements of *Sphaerotilus* revealed that while it tended to grow best on certain soluble organic substrates of low M.W. a wide range of organic substrates would support its growth including the constituents of well purified effluents which probably consist mainly of cellular material and humic substances of large M.W. Only very low levels of inorganic N and P seemed necessary for growth. On the whole, growth in rivers correlated best with the BOD or concentration of soluble organic carbon in the river water. Where growths were higher than might have been expected from the levels of these 2 criteria it was frequently found that the water contained relatively easily degradable soluble organic compounds released in poorly treated effluents.

On the whole, however, it does not seem possible to specify the growth promoting potential of mixtures of effluents (and river waters) with quite sufficient certainty for fixing effluent standards from measurements solely of their chemical characteristics. A useful supplementary guide can be obtained, however, by observing the tendency of the mixture to produce slimes under standard conditions in a laboratory scale recirculating channel.

(ii) Contamination by pathogens

The fate of pathogens in natural waters is naturally of interest in relation to public health. The fact that pathogens can be recovered at times from effluents, and that most remain viable for a time in natural waters to which they are discharged, has often given rise to questions as to whether any advantage could be gained by disinfecting effluents. Leaving aside the question as to whether or not any risk of bathing in rivers is sufficiently different in those contaminated with effluent from those that are not to make such a practice worthwhile, it would suffer from any number of practical drawbacks including that:

(i) with effluents typical of those produced at conventional sewage works the dosage of chlorine required, and thus the costs involved, would be comparatively high, especially in the case of effluents containing substantial concentrations of ammonia;

(ii) because of the acute toxicity of chlorine to fish, facilities would have to be provided to remove any excess and the possibilities of accidental failures would need to be considered;

(iii) disinfection of effluents would not necessarily eliminate the possibility of contamination of the water with pathogens from other sources, whose growth (or survival) would conceivably be supported by organic matter not fully eliminated by chlorination;

(iv) elimination of nonpathogenic organisms might reduce the rate of those potentially useful processes of self purification which one might wish to encourage.

(b) Marine pollution

Because the volume of the oceans while immense is nevertheless finite, and because they represent the terminal point of water carriage systems of waste disposal, it is perhaps reasonable to question whether one should adopt a different attitude to them than to, say, rivers from which the polluting material tends to be continually flushed away. In the case of materials which are readily decomposed biologically or chemically to innocuous products there seems to be no obvious reason to depart from the belief that the oceans have a certain self purifying capacity, available for the use of society, and that, providing one has determined how far this capacity can be taken up without producing unacceptable side effects and arranged the system of disposal so that this is not exceeded, the situation would be perfectly satisfactory. With materials that are not biodegradable, practice in the past has been based on the belief that the dilution ultimately available is enormous, that huge quantities of most such materials are naturally present, and that the amounts in wastes released by man are far too small to alter the situation to a significant extent.

Opinions of course can differ about the levels of contamination that could be regarded as acceptable, and there has been a number of instances in which engineering design or method of operation of disposal schemes has been inadequate to prevent development of aesthetically unsatisfactory conditions in localized areas.

The practice of discharging sewage into the sea with, when compared with inland waters, relatively little treatment has provoked much discussion. The aspect which is perhaps most relevant in the present context concerns the issue that has often been debated as to the desirability or otherwise of the adoption of bacterial standards as is the practice in some countries. There are many reasons why this has not been done in the U.K., among the most powerful being the absence of any statistical evidence of health risk to bathers in situations where relatively straightforward engineering measures have been taken to ensure absence of gross visible pollution (Medical Research Council, 1959).

Similarly, while there is no doubt that marine organisms, such as shellfish, can become dangerous to eat as a result of accumulation of pathogenic organisms, adoption of practical safeguards such as locating outfalls at sites remote from shell fisheries and the cleansing of these organisms before they are sold in the open market, appears to have eliminated much of the risk. But the medical aspects of these types of problems are outside the scope of this paper.

From an engineering standpoint an argument against the adoption of bacterial standards has been the fact that the fate of micro-organisms derived from sewage is not yet sufficiently understood for it to be possible to design a scheme so that any proposed standard could be met with any given safety margin.

One of the objects of current research is to eliminate at least some of the areas of uncertainty, and if possible to develop methods of predicting the degree of bacterial contamination (or at least that due to coliforms) that result from discharging sewage into the sea at any given point after any given degree of pretreatment. Attention has been concentrated, though not exclusively, on coliforms, principally because they are present in large numbers and so can be detected relatively easily with high sensitivity, and also because there is a background of satisfactory traditional practice relating to their use as an indicator organism in assessment of the suitability of water for drinking. Progress in this field has been reviewed elsewhere (Gameson, Munro & Pike, 1970). It is perhaps sufficient to note here that an approach under examination is based on the premise that, apart from the rate of injection of organisms, the 2 main factors determining numbers from a particular source are their mortality in the sea and the rate at which they are dispersed by physical processes. Bacterial mortality is affected by many environmental factors including predatory or infective attack by other organisms. Certainly one of the most important is light intensity. Generally, numbers of bacteria in sea water under constant conditions decline roughly logarithmically and it is customary to express the rate of decline in terms of T_{90}, the time taken for them to fall by 90%. In bright sunlight in shallow water T_{90} for coliforms may be 1 h but in the dark the figure is increased to 2 or 3 days. The complexities of the problem have been such that attempts at prediction have been largely confined to the mean situation.

Some encouraging success has been achieved in the sense that it has been possible to show that predicted numbers of coliforms present at particular sites are of the same order as those observed when figures are assumed for average mortality rates reasonably consistent with those estimated from independent experiments. However it is still uncertain whether it will be possible to provide generalized relations from which numbers at any site can be forecast with acceptable accuracy for engineering design from relatively simple basic measurements.

In such research profitable use has been made of particular readily identifiable organisms, easily produced by techniques of mass culture, for examination both of bacterial mortality and physical dispersion. One such is *Serratia marcescens* and another the spores of *Bacillus subtilis*. The former serves as a tracer which appears to suffer mortality at roughly the same sort of rate as coliforms and can be useful in determining the proportion of contamination arising from particular sources when there is more than one in the area of interest, the latter is also useful in this connection and in addition since its rate of mortality is extremely low it affords an alternative to chemicals or radioisotopes as a means of estimating the maximum contamination that might arise if an organism did not suffer mortality.

Among a number of other microbiological aspects related to marine pollution on which further information is needed are the extent to which release of nutrient materials containing N and P may be responsible for the excessive growth of algae; the role of marine organisms in determining the fate of slowly degradable or conservative toxic substances (released in effluents or dumped as sludges) such as chlorinated pesticides, polychlorinated biphenyls and heavy metals such as cadmium, lead, and mercury and the scope for dealing with problems of oil pollution by utilizing the ability of certain organisms to metabolize oils.

The second category of problem seems especially likely to require increased attention in the immediate future, not least because there have been a number of events in the past few years indicative of the dangers of assuming that a scheme of disposal will necessarily be satisfactory if it simply ensures that no immediately overt deleterious effects are caused. The problems arising from the accumulation in higher species of the food web, of substances such as chlorinated pesticides have received world wide publicity and there is a growing awareness that somewhat similar difficulties may arise in regard to the other types of substance mentioned. The fate of mercury is of particular interest in the present context since it has been reported that even when released in the form of mercuric sulphide, one of the most insoluble substances known, it can nevertheless be converted by aerobic bacteria to the soluble methyl form which, as experience in Japan has shown, can be taken up by fish and render the flesh too dangerous for human consumption even when the degree of contamination is insufficient to produce obvious damage to the fish themselves.

5. References

Curtis, E. J. C., & Harrington, D. W. (1971). The occurrence of sewage fungus in rivers in the United Kingdom. *Wat. Res.* **5**, 281.

Gameson, A. L. H., Munro, D., & Pike, E. B. (1970). Effects of certain parameters on bacterial pollution at a coastal site. *Institute of Water Pollution Control, Symposium on Water Pollution in Coastal Areas* p. 34.

Hutchinson, M. (1971). The disinfection of new water mains. *Chemy Ind.* p. 139.

Medical Research Council (1959). Sewage contamination of bathing beaches in England and Wales. Memorandum No. 37. London: H.M.S.O.

Pike, E. B., & Curds, C. R. (1971). The microbial ecology of the activated sludge process. In *Microbial Aspects of Pollution,* Soc. appl. Bact. symposium. London: Academic Press.

Vuuren, L. R. J. van, Heren, Mr R., Stander, G. J. & Clayton, A. J. (1970). The full-scale reclamation of purified sewage effluent for the augmentation of the domestic supplies of the City of Windhoek. *5th International Water Pollution Research Conference, San Francisco.*

Microbial Aspects of Pollution in the Food and Dairy Industries

E. G. MULDER, J. ANTHEUNISSE AND W. H. J. CROMBACH

Laboratorium voor Microbiologie der Landbouwhogeschool,
Wageningen, The Netherlands

CONTENTS

1. Introduction

VARIOUS aspects of the microbial degradation of wastes from the food and dairy industries have been studied by the authors and their colleagues during the last 10 years. Investigations have been carried out with activated sludge grown under laboratory conditions, or obtained from oxidation ditches or conventional aeration tanks, and with suspensions and flocs of pure cultures of bacteria isolated from various types of activated sludge or from waste water.

In addition to the formation and the microbial composition of activated sludge we have considered its metabolic activity. In addition, studies have been made of the separation of sludge and effluent in connection with the phenomenon of bulking. In the present paper a survey is given of the main results of these investigations.

2. Microbial Composition of Dairy Waste Activated Sludge

The building up of a dairy waste activated sludge was studied by Adamse (1966, 1968a) using an oxidation ditch (Pasveer, 1960) and also on the laboratory scale. In both cases the experiment was started by aerating highly diluted dairy waste water (equivalent to a supply of 20 mg of BOD_5/l). This loading was gradually increased to 60 mg $BOD_5/l/5$ days after 2 weeks and to 250 mg/l/day after 8 weeks, when *c.* 4 g (dry wt) of sludge was present/l. In the laboratory experiments the stage of full operation had been reached when the load/g of sludge/day was *c.*60 mg BOD_5. Some vessels of the laboratory apparatus received the same substrate as the oxidation ditch, others received an artificial dairy waste, resembling as much as possible the average waste water of the oxidation ditch. This artificial waste water consisted of 1 g of a 1:3 mixture of skim milk powder and whey powder with a BOD_5 of *c.* 660 mg/g of waste dry material/l of distilled water. In the laboratory experiment, the substrate was added once a day after settling of the sludge and removal of the effluent. In the oxidation ditch the dairy waste water was added during a daily period of 8 h.

Neither the oxidation ditch nor the laboratory vessels was started with sludge from an existing activated sludge plant. Because the oxidation ditch had been built in sand, the microflora of soil suspensions and that of dairy waste must be considered as the inoculation material of the ditch. In some of the laboratory experiments the same inoculation material was used, in others a different soil suspension was introduced.

The oxidation ditch used by Adamse had a volume of 200 m^3. When in full operation it was daily supplied with 70 m^3 of fresh waste water which, after dilution with the 130 m^3 of purified waste water present in the ditch, gave a BOD_5 value of *c.* 250 mg/l, corresponding to *c.* 60 mg of BOD_5/g of sludge. Several characteristics of oxidation ditches differ from those of conventional aeration tanks (Table 1). Thus the sludge of oxidation ditches has a considerably

Table 1

Comparison of the activated sludge process in aeration tanks
and in oxidation ditches

Characteristic	Aeration tank	Oxidation ditch
Sludge content (g dry wt/l)	1 – 1.5	1
Load (mg BOD_5/g of sludge/day)	700	50
Approximate detention time of waste water	8h	3d
Approximate age of sludge (days)	2	40 – 60

lower loading rate than that of conventional oxidation tanks. The detention time of the waste water is much longer than that of an aeration tank, and this results in the growth of the sludge in an oxidation ditch being slow, the age of the sludge being high and the sludge production low.

To study the microbial composition of the dairy waste activated sludge at different stages of development, Adamse (1966, 1968*a*) sampled at 9, 28 and 56 days and made plate counts on tryptone-glucose-beef extract agar supplemented with skim milk. Approximately 1200 organisms were isolated and identified (Table 2), and the outstanding features in both systems were: (a) 9 days after starting the aeration of the dairy waste water, the majority of bacteria in the young flocs were pseudomonads and achromobacters, (b) some 3 weeks later, the relative numbers of pseudomonads and achromobacters had decreased, but strains of the corynebacteria had considerably increased, (c) this tendency was maintained during the subsequent 4 weeks so that in the fully grown sludge the coryneform bacteria were by far the dominating group, comprising 60–70% of the isolates.

Table 2

Micro-organisms isolated from developing activated sludge at different stages (Adamse, 1966)

Organism	Number (% of total) isolated from experiment in					
	oxidation ditch after (days)			laboratory experiment after (days)		
	9	28	56	9	28	56
Pseudomonads	30·5	20·0	15·4	41·3	11·4	5·1
Chromobacteria	19·1	–	–	1.0	–	–
Achromobacters	41·5	22·5	23·1	40·3	43·0	15·2
Corynebacteria	4·5	40·0	61·5	13·4	40·4	66·6
Others*	4·4	17·5	–	4·0	5·2	13·1

* This group includes enterobacteria, micrococci, brevibacteria, caulobacters yeasts and moulds.

A more detailed study of the coryneform bacteria by Adamse (1966, 1968a) revealed that the majority of the strains belonged to the genus *Arthrobacter,* a type of pleomorphic bacterium which display a growth cycle (Mulder et al., 1966). Coccoid coryneform bacteria, on transfer to a fresh nutrient medium, germinate with 1, 2 or 3 germination tubes, giving rise to more or less branched rods. Subsequently, relatively long irregular rods are formed, often showing "snapping" division (V formation). With increasing age of the culture, the rods become shorter and ultimately the cells are again in the coccus form (Plate 1). The last stage usually occurs during the stationary phase of growth. Although the coccoid cells are often thought to be resting forms ("arthrospores"), this is only partly true as they are metabolically very active. Washed cocci absorb and oxidize various substrates as readily as do rods. Multiplication and growth of coccoid cells is possible, and it is highly probable that in soil and in activated sludge, arthrobacters often occur as slowly growing cocci. This conclusion was confirmed by Adamse by direct microscopic observation of the dairy waste activated sludge which revealed the presence of large numbers of coccoid cells. Although no record has been made in the literature of the occurrence of arthrobacters as the predominating bacteria in dairy waste activated sludge, some authors have reported the isolation of micrococci and brevibacteria (Jasevicz & Porges, 1956; Porges, 1960). In activated sludge they morphologically resemble the coccoid cells of *Arthrobacter,* and they may be identical.

An extensive study by the present authors and their colleagues at the Laboratory of Microbiology revealed (Table 3) that the arthrobacters of dairy waste activated sludge are very similar to soil arthrobacters and clearly different from the cheese types (Mulder & Antheunisse, 1963; Mulder et al., 1966; Adamse, 1970; Kortstee, 1970; Antheunisse, 1971; Crombach, 1971).

Table 3

Characteristics of arthrobacters from different sources

Characteristic	% of strains tested giving positive results and, in parentheses, total number tested from			
	Soil	Dairy Waste	Cheese	
			Orange strains	Non-orange strains
Proteolytic activity	76(96)	89(71)	95(38)	20(51)
Growth with inorganic N	95(112)	100(71)	6(31)	33(51)
without vitamins	38	45	3	25
with vitamins	57	55	3	8
Requirements for organic N	5	0	80	63
without vitamins	1	0	13	22
with vitamins	4	0	68	41
Polysaccharides as reserve material	82(11)	88(71)	0(33)	0(13)
Uric acid decomposition	81(102)	90(10)	96(52)	15(51)
Growth with choline as C and N source	96(55)	96(52)	–	4(52)
Salt tolerance (8% NaCl)	0(13)	7(71)	100(8)	100(12)
DNA base composition (G + C%)	65–67(12)	64–67(3)	62·5–64(9)	65·5–67(10)

To imitate the shift in microflora of his dairy waste activated sludge, Adamse (1966) inoculated aerated and sterilized artificial waste water in a laboratory vessel with a mixture of 3 different strains of each of the 3 groups of isolates, respectively. There was a steady increase in number of the arthrobacters as compared with both other types of bacteria (Fig. 1) so that at the end of the experiment *c.* 80% of the flora consisted of arthrobacters.

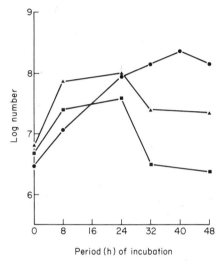

Fig. 1. Growths of *Arthrobacter, Achromobacter* and *Pseudomonas* sp. in mixed culture (Adamse, 1966). Closed circles, *Arthrobacter* sp.; closed triangles, *Achromobacter* sp.; closed squares, *Pseudomonas* sp.

The microbial composition of sludge samples taken during the assimilative phase (4 h after feeding) did not differ from that of sludge in the endogenous phase (4 h before feeding). This was the case both with sludge of 28 days and that of 56 days (Adamse, 1966).

3. Principles of Floc Formation

The oldest theory on the formation of activated sludge attributed an important role to certain slime-forming pseudomonads, known as *Zoogloea ramigera* (Butterfield, 1935). Subsequent authors were of the opinion that many types of bacteria, when occurring in the stationary phase, have a tendency to flocculate. This was thought to be due to the low energy level of the bacteria when the substrate is reaching exhaustion (McKinney & Weichlein, 1953; McKinney, 1962). More recently, several authors have again stressed the importance of zoogloea as a floc former (e.g. Dugan & Lundgren, 1960; Unz & Dondero, 1967). According to Crabtree *et al.* (1965), formation of the reserve material poly-β-hydroxybutyrate is responsible for the floc formation of activated sludge. Friedman *et al.* (1968, 1969) were able to dissolve bacterial flocs, which contained many fibrils, by treatment with cellulase, thus indicating that the fibrils are responsible for the aggregated growth of the bacteria and that they might consist of cellulose.

Recent investigations on floc formation in the authors' laboratory by Dienema & Zevenhuizen (1971) have involved testing bacterial strains, many isolated from activated sludge and some obtained from culture collections, for floc formation and slime production. Several of the slime-forming isolates were unable to flocculate during the exponential growth phase, and several of the floc-forming isolates produced no extra cellular slime or poly-β-hydroxy-butyrate.

Electromicroscope investigations of the flocculated bacteria revealed that many of them possess fibrils which connect cells (Plate 2*a*), and X-ray, I.R. and chemical analyses showed them to be cellulosic in nature. Addition of cellulase dissolved the cellulose as a result of which the cells lost their cohesion and went into suspension. In a floc the cellulose content is relatively low, so that the weight of the flocs is nearly the same as that of the flocculated cells.

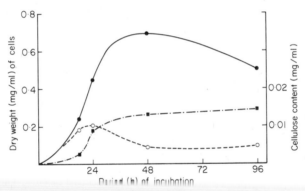

Fig. 2. Growth of *Pseudomonas* L-8 at 30° in casitone glucose medium (after Deinema & Zevenhuizen, 1971). Dry weight of bacteria in flocs, closed circles; in suspension, open circles; cellulose content of the culture, closed squares.

Floc formation due to cellulose fibrils starts at an early stage of the exponential growth phase, but there is always a proportion of cells in suspended form (Fig. 2). No data are yet available concerning the characteristics of suspended and flocculated cells, but when the flocs were separated by centrifugation and resuspended in a fresh nutrient medium, some of the cells separated from the flocs and resumed the growth of suspended cells. A possible way of deciding whether flocculated cells grow as fast as those in suspension would be to add ciliates which should consume the suspended cells only.

So far, it is unknown what part cellulose-forming bacteria play in the

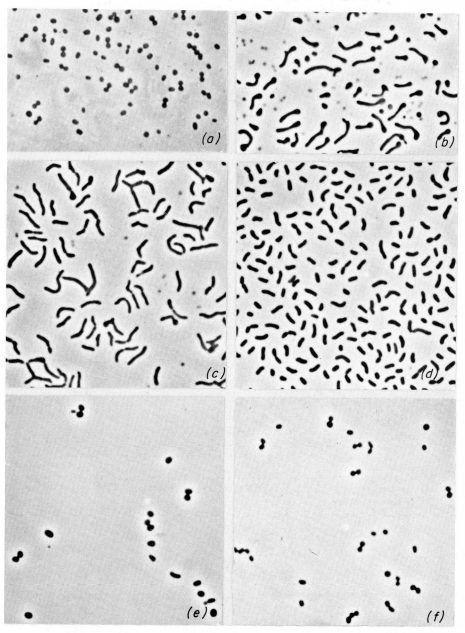

Plate 1. Life cycle of *Arthrobacter* strain AC 793. The culture was isolated from dairy waste activated sludge, grown for different periods of time on yeast extract-glucose agar at 25°. (× 1625)

(*a*), Cocci used for inoculations; (*b*), germinating cocci after 6h; (*c*), 12h; (*d*), 24h; (*e*), 4 days; (*f*), 7 days. (× 1625)

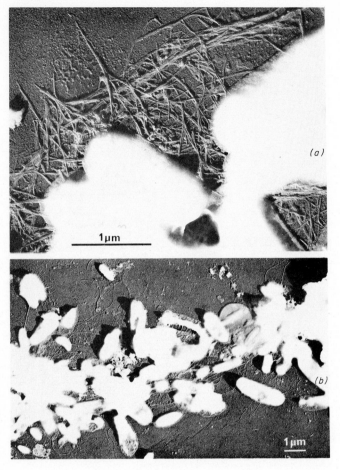

Plate 2. Shadowed preparations of bacterial flocs.
(a) *Pseudomonas* strain X-3; cellulose fibrils in small bundles, seldom single; (b) Fibrils in activated sludge.

Plate 3. *Sphaerotilus natans.*
 (*a*) Strain 3 in basal salts, solution + glucose and peptone. (× 1625)
 (*b*) In bulking activated sludge. (× 650)

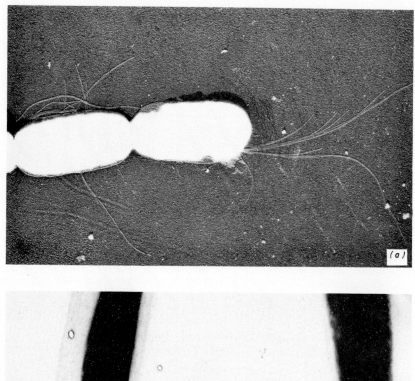

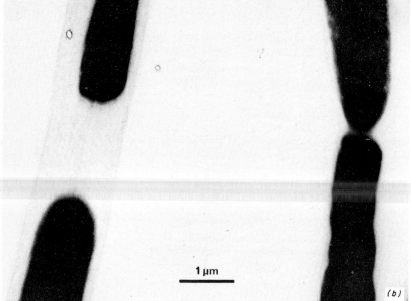

1 µm

Plate 4. *Sphaerotilus natans.*
  (*a*) Strain 7, showing intertwined bundles of subpolar flagella, shadowed with palladium. (× 15,000)
  (*b*) Strain 52, rods within sheaths. (× 15,000)

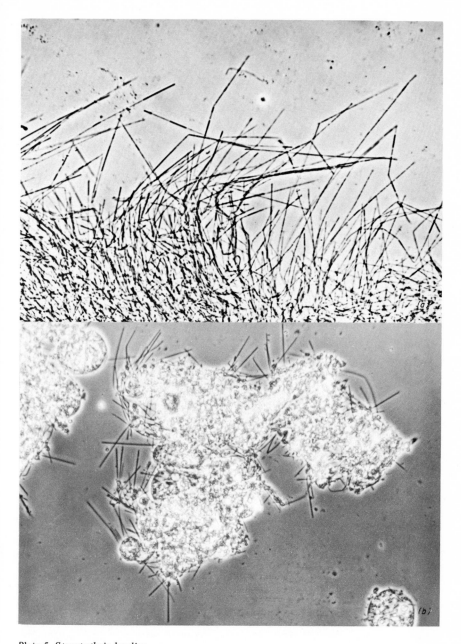

Plate 5. *Streptothrix hyalina.*
(*a*) Pure culture in a synthetic medium. (× 1625); (*b*) In bulking activated sludge. (× 650)

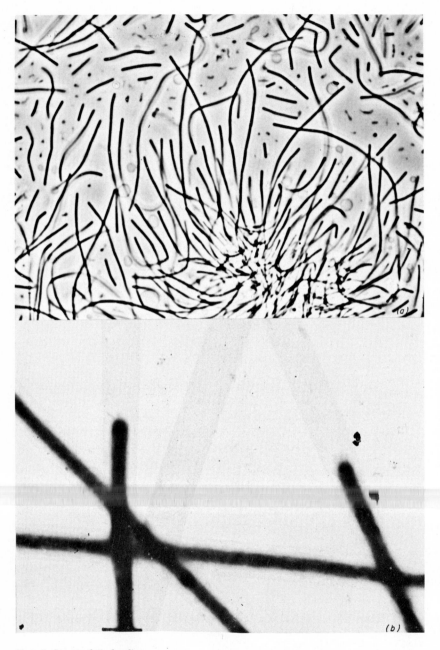

Plate 6. *Streptothrix hyalina.*
   (*a*) Growth on a starch-glutamate—soil extract agar. (× 1650)
   (*b*) Rods within sheaths. (× 15,000)

formation of activated sludge under natural conditions. Electronmicrographs revealed a certain number of cells with fibrils (Plate 2b). However, addition of cellulase only had a slight effect on the structure of the flocs.

4. Metabolic Activities of Activated Sludge

If it is assumed that activated sludge consists of 10–25% by weight of living bacteria (this value may vary considerably for different sludges), it will be easily understood that the substrate supply is far too low to enable exponential growth to take place of more than a small portion of the organisms present. This is not only true of oxidation ditches (50 mg BOD_5/g of sludge dry wt/day) but also of conventional aeration tanks with a 10–15 fold higher load. This means that the growth rate of sludge organisms generally is low and often approaching zero. However, the ability of these organisms to absorb and to oxidize substrates and to synthesize and accumulate reserve materials is often very pronounced. Adamse (1966) showed this for dairy waste activated sludge supplied with artificial dairy waste (containing lactose and casein in the ratio 7:2).

The large proportion of arthrobacters found by Adamse in dairy waste activated sludge may fit well with the tendency of this type of sludge to synthesize large amounts of polysaccharides. When present in the coccoid form, arthrobacters grow slowly or not at all. However, they have a marked ability to synthesize and accumulate polysaccharides (Mulder & Antheunisse, 1963), and Zevenhuizen (1966a) showed the intracellular polysaccharide of arthrobacters to be glycogen. Adamse (1966) grew one representative strain of each of his 3 main groups of isolated sludge bacteria in shaken culture in a nutrient solution containing 1% of lactose and 0·4% of casitone and after 48 h at 25° the *Arthrobacter* cells contained 52% of carbohydrate, the *Achromobacter* cells 8% and the *Pseudomonas* cells 13%. By means of Warburg experiments, van Gils (1964) in our laboratory has shown that synthesis of large amounts of intracellular polysaccharides from added glucose occurs in many activated sludges grown on domestic sewage. Owing to the ready synthesis of polysaccharides, a much lower proportion of glucose was oxidized than would have been the case if protein and nucleic acids had been formed, and this was reflected in the much lower oxygen uptake (Fig. 3). In these experiments the domestic sludge had to be pre-cultivated in the laboratory in the presence of glucose and ammonium sulphate: acetate did not need this pretreatment.

The synthesis and accumulation of carbonaceous reserve materials, like polysaccharides, by suspended *Arthrobacter* cells are promoted by adverse growth conditions (Zevenhuizen, 1966a, b). This is particularly true of nitrogen deficiency, as nitrogen is required in large amounts for the synthesis of protein and nucleic acids.

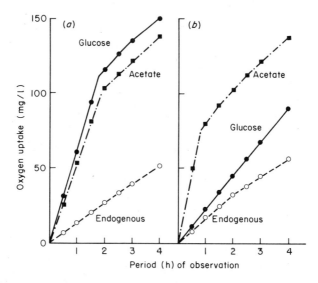

Fig. 3. Oxygen uptake of washed suspensions of activated sludge, pre-cultivated in media with glucose (*a*) or sodium acetate (*b*). The sludge content in (*a*) was 1320 mg/l and in (*b*) 1620 mg/l (dry wt). The concentration of added glucose was 470 mg/l and of sodium acetate 530 mg/l in both experiments.

The synthesis of polysaccharides by suspensions of arthrobacters or by dairy waste activated sludge when supplied with artificial dairy waste, may be because sugar is taken up by the bacteria more readily than protein which must first be degraded to molecules of smaller size such as amino acids and ammonia.

The addition of nitrogen in the form of $(NH_4)_2SO_4$, however, gave only a small increase in the dissimilation level of glucose (Fig. 4), 10% of the amount taken up the sludge being oxidized. The small increase in O_2 uptake brought about by adding $(NH_4)_2SO_4$, particularly at the highest glucose concentration used, was the result of the utilization of some of the carbohydrate in synthesizing nitrogen-containing cell components. The fact that even in the presence of adequate amounts of available nitrogen, a large part of the added glucose was synthesized to reserve carbohydrates demonstrates that the tendency to form new cells in this sludge was only slight. This may have been due to a particular bacterial type being the predominating organism of the floc or to particular nutritional conditions or it may be a general characteristic of activated sludge.

That nitrogen added to bacteria with a high content of intracellular reserve material under certain conditions may cause a ready utilization of this material was shown by Zevenhuizen (1966*a, b*) in an experiment with *Arthrobacter globiformis* (Fig. 5). This organism was grown in a medium with a relatively high C/N ratio resulting in cells with a carbohydrate content of *c*. 70%.

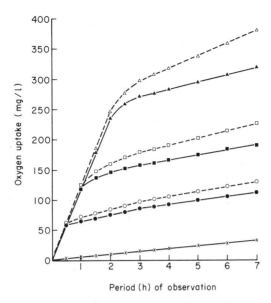

Period (h) of observation

Fig. 4. Effect of glucose concentration and $(NH_4)_2SO_4$ on the oxygen consumption of washed suspensions of activated sludge, grown in a basal salts medium with $(NH_4)_2SO_4$ and glucose (van Gils, 1964). The activated sludge content was 2540 mg/l (dry wt). Closed symbols, with glucose alone; open symbols, with glucose and $(NH_4)_2SO_4$ (300 mg/l) added. Crosses, endogenous respiration; circles, with 600 mg/l; squares, with 1200 mg/l glucose; triangles, with 2400 mg/l.

Transfer of the collected and washed bacteria to a basal salts medium containing 0·25% of $(NH_4)_2SO_4$, but no organic compound, brought about a ready drop in glycogen content but resulted in a synthesis of protein and other organic nitrogenous compounds. This gave an initial increase in the numbers of viable arthrobacter cells, which, on further aeration, declined significantly due to the lack of endogenous substrate. A control culture without added nitrogen decreased only slowly in glycogen content. The cells were not deprived of intracellular substrate which supplied energy of maintenance so that most of them survived the aeration period of 2 weeks.

5. Bulking of Activated Sludge

One of the most important aspects of the activated sludge process is the sedimentation of the sludge solids. It follows the period of agitation and aeration during which the substrates of the waste water are being removed by the sludge organisms. The sedimentation should be completed in a relatively short period of time so that sludge and effluent can be easily separated.

The parameter generally used for describing the settling ability of activated sludge is the sludge volume index (SVI). This is the sludge volume (ml) of 1 g

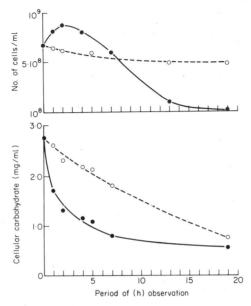

Fig. 5. Number of viable cells and intracellular carbohydrate of a washed suspension of *Arthrobacter globiformis,* aerated with and without nitrogen compound. Open circles, no added nitrogen; closed circles, 2.5 g/l $(NH_4)_2SO_4$ added.

dry wt of sludge after a settling period of 30 min. Sludge with a good settling ability should have a SVI < 150. When the SVI is high, for instance 500 or 800, settling is poor and this is called "bulking". The effluent may be clear but the volume is much too small.

Bulking is a microbial phenomenon. Although slime-forming bacteria have been stated to be involved (Heukelekian & Weisberg, 1956), filamentous bacteria are the main cause. Until the last few years, *Sphaerotilus natans* has been thought to be largely responsible for the poor settling, and indeed the predominant occurrence of this organism often coincides with bulking. However, a number of other filamentous micro-organisms may be found in activated sludge and in some cases, when occurring in large quantities, they also bring about bulking. Micro-organisms causing bulking have been reviewed by Pipes (1967, 1969), Hünerberg *et al.* (1970) and Farquhar & Boyle (1971).

An extensive study of the subject has also been made in the authors' laboratory. As far as the aeration tanks of the conventional type and of oxidation ditches in the Netherlands are concerned, 2 types of bulking can be distinguished, viz. (1) a *Sphaerotilus natans* type in overloaded sludge which may be found in aeration tanks and sometimes in oxidation ditches, and (b) a non-*Sphaerotilus* type which is very often encountered in underloaded oxidation ditches. In the latter, which from the nutritional point of view is quite different from the former, various micro-organisms have been found to be responsible for bulking. In addition to the sheath-forming *Streptothrix hyalina*, a number of Gram positive as well as Gram negative, non-sheath forming, filamentous bacteria have been studied by phase contrast microscopy and by electron microscopy. Some of them have been isolated by a procedure for selecting filamentous bacteria (Eikelboom & van Veen, unpublished). Pasveer's (1969) filamentous coliform organism might be another type of organism causing poorly settling sludge in underloaded oxidation ditches.

Sphaerotilus natans is a large Gram negative bacterium (1·2—2·4 $\times$ 3—10 μm) occurring in chains within a sheath of uniform width (Plates 3*a* and 4*b*). Single cells outside the sheath are motile by means of a bundle of subpolar flagella which are often intertwined and give the impression of a single large flagellum (Plate 4*a*). Under natural conditions, "false branching" is a common feature. The organism is adapted to the growth conditions occurring in overloaded activated sludge. It grows relatively readily, is able to decompose large numbers of sugars and other organic compounds, and synthesizes considerable amounts of cell material. It can form and accumulate large amounts of reserve material like poly-β-hydroxybutyrate and glycogen (Mulder & van Veen, 1963; Mulder, 1964). *Sphaerotilus* can also assimilate inorganic nitrogenous compounds, but it grows better with nitrogen in an organic form such as glutamic or aspartic acid. It requires vitamin B_{12} which can be replaced by methionine.

Although it is an aerobic micro-organism, *Sph. natans* is able to grow at low oxygen concentrations. This is undoubtedly one of the reasons for its occurrence in large quantities in overloaded sludge, where the oxygen supply cannot meet the oxygen demand. Within the flocs, *Sphaerotilus* is found in masses of long filaments which apparently prevent the settling of the sludge to a compact sediment (Plate 3*b*).

In a number of laboratory experiments with overloaded bulking activated sludge, containing excessive amounts of *Sph. natans,* elimination of this organism was achieved by altering the continuous feeding system into a discontinuous one (shock loading). These experiments were carried out by Kerstens and by Berns (unpublished). Similar results have been obtained with an overloaded oxidation ditch (R.A.A.D., 1966). In both instances the total amount of substrate supplied was the same. Shock loading undoubtedly has an adverse effect on the dissolved oxygen content of the sludge suspension during the feeding phase (e.g. Adamse, 1966, 1968*b,c*) (Fig. 6) and so might favour the development of *Sph. natans* in the sludge. However, the prolonged period of decreased substrate supply (endogenous phase) which in the above mentioned experiments followed the short period of excessive feeding, apparently affected the non-filamentous bacteria more beneficially than *Sph. natans.* The latter organism metabolizes the reserve material accumulated during the feeding

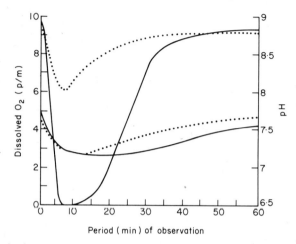

Period (min) of observation

Fig. 6. Dissolved oxygen and pH value of activated sludge suspensions at 25° after adding artificial dairy waste (powder mixture) (Adamse, 1966). Upper curves, dissolved oxygen; lower curves, pH value. Broken lines, 0·4 g/l and solid lines, 0·75 g/l of powder mixture added.

period, presumably more readily and much less economically than do the non-filamentous bacteria like *Arthrobacter* (Mulder, 1964) and as a result *Sph. natans* is gradually overgrown.

One of the filamentous bacteria of the non-*Sphaerotilus* type, *Streptothrix hyalina,* has been studied in the authors' laboratory by van Veen, van der Kooij Geuze, & van der Vlies (1971). It has thin, Gram negative rods of considerable length (0·35 − 0·45 x 3·2 − 5·0 μm) which occur in chains within hardly visible hyaline sheaths 0·5 − 0·8 μm wide (Plates 5a, 6a, b). Free cells outside the sheaths were only sporadically observed, and no motility or flagella formation was evident (Mulder & van Veen, in press).

The growth rate of pure cultures is considerably lower than that of *Sphaerotilus.* It utilizes sugars as the carbon and energy source, but synthesizes and accumulates no poly-β-hydroxybutyrate. Inorganic nitrogenous compounds may serve as the nitrogen source, but the organism prefers organic compounds like glutamate or peptone. In addition to vitamin B_{12}, thiamine is required as a vitamin.

In bulked underloaded activated sludge from oxidation ditches, many straight, sheathed chains of bacteria projecting from the flocs may be seen (Plate 5b).

The development of *Streptothrix* and other non-*Sphaerotilus*, non-sheath forming, filamentous bacteria in this type of sludge is thought to be due to the low concentration of nutrients. This reduces the growth of non-filamentous sludge organisms more seriously than that of the filamentous ones which, because of their relatively large surface areas, are in a better position to absorb nutrients from highly diluted solutions. Evidence for this hypothesis is provided by the fact that increased loading of the sludge, or shock loading, which temporarily increases the supply of substrate, may eliminate this type of bulking. This is apparently due to the fact that non-filamentous sludge organisms are more favourably affected than the filamentous types which gradually become overgrown by the former.

There is an indication that *Streptothrix* grows more readily in sludge derived from proteinaceous wastes.

6. Waste Water of the Potato Flour and Straw Board Industries

The production of potato flour and of straw board in The Netherlands is concentrated in the north-eastern part of the country. In the last few years the production of potato starch and of secondary products made from starch has risen considerably, due to a larger proportion of the area of reclaimed peat soil being used for the growing of potatoes. The production of board from straw is constantly declining because the straw is used for other purposes.

The potato mills handle $>2 \times 10^6$ tons of potatoes annually. When only potato flour is produced, *c.* 20,000 tons of protein, 20,000 tons of amino acids and other soluble organic nitrogenous compounds, 6000 tons of sugar and 20,000 tons of K_2HPO_4 are discarded as waste products (Peters, 1960). The protein may be precipitated by heat and then collected and sold as cattle fodder. This is at present being done with only a minor portion of the protein available. Removal of the nonprotein, soluble, organic compounds by the use of yeasts or of activated sludge in an oxidation ditch has been tested in pilot plants; it has never been put into practice.

Most of the waste of these agricultural industries is disposed into a system of canals which contain stagnant water. Here it causes an enormous pollution, estimated as equivalent to a population of 20×10^6, during the autumn months when the processing of the potatoes is in progress. Elimination of the wastes takes place partly by anaerobic microbial processes in the canals, giving rise to an offensive smell, and partly by aerobic digestion in the estuary of the river Eems, where the excess canal water is discharged during the wet late autumn and winter months.

Plans have been made for a more direct disposal of these wastes to the Eems estuary by means of a system of pipe lines. There is much opposition to this project by several biologists who fear that the supply of such large amounts of organic wastes may spoil the water quality of the estuary which is unique for its

MAP—4

natural life. Removal of the protein and possibly the amino acids and sugars by the factories, in combination with a direct disposal of the remaining wastes to the Eems estuary, would probably meet the demands of the population of the industrial areas, of the biologists and of those people who think that discharge of such large amounts of valuable nutrients would be inadmissible.

Eggink (1965) has investigated extensively the pollution of the canals and of the Eems estuary by the wastes of the potato flour and straw board industries. Through studies of the relationship between the existing polluting load and the distribution of BOD and oxygen content of the estuary for >5 years, he was able to calculate the future oxygen content, assuming that all the wastes of the agricultural industries would be discharged to the estuary by means of the planned pipe line. It should be stressed, however, that after his calculations were made the area cropped with potatoes has considerably increased, so that the total yield of potatoes and also the amount of wastes have markedly risen.

Eggink (1965) also studied the biochemical oxygen consumption of different types of waste water of the agricultural industries under different conditions. This was done in 2 ways: (a) using the BOD dilution test and (b) using a modified Sierp apparatus in which the oxygen consumption was measured manometrically, after absorption of CO_2.

Figure 7 illustrates a typical oxygen consumption curve of a synthetic waste

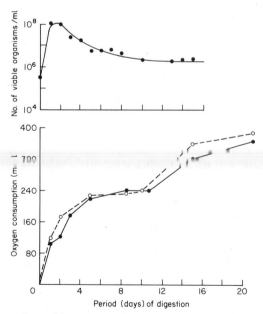

Fig. 7. The numbers of bacteria and the oxygen consumption in a BOD dilution test in a synthetic waste water inoculated with sewage. Closed circles, sewage used immediately after filtration; open circles, sewage used after a pre-aeration period of 48h (Eggink, 1965).

water containing 150 mg of glucose + 150 mg of glutamic acid/l of buffered basal salts medium when inoculated with a small amount of filtered sewage. The oxygen consumption curve had 2 breaks, one after *c*.1 day, and the second after *c*. 10 days.

The first pause, which was even more clearly obtained with the Sierp apparatus (Fig. 8), is known in the literature as the plateau of Busch (Busch, 1958; McWhorter & Heukelekian, 1964; Gaudy *et al.*, 1965). This plateau coincided with the maximum bacterial count, as shown by Eggink (1965). This pause in oxygen uptake was undoubtedly due to exhaustion of one of the substrates, viz. glucose. During this pause the microflora apparently became adapted to a second substrate which was subsequently oxidized at about the same rate. Whether this second substrate was glutamic acid or cell material from dead bacteria is unknown. The fact that the resumed oxygen uptake coincided with a fall in the numbers of viable bacteria suggests that dead bacteria were involved. This is in agreement with the observation that no plateau of Busch occurred when the filtered sewage, which was used as inoculation material, had been pre-incubated for 1 day. During this period many protozoa had developed and they were presumably responsible for eliminating the first pause in oxygen uptake. Further evidence of oxidation of cell material from dead bacteria being the cause of the resumed oxygen uptake after the first pause may be derived from similar investigations by McWhorter & Heukelekian (1964) and by Gaudy *et al.* (1965) in which the plateau of Busch was shown to occur when only a single compound had been supplied as the substrate.

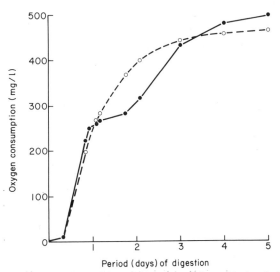

Period (days) of digestion

Fig. 8. Effect of pre-aeration of inoculation material on biochemical oxygen consumption of synthetic waste water (Eggink, 1965). Closed circles, no pre-aeration; open circles, inoculated material pre-aerated for 24h.

The second resumption of the oxygen uptake after a prolonged pause, *c.* 2 weeks after starting the experiment (Fig. 7), undoubtedly was due to nitrification. This may be concluded from the fact that the retarded rise in oxygen uptake did not occur when the Sierp apparatus was used. Here, inadequate amounts of CO_2 were available for the autotrophic nitrifiers.

The delay in development of nitrifying bacteria, which was also observed in decaying potato flour-processing waste water, and which has also been reported in the literature (e.g. Leclerc, 1960) was apparently due to a toxic effect of traces of organic compounds on the development of these bacteria. Evidence concerning this assumption derives from a comparison between the oxygen uptake by fresh and by decaying potato flour-processing waste water (Fig. 9). Although in both cases an inoculation with a small amount of sewage had been applied, nitrification in the decaying waste water started much earlier than in the fresh waste water, apparently as a result of an earlier elimination of the organic waste products due to the presence of an adapted microflora.

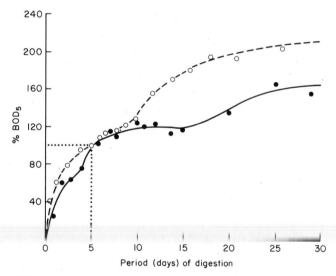

Fig. 9. BOD curves of fresh (closed circles) and decaying (open circles) potato flour-processing waste water (Eggink, 1965).

The retarding effect of organic wastes on nitrification, as reported above, is more or less in contrast with the results obtained with certain activated sludge systems in which large amounts of nitrogen may be driven off by a combination of nitrification and denitrification. It is not known whether this discrepancy depends on differences in types of nitrifying organisms or on differences in type of organic compounds involved.

The oxygen consumption due to nitrification constitutes an important portion of the amount of oxygen required for the complete oxidation of proteinaceous material like potato flour-processing wastes as can be seen from Fig. 9. It depends on the large amount of oxygen required for nitrification (2 moles of oxygen for 1 mole of ammonia). In contrast with the aerobic breakdown of organic substrates, of which $> 40\%$ may be used for the synthesis of cellular components, nitrifying bacteria utilize only a very small part of their ammonia substrate for cell synthesis. This means that BOD during the growing phase of the autotrophic bacteria is nearly equal to total oxygen demand for the oxidation of ammonia, whereas in the case of organic substrates there is a wide gap between the BOD during the growth phase of the heterotrophic bacteria and the amount of oxygen required for complete oxidation of the substrates.

As a result of the high oxygen demand for nitrification, the BOD_{30} of fresh potato flour wastewater was $c.160\%$ of the BOD_5, and the BOD_{30} of decaying potato wastewater was $>200\%$ of BOD_5 of the same wastewater (Fig. 9). The higher value of the latter was because part of the organic wastes already had been eliminated when the experiment started.

7. Acknowledgments

We are much indebted to Mr H. G. Elerie of the Technical and Physical Engineering Research Service at Wageningen for taking the electron micrographs.

8. References

Adamse, A. D. (1966). Bateriological studies on dairy waste activated sludge. Thesis, Wageningen University.

Adamse, A. D. (1968a). Formation and final composition of the bacterial flora of a dairy waste activated sludge. Wat. Res. 2, 665.

Adamse, A. D. (1968b). Response of dairy waste activated sludge to experimental conditions affecting pH and dissolved oxygen concentration. Wat. Res. 2, 703.

Adamse, A. D. (1968c). Bulking of dairy waste activated sludge. Wat. Res. 2, 715.

Adamse, A. D. (1970). Some characteristics of arthrobacters from a dairy waste activated sludge. Wat. Res. 4, 797.

Antheunisse, J. (1971). DNA decomposition by soil micro-organisms. Antonie van Leeuwenhoek 37, 258.

Busch, A. W. (1958). BOD progression in soluble substrates. Sewage ind. Wastes 30, 1336.

Butterfield, C. T. (1935). Studies on sewage purification II. A zoogloea forming bacterium isolated from an activated sludge. Publ. Hlth Rep. Washington 50, 671.

Crabtree, K., McCoy, E., Boyle, W. C. & Rohlich, G. A. (1965). Isolation, identification, and metabolic role of the sudanophilic granules of Zoogloea ramigeru. Appl. Microbiol. 13, 218.

Crabtree, K. Boyle, W., McCoy, E. & Rohlich, G. A. (1966). A mechanism of floc formation by Zoogloea ramigera. J. Wat. Pollut. Control. Fed. 38, 1968.

Crombach, W. H. J. (1971). Comparison of deoxyribonucleic acid base composition of soil arthobacters and coryneforms from cheese and sea fish. Antonie van Leewenhoek. In press.

Deinema, Maria H. & Zevenhuizen, L. P. T. M. (1971). Formation of cellulose fibrils by Gram-negative bacteria and their role in bacterial flocculation. Arch. Mikrobiol. 78, 42.

88   E. G. MULDER, J. ANTHEUNISSE AND W. H. J. CROMBACH

Dugan, P. R. & Lundgren, D. G. (1960). Isolation of the floc-forming organism *Zoogloea ramigera* and its culture in complex and synthetic media. *Appl. Microbiol.* **8**, 357.
Eggink, H. J. (1965). Het estuarium als ontvangend water van grote hoeveelheden afvalstoffen. Thesis, Wageningen University. (In Dutch with English summary).
Farquhar, G. J. & Boyle, W. C. (1971). Occurrence of filamentous micro-organisms in activated sludge. *J. Wat. Pollut. Control Fed.* **43**, 779.
Friedman, B. A., Dugan, P. R., Pfister, R. M. & Remsen, C. C. (1968). Fine structure and composition of the zoogloeal matrix surrounding *Zoogloea ramigera. J. Bact.* **96**, 2144.
Friedman, B. A., Dugan, P. R., Pfister, R. M. & Remsen, C. C. (1969). Structure of exocellular polymers and their relationship to bacterial flocculation. *J. Bact.* **98**, 1328.
Gaudy, Jr., A. F., Bhatla, M. N., Follett, R. H. & Abu-Niaaj, F. (1965). Factors affecting the existence of the plateau during the exertion of BOD. *J. Wat. Pollut. Control Fed.* **37**, 444.
Gils, H. W. van (1964). Bacteriology of activated sludge. Thesis, Wageningen University.
Heukelekian, H. & Weisberg, E. (1956). Bound water and activated sludge bulking. *Sewage ind. Wastes* **28**, 558.
Hünerberg, K., Sarfert, F. & Frenzel, H. J. (1970). Ein Beitrag zum Problem 'Blähschlamm'. *Gas Wass.* **111**, 7.
Jasewicz, Lenore & Porges, N. (1956). Biochemical oxidation of dairy wastes. VI. Isolation and study of sludge micro-organisms. *Sewage ind. Wastes* **28**, 1130.
Kortstee, G. J. J. (1970). The aerobic decomposition of choline by micro-organisms. I. The ability of aerobic organisms, particularly coryneform bacteria, to utilize choline as the sole carbon and nitrogen source. *Arch. Mikrobiol.* **71**, 235.
Leclerc, E. (1960). The self-purification of streams and the relationship between chemical and biological tests. In *Waste Treatment.* Ed. P. C. G. Isaac. Oxford: Pergamon Press.
McKinney, R. E. (1962). *Microbiology for Sanitary Engineers.* New York: McGraw-Hill.
McKinney, R. E. & Weichlein, R. G. (1953). Isolation of floc-producing bacteria from activated sludge. *Appl. Microbiol.* **1**, 259.
McWhorter, T. R. & Heukelekian, H. (1964). Growth and endogenous phases in the oxidation of glucose. In *Advances in Water Pollution Research, Proc. Intern. Conf., London.* Ed. W. W. Eckenfelder. Oxford: Pergamon Press.
Mulder, E. G. (1964). Iron bacteria, particularly those of the *Sphaerotilus-Leptothrix* group, and industrial problems. *J.appl.Bact.* **27**, 151.
Mulder, E. G., Adamse, A. D., Antheunisse, J., Deinema, Maria, H., Woldendorp, J. W. & Zevenhuizen, L. P. T. M. (1966). The relationship between *Brevibacterium linens* and bacteria of the genus *Arthrobacter. J. appl. Bact.* **29**, 44.
Mulder, E. G. & Antheunisse, J. (1963). Morphologie, physiologie et écologie des *Arthrobacter. Annls Inst. Pasteur, Paris* **105**, 46.
Mulder, E. G. & Veen, W. L. van (1963). Investigations on the *Sphaerotilus-Leptothrix* group. *Antonie van Leeuwenhoek* **29**, 121.
Mulder, E. G. & Veen, W. L. van (1971). *Genus Sphaerotilix* (Migula) In *Bergey's Manual of Determinative Bacteriology*, 8th ed. In press.
Pasveer, A. (1960). New developments in the application of Kessener brushes (aeration rotors) in the activated-sludge treatment of trade-waste waters. In *Waste Treatment,* Ed. P. C. G. Isaac. Oxford: Pergamon Press.
Pasveer, A. (1969). A case of filamentous activated sludge. *J. Wat. Pollut. Control Fed.* **41**, 1340.
Peters, H. (1960). De winning van bijprodukten van de aardeppelmeelbereiding. *De Ingenieur* **72**, 47.
Pipes, W. O. (1966). The ecological approach to the study of activated sludge. *Adv. appl. Microbiol.* **8**, 77.
Pipes, W. O. (1967). Bulking of activated sludge. *Adv. appl. Microbiol.* **9**, 185.
Pipes, W. O. (1969). Types of activated sludge which separate poorly. *J. Wat. Pollut. Control Fed.* **41**, 714.
Porges, N. (1960). Newer aspects of waste treatment. *Adv. appl. Microbiol.* **2**, 1.

R.A.A.D.   (Rijks   Zuivel-Agrarische   Afvalwaterdienst,   Arnhem,   1966).   Verslag proef-oxydatiesloten over de periode 1963 tot 1966. (In Dutch with English review: Report of tests on oxidation ditches in 1963, 1964 and 1965).

Unz, R. F. & Dondero, N. C. (1967). The predominant bacteria in natural zoogloeal colonies. I. Isolation and identification. *Can J. Microbiol.* **13**, 1671.

Veen, W. L. van, Kooij, D. van der, Geuze, E. C. W. A. & Vlies, A. W. van der (1971). The classification of a multicellular bacterium isolated from activated sludge. *J. gen. Microbiol.* In press.

Zevenhuizen, L. P. T. M. (1966a). Function, structure and metabolism of the intracellular polysaccharide of *Arthrobacter*. Thesis, Amsterdam University.

Zevenhuizen, L. P. T. M. (1966b). Formation and function of the glycogen-like polysaccharide of *Arthrobacter. Antonie van Leeuwenhoek* **32**, 356.

Aerobic Treatment of Agricultural Wastes

K. ROBINSON

*Bacteriology Division, School of Agriculture,
Aberdeen, Scotland*

CONTENTS

1. Introduction

WHY IS IT necessary to consider treatment of agricultural wastes? The need for increased efficiency, particularly in livestock production, has led to "intensification", often to such a degree that it is no longer possible to associate some intensive units with the most advanced traditional farm in which the balance of crops and animals allows recycling of the wastes as fertilizer. The development of fringe industries, e.g. poultry and vegetable processing, which are giving increased attention to the production of deep frozen, ready prepared foods has led to localized production of waste material which previously had access via the kitchen sink to a sewerage system. Legislation (e.g. Rivers, Prevention of Pollution Act, 1961) controls the quality of waste discharges or may completely prevent discharge.

The production and disposal of raw, semisolid or liquid wastes which are in excess of fertilizer demand may result in direct or indirect pollution of adjacent watercourses with their consequent loss as a social amenity or as a source of potable water. Deaths of animals and humans due to gases, e.g. CO_2 and H_2S, produced by anaerobic degradation of wastes collected and stored in pits beneath animal houses, have been recorded (Taiganides & White, 1969).

Storage of anaerobically decomposing waste or its use for irrigation may release odours which are unacceptable to the local population. Employees are reluctant to work with raw wastes or in highly odorous atmospheres.

Waste must be accepted as an inevitable end product of agriculture but a reduction or elimination of some or all of the disadvantages or hazards arising from its production may be obtained by the use of aerobic microbiological treatment in association with other chemical or physical processes e.g. floculation and mechanical separation.

The essential features of a treatment system suitable for use on a farm are that it should be robust and simple, form an integral part of the agricultural enterprise and be automatic, since it is unlikely that suitably qualified personnel will be available for its operation. A variety of small scale aerobic treatment systems will be described and the ability of these systems to provide an environment suitable for efficient microbial purification will be discussed.

2. Possible Systems

Several of the nonmechanical and mechanical systems currently in use, either commercially or experimentally, for the treatment of agricultural wastes are ostensibly "aerobic", i.e. their purpose is to encourage the development of aerobic micro-organisms using raw waste as a substrate. The principal difference between these systems is the method of oxygen supply, e.g. surface absorption/diffusion or mechanical aeration.

(a) Nonmechanical systems

The barrier ditch (Fig. 1) is a very simple continuous flow system with a primary stage, preferably with baffles to control the flow, in which settling of the gross solids occurs followed by a series of shallow stages separated by barriers. Some settling of solids occurs in these stages and reduction in the flow rate increases the treatment period. Oxygenation takes place at the surface of the liquid in each stage and also while the liquid flows in thin layers over each barrier.

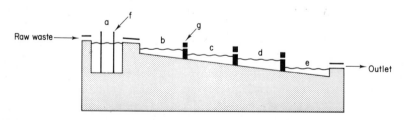

Fig. 1. The barrier ditch. a, Gross solid separation; b,c,d,e, stages of aerobic treatment; f, flow baffles; g, solids retention barriers.

The "lagoon" or "pond" is a storage treatment system favoured when liquid cannot be discharged to a watercourse or for conservation of waste water for irrigation (Jones, Day & Dale, 1970). The larger lagoons are 10—15 ft deep and may be used with shallower forms 3—5 ft deep which receive the overflow from the deep lagoon. Oxygenation in both is limited to the surface but in a suitable climate additional oxygen may be provided by the growth of algae.

Biological filtration in towers packed with stones, corrugated plastic sheeting or plastic filter media (Truesdale, Wilkinson & Jones, 1964; Hepherd, 1970) allows wastes to flow in thin films and thus increases the surface area for oxygen supply; treatment is effected by a film of organisms which develops on the surface of the filter media.

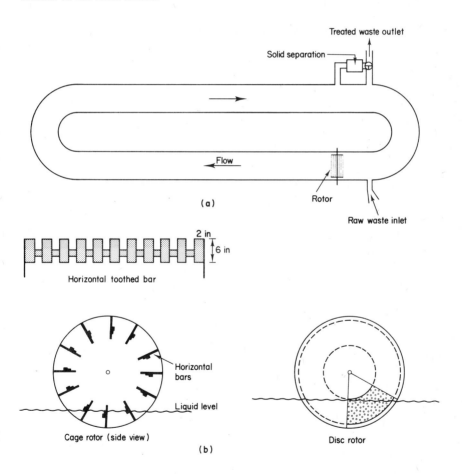

Fig. 2. (a) The oxidation ditch (b) oxidation ditch aerators/agitators.

(b) *Mechanical systems*

Mechanical systems, with their increased rate of oxygen supply, are potentially capable of preventing oxygen limitation during treatment, and their ability to maintain solids in suspension prevents the development of anaerobic conditions which occur in the solids which settle in nonmechanical systems. The oxidation ditch (Fig. 2) (Pasveer, 1959) has been used and examined for its ability to provide controlled treatment of wastes from pigs (Jones, Day & Converse, 1969; Robinson, Saxon & Baxter, 1971), poultry (Stewart & McIlwain, 1971) and cattle.

The contents of the ditch, which are maintained at a depth of 0·3–0·6 m, are circulated around the ditch at a speed sufficiently great to maintain solids in suspension, and aerated by a rotor. Cage rotors comprising horizontal toothed bars are frequently used (Baars & Muskat, 1959) but other types are equally suitable (South African Inventions Development Corporation, 1968). The oxidation ditch and other mechanical systems are not unlike a continuous culture apparatus, in that additions of substrate are small compared with the total volume of the treatment vessel which contains an established microflora capable of metabolizing the substrate. Additions of fresh substrate can be arranged to maintain metabolism at an efficient level; treated liquor can be removed to compensate for the addition of fresh substrate and to maintain the microbial population at an optimum level. Alternatives to the oxidation ditch are: (i) the surface aerator system (Fig. 3) in which a conical disc or stirrer blade revolving at or just below the surface creates a vortex and turbulence and (ii) the aerohydraulics system (Fig. 4) which uses an "air-gun" to achieve aeration and agitation (Bryan, 1964). The gun is placed at the bottom of the treatment vessel and an air bubble generated at the base of the gun moves up the barrel and draws waste liquor into the barrel. The large bubbles burst after exit from the barrel and oxygen transfer occurs between each bubble and the body of liquid in the vessel. The aeration capacity of this system may be increased by supplying air through perforated pipes beneath the surface of the liquor.

3. The Environment

Ideal microbial treatment or "purification" may be regarded as a process by which the pollutants in the raw waste are converted to microbial cells or insoluble substances which can be separated from water, the final endproduct. It is preferable that such an achievement should provide equally satisfactory results to other aspects of pollution associated with agricultural wastes. The efficiency of treatment depends on the ability of the treatment system to provide an environment which supports the growth and activity of a treatment microflora along with factors such as the balance between oxygen (O_2) and substrate supply, the chemical nature and biodegradability of the waste; pH value also has

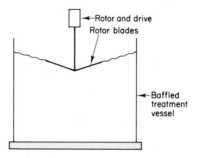

Fig. 3. The surface aerator system.

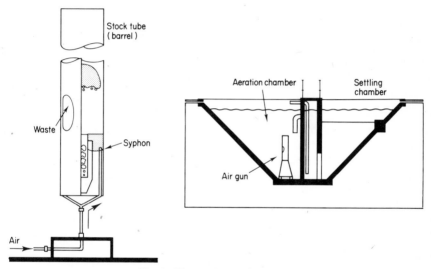

Fig. 4. The aerohydraulics system.

a profound influence on the selection, balance and activity of the various types of micro-organism comprizing the treatment microflora.

(a) *Oxygen supply*

Micro-organisms obtain their O_2 from dissolved gas in the liquid phase and replacement of dissolved O_2 from the atmosphere occurs via the gas/liquid interface which exists at the surface of a body of liquid or at the surface of a gas bubble, and the rate at which the transfer of O_2 from the gaseous to the dissolved state occurs is expressed by the equation

$$\frac{dc}{dt} = K(C_s - C_1)$$

where $K = O_2$ transfer rate; C_s = concn of dissolved O_2 at saturation; C_1 = concn of dissolved O_2 at time t.

The rate of O_2 supply and the buffering capacity provided by a body of liquid containing dissolved O_2 must be in excess of the O_2 required for microbial metabolism if the development of anaerobic conditions is to be prevented.

In nonmechanical systems, e.g. barrier ditches and lagoons, O_2 supply is limited to transfer at the liquid surface and by diffusion throughout the body of the liquid. Although values of transfer rates for these systems are not known, they can be expected to be similar to but lower than, those for natural waters because of limited mixing (Downing & Truesdale, 1955). Limitation of treatment by the low rate of O_2 transfer can be compensated for either by an increase in surface area or by a reduction in the substrate load.

The supply of O_2 by diffusion (Wuhrmann, 1964) to thin films of micro-organisms developing on the surface of biofiltration media has been investigated by Tomlinson & Snaddon (1966) who found the rate of diffusion to be $c.2/3$ of that for O_2 in water. The size of the filter and the flow rate of the liquid waste can be adjusted relatively easily, unlike the previous systems, so that O_2 does not become limiting.

Increases in the rate of O_2 transfer to large volumes of liquids or liquid-solid waste mixtures can be obtained by the use of mechanical aerators which increase the surface area of liquid exposed to the air, generate air bubbles in the liquid increasing the surface area of air to liquid, the agitation increasing the residence time of gas bubbles within the liquid during which transfer can occur. Factors affecting the rate of O_2 transfer by rotors, e.g. depth of immersion, are illustrated in Fig. 5. Determinations of transfer rate in oxidation ditches are of the order of 1% of O_2 saturation/min (Robinson, Saxon & Baxter, 1971), much lower than those which can be obtained in a laboratory continuous culture apparatus.

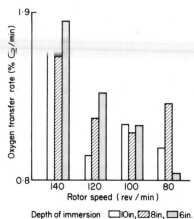

Fig. 5. Factors affecting the rate of oxygen transfer by a disc rotor.

In view of the low rates of O_2 supply to both nonmechanical and mechanical systems it can be anticipated that oxygen may become limiting very quickly if suitable allowance is not made in the design of the plant for a reserve supply of O_2 and/or a large volume of liquid containing dissolved O_2 which will buffer the system during period of peak O_2 demand.

(b) Substrate

The substrate must be considered in terms of its quantity and quality, i.e. the treatment system must be sufficiently large to permit adequate time for treatment to take place and the chemical nature and biodegradability of the waste should neither inhibit the activity of treatment micro-organisms nor be subject to sudden wide variations which would impair the maintenance of a stable microbial population.

The polluting capacity of a waste or its O_2 requirement during treatment can be determined by tests for biochemical oxygen demand (BOD) and chemical oxygen demand (COD). Values of the quantities of different types of waste and their oxygen demand are shown in Table 1. Data exist for the value of animal wastes as fertilizer (Table 2) but there is little detailed information on the chemical identity of the organic compounds or their relative biodegradability, knowledge of the latter being particularly important if its value as a substrate is to be assessed correctly and the discharge of nondegraded material in the final effluent is to be avoided.

Fibrous material and insoluble faecal solids in pig faeces appear to be nonbiodegradable even after prolonged aeration, whereas the soluble nitrogen (N) compounds in pig urine are used rapidly but at quite different rates, except in the presence of copper which is excreted in high concentrations when included in mineral supplements added to the diet (Robinson, Draper & Gelman, 1971).

The fact that certain insoluble components of the substrate are nonbiodegradable means that they may be separated prior to or during treatment and that their contribution to the BOD or COD of the substrate does not need to be included as part of the biological load on the treatment system. The knowledge that toxic substances are present in the waste indicates the need to exclude such substances from the system. The quantity and nature of different substrates in a waste exert a strong influence on the initial selection of the treatment microflora and subsequently on the changes which occur in the organisms comprising that flora. Differences in the microflora of activated sludge (Pike & Curds, 1971) and in dairy wastes (Mulder, 1971) illustrate these selective influences.

(c) pH value

No attempt has been made to control the pH value of aerobic treatment systems used for agricultural wastes. The initial pH of the system depends on that of the

Table 1

The output of livestock wastes and their oxygen demand

Waste form	Output (l/day)	Total solids (kg/day)	BOD (g O_2/day)	COD (g O_2/day)
Poultry	0·54	0·18	90·7	181·5
Pigs	7·3	0·5	172·4	324·3
Cattle	47·1	3·8	646·5	3221·0

Table 2

The analysis of pig faeces

Constituent	Amount (p/m)
Crude fibre	165,000
Ash	178,000
Total N	28,000
" P	21,000
" K	4000
" *Cu	1421
" Mn	247
" *Zn	1821

*Added as a supplement to feed ration.

Table 3

The effect of pH oxygen demand

pH value	Oxygen demand (% dissolved O_2/min)
6·0	1·41
6·6	2·33
7·2	2·86
7·8	3·13

waste, but once a microflora has become established changes in pH value occur in response to the metabolic activity of the microflora. In pig waste treatment it has been found that an alkaline pH value (8·5–9·0) can be maintained when the substrate has a high N content and the maintenance of such pH levels corresponds with a high rate of reduction of O_2 demand of the substrate; lower rates of substrate supply lead to the production of acid conditions (pH $5\cdot5-6\cdot0$). Differences in the microflora can be recognized in these circumstances, and influences of pH on the metabolic activity can be measured (Table 3).

The changes in pH value which occur in a system in response to the rate of substrate addition illustrate that a multistage treatment process which permits the development of a different microflora in each stage may be preferable to a single stage system in which the established pH value may not favour the development and activity of micro-organisms degrading other substrates in the waste.

(d) Oxygen demand

The O_2 demand of a system is a function of the number and types of micro-organisms present and the quantity, rate of addition and nature of the substrates; whether this demand is satisfied or not will depend on a balance existing between the fixed rate of O_2 supply and the variable rate of demand. The effect of adding substrate, i.e. the depression of dissolved O_2 concentration and an increase in the rate of uptake by the micro-organisms, is shown in Fig. 6.

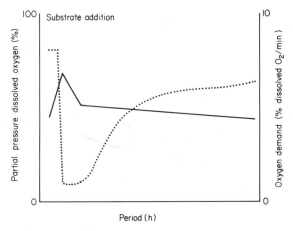

Fig. 6. The effect of substrate addition on oxygen demand and availability. Dotted line, dissolved oxygen; unbroken line, oxygen demand.

Oxygen demand falls sharply again when the dissolved O_2 reaches very low levels and continues to fall for several hours, even though the dissolved O_2 level rises considerably. It can be assumed that treatment efficiency falls during such periods and continued addition of substrate prevents restoration of a more normal O_2 demand/supply relationship. Many mechanical aeration systems contain only 20% of dissolved O_2 in relation to the saturation value and are therefore very susceptible to O_2 depletion by the application of a single large daily addition of substrate. The activity of the microflora and the rate of purification of substrate can be increased or maintained at a more satisfactory level by the use of regular small additions of substrate; this also prevents continual fluctuations in the different species of the microflora as a result of alternate periods of starvation and overfeeding.

(e) *Temperature*

Treatment systems are normally allowed to operate at atmospheric temperature, and this may be subject to wide variations. Unprotected systems may freeze over completely during winter or have large losses of moisture by evaporation in summer, and these create problems in maintaining controlled input of raw waste and output of treated waste. One of the major effects of temperature is its influence on the selection of treatment micro-organisms. An almost continuous inoculum of organisms derived from the intestine is added to systems associated with livestock units but these do not appear to develop as numerically significant members of the microflora growing at treatment temperatures (15–20°). Optimum temperatures for strains of *Acinetobacter*, which occur constantly in aerated pig waste, are 20–30° (Robinson, Baxter & Saxon, 1970).

(f) *The microflora*

Although a great deal is known about the microbial ecology of aerobic sewage treatment processes (Harkness, 1966) examination of agricultural wastes has been restricted to the fate of pollutant bacteria, e.g. coli-aerogenes and enterococci. In preliminary laboratory studies of the treatment microflora of aerated pig urine, *Acinetobacter* developed as the dominant flora (Robinson, Saxon & Baxter, 1971) and similar organisms were subsequently isolated from several operational treatment systems. More extensive investigations followed changes in the microflora in response to variations in the quantity and nature of the substrate (Saxon & Robinson, 1971). It is unfortunate that the poor response of many of the isolates to laboratory tests and lack of knowledge on the substrates present in the waste does not permit the significance of these organisms in the treatment process to be established. The types of organism found in aerated pig waste may not necessarily develop in systems treating other types of waste. The number of pollutant organisms, e.g. faecal coliforms, in the primary effluent from an oxidation ditch is high ($> 1800/100$ ml) and tests of intestinal pathogens show that they are able to survive aerobic treatment.

4. Discussion

It would be unreasonable to expect simple systems like the barrier ditch to purify waste at a high rate because of factors such as low O_2 supply and uncontrolled liquid flow through the system; and in addition they do not begin to treat much of the solid material. However they can be rather successful in situations where the effluent is allowed to "disappear" rather than be discharged directly into a watercourse, and solids can be disposed of on the land.

Lagoons must be considered as systems of storage rather than of aerobic treatment. Most lagoons are designed to give an overcapacity, and with the removal of liquid for irrigation and loss by evaporation it may be many years before the accumulation of solids creates a practical problem. The lack of adequate aerobic conditions allows the production and escape of obnoxious gases so that problems of nuisance can arise if the lagoon is sited adjacent to an urban area.

Credit for the development of mechanical aeration systems for agricultural wastes must go to engineers who, by processes of trial and error, have modified similar systems used in large scale sewage purification with reasonable success. Greater efficiency of treatment can be expected when the types of micro-organism responsible for treatment are known and their optimum environmental conditions have been established. Studies on the aeration of pig wastes indicate that insufficient attention has been paid to the establishment of a treatment microflora and its maintenance in an active state, and results show that microbiological treatment *per se* does not produce an effluent with an acceptable BOD even though the residual soluble substrates are only slowly biodegradable. The use of post-treatment processes to remove fine suspended solids and microbial cells is essential not only to reduce the BOD but also to eliminate intestinal and pathogenic micro-organisms. When complete treatment is the ultimate aim no major difficulties are experienced with other aspects of waste handling and disposal. Treated liquors have only a slight musty odour, clean odourless solids are produced and, under suitable operating conditions nitrogen can be removed as ammonia rather than nitrate.

The selection of a suitable system of treatment should depend in the first instance on its ability to control or eliminate pollution from any particular agricultural enterprise, but the economics of the system carry almost equal weight unless it can be shown that treatment is the only solution to a particular problem or that the system is as cheap and more convenient that the existing method of waste handling and disposal.

5. References

Baars, J. K. & Muskat, J. (1959). Oxygenation of water by bladed rotors. *Res. Inst. Publ. Hlth. Eng.* Rep. No. 28. Delft, Holland: TNO.

Bryan, J. G. (1964). Physical control of water quality. *J. Br. Wat. Wks Ass.*

Downing, A. L. & Truesdale, G. A. (1955). Some factors affecting the rate of solution of oxygen in water. *J. appl. Chem.* **5**, 570.

Harkness, J. (1966). Bacteria in sewage treatment processes. *J. Proc. Inst. Sew. Purif.* **6**, 542.

Hepherd, R. Q. (1970). Easing the slurry load. *Pig Fmg* **12**, 52.

Jones, D. D., Day, D. L. & Converse, J. C. (1969). Field tests of oxidation ditches in confinement swine buildings. In *Proc. Animal Waste Management.* Cornell Univ. Conf. Am. Soc. Agric. Engrs.

Jones, D. D., Day, D. L. & Dale, A. C. (1970). Aerobic treatment of livestock wastes. *Univ. Illinois Bull. No. 737.*

Mulder, E. G. (1971). Microbial aspects of pollution in the food and dairy industries. In Soc. appl. Bact. Symp. on *Microbial Aspects of Pollution.* London: Academic Press.

Pasveer, A. (1959). A contribution to the development of the activated sludge process. *J. Proc. Inst. Sew. Purif.* **4**, 436.

Pike, E. B. & Curds, C. R. (1971). Microbial ecology of the activated sludge process. In Soc. appl. Bact. Symp on *Microbial Aspects of Pollution.* London: Academic Press.

Robinson, K., Baxter, S. H. & Saxon, J. R. (1970). Aerobic treatment of farm wastes. In Symposium on Farm Wastes. *J. Wat. Pollut. Control Fed.* 122–131.

Robinson, K., Draper, S. R. & Gelman, A. L. (1971). Biodegradation of pig waste: breakdown of soluble nitrogen compounds and the effect of copper. *Envir. Pollut.* **2**, 69.

Robinson, K. Saxon, J. R. & Baxter, S. H. (1971). Microbiological aspects of aerobically treated swine waste. In Symposium on Livestock Wastes. *Am. Soc. Agric. Engrs,* in press.

Saxon, J. R. & Robinson, K. (1971). The microflora of aerated pig waste. (In preparation).

South African Inventions Development Corporation (1968). A guide to the use of the Huisman-Orbal activated sludge system for sewage purification. Council Sci. Indust. Res. Pretoria, South Africa.

Stewart, T. A. & McIlwain, R. (1971). The aerobic storage of poultry manure. In Symposium on Livestock Wastes. *Am. Soc. agric. Engrs* **14**, in press.

Taiganides, E. P. & White, R. K. (1969). The menance of noxious gases in animal units. *Am. Soc. agric. Eng.* **12**, 359.

Tomlinson, T. G. & Snaddon, Dorothy H. M. (1966). Biological oxidation of sewage by films of micro-organisms. *Air. Wat. Pollut.* **10**, 865.

Wuhrmann, K. (1964). Microbial aspects of water pollution control. *Adv. appl. Microbiol.* **6**, 119.

The Role of Strict Anaerobes in the Digestion of Organic Material

P. N. HOBSON AND B. G. SHAW

The Rowett Research Institute,
Bucksburn, Aberdeen, Scotland

CONTENTS

1. Introduction

THIS CONTRIBUTION is not intended to be a detailed review of work on effluent disposal or anaerobic digestion, but only to give a background to our bacteriological results. A detailed review of anaerobic digestion has recently been made by Kirsch & Sykes (1971). The problems of microbial interactions and control in mixed bacterial and multisubstrate systems have also been discussed previously by one of us (Hobson, 1969, 1971).

In the biological treatment of effluents anaerobic digestion has long had a place, usually as part of a 2 stage aerobic and anaerobic process. In the treatment of domestic sewage the solid material is separated from the liquid waste. The liquid is treated aerobically and the solids, together with excess sludge from activated sludge plants, are digested anaerobically. The anaerobic digestion removes part of the organic matter of the sewage and leaves a residue which is comparatively innocuous and more easily dried than the original material.

In its usual form in domestic systems the anaerobic digester takes the form of a single stage, continuous fermenter, heated by gas from the digestion and stirred slowly so as to make the contents reasonably homogeneous. The usual turnover time is *c.* 30 days. The anaerobic 'contact process' provides for a secondary stage where settling takes place and some of the solids are returned to the primary digester, thus providing a feedback of bacteria. This process has been used to increase the efficiency of anaerobic digestion of dilute liquid wastes produced in large quantities, but would not seem to be suitable for wastes with high solids contents. Another process for anaerobic treatment of low solids waste is the upflow filter. This has been used so far mainly on an experimental scale, but a similar upflow contact process has been used on a small scale for some domestic and other wastes.

The adequate treatment of farm effluent, particularly that from intensive farming units is now becoming a problem of importance. It seemed to us that anaerobic digestion offered possibilities for treatment of the strong, high solids content farm effluent, as this is akin to the solids fraction of domestic sewage. The bacteriological work reported here was done during experiments on the treatment of piggery effluent by a single stage, anaerobic digestion, similar to the standard domestic sewage anaerobic digester.

2. The Overall Process of Anaerobic Digestion

Although anaerobic digestion has been in use for many years, and some practical "rules" for running of digesters have been formulated, relatively little is known about the biochemistry or bacteriology of the process. This is partly because the strict anaerobes constituting a large part of the digester flora are not easy to cultivate. Application of the methods used in rumen bacteriology has made the analysis of digester flora a much more practical possibility.

Generally the efficiency of sewage digestion processes is measured in terms of the sewage engineer's parameters, BOD, COD, etc, and there have been few meaningful analyses on the composition of wastes before and after treatment. A detailed analysis of waste would take a long time and might not be of much advantage, as the composition of the waste varies from works to works and from time to time even in the same sewage works. Also such material as "cellulose" varies from the treated cellulose of domestic papers to the lignin coated, highly crystalline cellulose of plant fibre residues.

However, one can say that domestic or farm wastes contain similar substances, but in varying amounts, whatever the source, and from the end products of the digestion and by analogy with the activities in the rumen, which is a well documented microbial habitat similar in many ways to the anaerobic digester, one might deduce that the overall processes of anaerobic digestion are as outlined below. Farm animal wastes contain, of course, higher proportions of plant residues than do domestic wastes.

Faeces consist of food residues, intestinal bacterial cells, residues of intestinal enzyme and bacterial action, and intestinal secretions in a salts solution. The urine contains salts and nitrogenous excretion products. Farm effluent may or may not be diluted with animal house washing water. It may also have been standing for some time, for instance in drainage channels or tanks under a pig house. The faeces-urine mixture also contains the products of the anaerobic fermentation which can occur here; these are mainly volatile fatty acids (VFA). The extent of this fermentation depends on the degree of anaerobiosis, i.e. the depth of the sewage mass, and the temperature at which it is held. This latter is usually fairly low, but in a tank of pig effluent held at 35° we have found the VFA concentration to rise from 2500 to 13,500 p/m (as acetic acid) in 14 days. In this primary fermentation any soluble sugars and some of the more easily degraded polysaccharides, such as starch from pig feeds, remaining in the excreta are fermented, and the pH of the material falls. Carbon dioxide is the principal gaseous product; in the test quoted above the composition of the gas was 16% of methane + 84% CO_2.

Anaerobic digestion proper is a 2 stage process. In the first stage bacteria grow by anaerobic fermentation using as energy sources principally the carbohydrates of the effluent. These consist of starch, cellulose, hemicellulose and pectic residues from the diet, especially of animals, glycerol from lipids, and bacterial polysaccharides and mucopolysaccharides from the intestines. This fermentation is carried on by a varied flora and can produce a mixture of VFA, ethanol, CO_2 and hydrogen. Lactic and succinic acids are also possible primary fermentation products, but these are most likely fermented by other bacteria to VFA. As the carbohydrates are mainly in the form of polysaccharides their breakdown requires bacteria with the necessary hydrolytic enzymes such as amylase, cellulase and pectinase. Bacterial lipases hydrolyse esters to glycerol and long chain fatty acids; the resultant glycerol is fermented and the fatty acids are probably converted to bacterial lipids.

The bacteria also require a source of nitrogen (N) supplied by the proteinaceous material of the slurry through the action of bacterial proteases, and ammonia (NH_3) either present in the slurry or produced by deaminative or ureolytic bacteria. More complex N compounds, such as purines, can be fermented as sources of energy by some anaerobic bacteria, or broken down, in each case the products being ammonia, VFA and CO_2. The bacteria also need a source of sulphur and this comes mainly from sulphur-containing amino acids, either directly or after their conversion to sulphide. Trace elements sufficient for bacterial growth should be present normally in the slurry.

A further possibility of degradation of amino acids is by their use as energy sources by Stickland-type reactions or other fermentations such as are carried out by certain clostridia or *Diplococcus glycinophilus*. Bacteria such as these occur only to a limited extent in the rumen, but they may be more active in digesters.

As this is a mixed microbial habitat one might expect, by analogy with the rumen, growth factors for some bacteria to be produced by other bacteria.

The main processes occurring in the first stage thus a fermentation of carbohydrates and accumulation of bacterial cells. This process serves to break down the solid material of the slurry, but the extent of the bacterial accumulation depends on many factors. The anaerobic fermentation is essentially less efficient than an aerobic one in converting substrate energy to bacterial cells, although a number of anaerobic rumen bacteria, and others, similar to those in digesters, have high growth yields (Hobson & Summers, 1967). Even with an anaerobic rumen bacterium using only ammonia and carbohydrate as a source of energy and cell mass, < 20% of the carbohydrate occurred as cell mass, the rest was fermented; and this is a high figure for conversion of carbohydrate to cells. In most anaerobes growing on complex media almost 100% of the energy source is fermented, and the cell mass comes from other constituents of the medium. Cell yields are also decreased by maintenance requirements of bacteria growing slowly and by uncoupled fermentations (Hobson & Summers, 1967; Henderson, Hobson & Summers, 1969), although the latter is unlikely unless carbohydrate is in excess. However, so far as breakdown of the carbohydrates of the slurry goes it does not matter whether fermentation is coupled to bacterial growth or not. What is affected by maintenance or uncoupled fermentation is the extent to which the nitrogenous materials of the slurry, especially the ammonia, are converted to bacterial cells. The extent of fermentation of some carbohydrates, such as cellulose, may also be limited by the concentration or rate of action of the bacterial hydrolases. If carbohydrate availability is the limiting growth factor one might suggest that a better overall conversion of nitrogenous material to relatively innocuous bacterial cells might be obtained by adding some cheap carbohydrate to the slurry before fermentation.

Lignin-type aromatic compounds are probably not degraded to any extent anaerobically as lignin degradation is an aerobic process, although there is a small possibility of methanogenesis, as the aromatic ring in benzoate can be degraded by methanogenic bacteria from sewage (Nottingham & Hungate, 1969).

All these processes are carried out by a mixed anaerobic flora, many of them strict anaerobes like the rumen flora (using 'strict' in the way defined by Loesche, 1969), and the methanogenic bacteria are strict anaerobes. There are also facultative anaerobes present in varying numbers, and while these aid in breakdown of the slurry they are also most likely responsible for reducing the digester contents, by absorbing the oxygen in any air in the digester, to an E_h value at which the anaerobes can grow. The numbers and types of bacteria involved are considered in the next section.

The primary fermentation of the slurry results largely in lignin, cellulose materials (as degradation of cellulose, being a slow process, is never complete), bacterial cells and residual solid nitrogenous materials, unused ammonia and possibly amino acids, together with fermentation acids in the liquid: there is also CO_2 and H_2 in the gas. The fermentation acids in the liquid represent not only a source of pollution if the material were run to waste at this stage, but also a possible source of endproduct inhibition of the primary fermentations. These acids, then, must be removed as far as possible.

In the second stage of the fermentation methanogenesis occurs. Here bacterial cells are produced with methane formation providing the energy for growth. The N requirements of the methanogenic bacteria are not completely known: it seems that they use NH_3 in the digester fluids and possibly some amino acids. The substrates for methane formation by different species of methanogenic bacteria have been given as hydrogen + CO_2, formic, acetic and butyric acids, ethanol and methanol. Hydrogen + CO_2 and formate are definitely substrates; the position regarding some of the other compound is confused owing to the difficulty of growing the bacteria and the possibility that some methanogenic species are actually mixed cultures, as has been shown for *Methanobacillus omelianskii* (Bryant *et al.*, 1967). All the possible substrates for methanogenesis occur in the primary anaerobic digestion, methanol being a product of demethylation of pectic materials in plant food residues. However, the action of the methanogenic bacteria lowers the VFA concentration of the digester fluid and converts some of the nitrogenous material to bacterial cells, although, again, as methanogenesis is an inefficient reaction for production of growth energy, the former aspect of the process is the most important in digester operation. Kirsch & Sykes (1971) have reviewed methanogenesis in some detail.

Methanogenesis has been said to be subject to substrate inhibition by VFA so that, as the methane bacteria are slow growing, removal of VFA by the methane bacteria is often the rate limiting process in anaerobic digestion. Too rapid a primary fermentation results in accumulation of acids, the cessation of methanogenesis and the breakdown of the digestion. Ammonia may also be an inhibitor of the methane bacteria. However, we must remember that the primary fermentations can also be slow. Farm wastes, especially, contain only the plant residues which have resisted degradation by intestinal enzymes and rumen or caecal bacteria, and cellulolysis of these residues is a slow process. All these factors combine to make the overall anaerobic digestion slow compared with many commercial microbial processes.

The rate of digestion is affected by the temperature of the fermentation, although there is also some evidence that different flora predominate at different process temperatures. Domestic sewage digesters are usually run in the mesophilic range, $c.35°$, and we selected this temperature for our farm sewage digesters as offering the best compromise between rate of digestion and heating of a farm scale digester.

3. Anaerobic Digestion of Piggery Wastes

In several instances chemical or bacteriological analysis has been done on samples from laboratory scale anaerobic digesters with artificial wastes made from laboratory materials or with sewage sludge fed with a carbohydrate such as glucose. It is uncertain how far such results are applicable to actual sewage digesters. However, there have been some investigations of digestion of actual sewages.

Sewages vary in compostion, so that the extent of breakdown of particular components depends on the balance between the various components, but there are reactions common to all digestions. Because the digesters are continually inoculated with excreta containing large numbers of viable bacteria suited to life in the comparatively anaerobic environment of the intestines and because they are also subject to contamination by air- or soil-borne organisms (and as a practical measure digesters are often started by inoculating one from another), it seems that, although the balance of the flora may vary from one digester to another, any bacteria isolated can be regarded as potentially capable of playing a role in any digester.

Our own work on the treatment of piggery wastes has been done on 15 l. laboratory digesters. These will be scaled up later, but as there are no problems involved in aeration it would seem valid to assume that the results obtained from laboratory scale digesters are closely related to those that would be obtained from a large scale digester. The experimental digesters were of the completely mixed, single stage type, without feedback, as this is the type generally used in domestic sewage plants and would be the simplest type for farm use. One was fabricated from a stainless steel vessel, the other from polypropylene. Both were fitted with low speed (80 rev/min) paddle stirrers, which kept the contents relatively homogeneous, gas outlets to a small gas meter and tubes for loading and unloading, and for flushing the digesters with CO_2. The digesters were heated in water baths at $35°$. Owing to the difficulties of obtaining small sludge pumps the digesters had to be run on a semicontinuous basis, with unloading of part of the contents and loading with fresh sludge once each 24 h. This was done by a suction and CO_2 pressure process that excluded air. We realize that this could be inefficient and that a continuous flow operation could probably increase the overall loading rate and degree of digestion of the sludge; and new digesters, working on this principle are now being brought into use. However, the digestion processes are the same in each case, therefore, and while small changes in the numbers of bacteria might be expected the types and relative proportions should be the same in each case. This also applies to different solids loading in the semicontinuous process.

The waste used was obtained from an intensive piggery unit, housing pigs fed on a barley ration containing 200 p/m of copper. It consisted only of the faeces-urine slurry accumulating for 10 days under a slatted floor as no washing water was used. This slurry contained 10–15% of total solids and represents the strongest form of piggery effluent of commercial practice. In initial experiments it was shown that this mixture did not attain a full anaerobic digestion when incubated at 35°; acid production eventually stopped the fermentation. The slurry was then diluted to 2% solids content at which dilution a complete digestion process could be obtained. Several such digestions have been run for varying periods, but 2 have been used for most of the work and have now been running continuously for 18 months. One was seeded with material from a domestic anaerobic sewage digester, the other was allowed to build up by regular addition of piggery waste to a digester initially filled with water. Both have attained essentially the same digestion rate and process. After 8 months running at the initial loading rate (Table 1) the loading rate was increased slowly,

Table 1

Anaerobic digestion of piggery waste at low loading rate

Sample	Input	Effluent	Reduction %
Whole sample (BOD$_5$)	5400	1300	76
Supernatant fluid (BOD$_5$)	4200	700	83
Whole sample (COD)	24,300	14,900	39
Supernatant fluid (COD)	8200	4300	48
Total solids	20,800	12,900	38
Suspended solids	19,500	10,900	44

Loading rate, 0·013 lb volatile suspended solids/ft^3/day. Turnover time, 37·5 days.

allowing at least 2 turnover times between increasing the input, until at the time of writing the loading rate was as shown in Table 2. This is probably not the maximum capacity of the digestion even with semicontinuous loading, and experiments are still in progress. The VFA concentration in the digesters has

Table 2

Anaerobic digestion of piggery waste at high loading rate

Sample	Input	Effluent	Reduction %
Whole sample (BOD$_5$)	9500	4200	56
Supernatant fluid (BOD$_5$)	7300	1500	79
Whole sample (COD)	42,500	29,400	31
Supernatant fluid (COD)	14,400	4400	70
Total solids	36,400	26,800	26
Suspended solids	34,100	25,800	24

Loading rate, 0·14 lb volatile suspended solids/ft^3/day. Turnover time, 14 days.

remained < 2000 p/m (as acetic acid) and some typical analyses of input and output are given in Tables 1 and 2 in the usual water purification terms. It is apparent that although the process is still working well it is not quite so good at the higher loading rates, and in practice a mean between purification and loading

Table 3

Main components of piggery waste before and after anaerobic digestion

	Amount (g/l) in	
	Input	Effluent
NH_3-N	0·5	0·6
N from other sources x 6·25	3·8	3·6
Crude fat	1·4	0·5
Neutral detergent fibre (essentially hemicellulose, cellulose, lignin)	9·8	5·9
Acid detergent fibre (essentially cellulose, lignin)	6·0	4·8
Lignin	1·8	1·7
Ash	3·2	3·1

Loading rate, 0·03 lb volatile suspended solids/ft³/day. Turnover time, 37·5 days.

rate (on which depends the volume of the digester needed for any particular farm) must be decided. A typical analysis of the changes in the main constituents of the material is shown in Table 3. The output from the digesters settles better and is less noxious than the input.

4. The Bacteriology of Anaerobic Digestion, with Special Reference to the Digestion of Piggery Wastes

(a) Non-methanogenic bacteria

Bacteriological work on anaerobic digestion has been reviewed by Kirsch & Sykes (1971) and more extensively by Toerien & Hattingh (1969), and these papers might be consulted for a complete list of references to work on anaerobic digestion.

As we have said, an anaerobic digester is subject to continuous inoculation by faecal bacteria and by the large number of bacteria in the surroundings of the digester. Even if many of these are aerobic types they may survive for long periods in the digester. Others, anaerobic or facultatively anaerobic ones, may be potentially able to grow but may in fact grow only to a small extent because of their inability to compete for nutrients, or because some growth factor is in a limiting concentration. In reality, these bacteria may be washed out of the digester, but their numbers are kept low by continued inoculation from the input. One might say, then, that almost any species of bacterium might be isolated from an anaerobic fermentor, so that lists of species isolated at different times are of doubtful value, especially if only one species was isolated by the use of one particular medium. The role of a species of bacterium in the digestion processes can only be assessed by knowledge of the proportion of the total population it represents and its fermentative and enzymic activities. The viable count of bacteria in a milieu like anaerobic sludge is difficult to assess because of the possibilities of clumping and attachment of bacteria to particles, and the work of Tempest, Herbert & Phipps (1967) has shown that in continuous pure cultures of long turnover time a large proportion of the cells may be nonviable; experience with rumen bacteria also shows that in this habitat, on most diets, only 10% or less of the total bacteria are viable. On the other hand one of us has argued elsewhere (Hobson, 1971) that cells, nonviable in the sense of inability to reproduce in an artificial medium, may still possess enzymic activities contributing to substrate degradation.

(i) *Methods*

In trying to assess the role of various bacteria we have used habitat-simulating media, and these were given higher counts than a number of conventional anaerobic media tested. All anaerobic media were prepared and inoculated by the strictly anaerobic techniques developed for rumen bacteriology (Hungate, 1950). Incubations were at 38°. Bulk samples were taken under CO_2 from digesters and these and other samples were diluted under CO_2, in 5 ml amounts, with an equal volume of diluting fluid. After mechanically blending on a Whirlimixer (Fisons, Loughborough) for 1 min decimal dilutions were prepared from which the culture media were inoculated in triplicate. The diluting fluids were similar in composition to the media, but without carbohydrates.

Total viable counts of anaerobic and facultatively anaerobic, nonmethanogenic bacteria were made in a medium, based on Medium 2 (Hobson, 1969a) containing a casein hydrolysate (Bacto Casitone), yeast extract and ammonium sulphate as sources of nitrogen and vitamins, digester fluid cleared by centrifuging as a source of growth factors, a mineral solution, cysteine as reducing agent and agar. A mixed energy source was added containing glucose, with lactate for bacteria such as the veillonellae, and, as some bacteria ferment only the disaccharide from polysaccharide hydrolysis, cellobiose and maltose. The medium was buffered with sodium bicarbonate and prepared and incubated under oxygen-free CO_2. Roll-tube culture were incubated for 7 days and 30–40 colonies picked from a suitable dilution for purification and further study. Amylolytic bacteria were counted in a similar medium containing maize starch in place of the mixed sugars. Likewise, total viable counts of aerobic and facultatively anaerobic bacteria were made, but in a medium omitting the digester fluid and cysteine, and with aerobic incubation in plates for 3 days.

Cellulolytic bacteria were counted anaerobically using a paper-strip medium (Mann, 1968) and a roll-tube medium, containing digester fluid, based on that of Maki (1954), but using a fine cellulose powder in which cellulolysis was indicated by clearing of the suspension. Incubation was for 4 weeks. Colonies were picked from roll tubes into a cellulose medium and tested, after purification, for cellulolytic activity in a paper-strip medium.

Proteolytic bacteria were counted anaerobically in roll tubes of a skim-milk medium based on that of Blackburn & Hobson (1962), but with digester fluid in place of rumen fluid. After 3 days, colonies surrounded by a clear zone were picked, purified and tested semiquantitatively for proteolytic activity in a medium containing acid-precipitated casein.

Pentosan-utilizing bacteria were initially cultured for 1 week in a medium containing wheat-flour pentosan (Butterworth, Bell & Garvock, 1960) and after isolation were retested for fermentation of pentosan.

Volatile fatty acid (C_2-C_6) fermentation products were determined by gas chromatography; formic, lactic and succinic acids by paper chromatographic methods.

(ii) *Bacteria present in the piggery waste*

As mentioned above, a complete count of bacteria in a slurry like piggery waste is almost impossible to obtain. We are most likely underestimating the numbers of bacteria. On the other hand the relative counts from the waste and digester effluent should be comparable and the types obtained representative of those in the material. But again it is probable that we are underestimating, for instance, the cellulolytic bacteria as some of them may be tightly bound to plant fibres.

However, in a series of counts from the pig waste diluted to 4% of settable solids the average total count on the anaerobic medium was c. 6.4×10^8/ml, and the aerobic count c. 2.4×10^8/ml. The anaerobic count would include facultative anaerobes capable of multiplication and performance of a useful role in the digester. Similarly, the aerobic count would include these and obligate aerobes, but neither the latter group, nor microaerophilic groups of aerobes, with requirements for traces of CO_2 etc., which could not grow in either medium, would be expected to play any significant part in the digestion. Of the bacteria growing in the anaerobic medium 3 main morphological groups were apparent: (a) facultatively anaerobic streptococci, which comprise 43–74% of the isolates in different batches of waste and which were the only facultative anaerobes, apart from staphylococci in one batch; (b) bacteroides, which constituted 20–80% of the strict anaerobes and (c) clostridia, which made up the remaining anaerobes, except for the occasional curved, Gram negative rod. Cellulolytic bacteria could not be detected in many repeated cultures using any cellulose medium. In several samples the amylolytic bacteria numbered $>$ 4×10^5/ml.

The bacteriodes were Gram negative, pleomorphic rods. All fermented a variety of mono- and di-saccharides (including cellobiose) and most were amylolytic; none was cellulolytic or proteolytic. The fermentation products were formic, acetic and butyric acids. The streptococci were the same as those found in the digesters and the clostridia were the group 1 type described below.

In comparing the results here with others reported for the bacterial content of pig faeces it must be remembered that this was not an exhaustive study of the types of bacteria present in pig faeces, that the faecal flora varies with the pig ration and that the faeces had stood in admixture with urine under fairly anaerobic conditions for some time. What is apparent is that though, as will be seen below, this material is capable of building up into a digester flora, changes in the predominant groups of bacteria occur, and some groups functional in digestion appear to be in very low (undetectable) numbers in the waste. The bacteria present in the waste have properties which would enable them to grow and to some extent break down the more easily degradable material of the waste during storage.

(iii) *The development of digestion of piggery waste*

The results (Fig. 1) show the development of the digestion process in a digester initially filled with water and to which was added quantities of piggery waste so that the water was replaced in *c.* 4 weeks. The digester then continued to run at a loading rate of 0·03 lb of volatile suspended solids/ft^3/day, with a turnover time of 37·5 days. The solids content of the digester was then *c.* 2%. It is evident that, in contrast to the incubation of the undiluted waste, if the waste is initially diluted so that a rapid fermentation (with acid production sufficient to stop bacterial activity) does not take place, then over a period of 6 or 7 weeks the balance of the flora will change to give a methane fermentation capable of keeping the acids at a low level. This is shown by the bacteriological tests which can be summarized as follows. The counts on the anaerobic and aerobic media remained much the same after the first week, averaging 5×10^6–5×10^7/ml. As might be expected from the bacterial composition of the pig waste, amylolytic bacteria were present from the beginning and remained at $>4 \times 10^4$/ml. A cellulolytic flora developed only at 4–5 weeks, when the counts reached 10^4 or

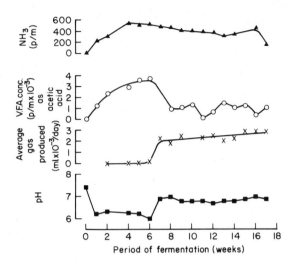

Fig. 1 The development of digestion of piggery waste.

10^5/ml, and by this time proteolytic counts were >4 x 10^4/ml. Methanogenic bacteria began to appear about the 3rd week in numbers of c. 10^3/ml and finally reached c. 10^6/ml by the 9th week. The composition of the digester gas varied as digestion developed. In the early stages the gas contained up to 25% of H_2 with 30–40% of CO_2 and 40–50% of methane. In the fully developed digestion H_2 was present only in trace amounts and the gas composition was c. 60% methane + 40% CO_2.

Bacteriologically the material from a domestic anaerobic digester used to inoculate the second digester was different from the untreated pig waste. As might be expected, the total count on the anaerobic medium was high (2·4 x 10^7/ml) and there was an amylolytic and cellulolytic flora. Counts of proteolytic bacteria were also high (>4 x 10^5/ml). The principal difference from the pig waste, though, was the large number of enterobacteria, mainly *Escherichia coli,* which made up c. 50% of the count on the anaerobic medium. The streptococci of the pig waste were not present (or were in undetectable numbers). *Escherichia coli* has been found before in large numbers in domestic sewage digesters. During the addition of pig slurry to the seeded digester the coli count decreased until they were absent, or present only in very small numbers. On the other hand the streptococci increased in number until they constituted, as in the digestion developed from water, to c. 50% of the anaerobic count. These facultatively anaerobic, streptococci, at present unclassified, like the coliforms, were not proteolytic, amylolytic or cellulolytic. As the coliforms of the original seeds disappeared when the pig waste was added one might suppose that they were not suited to conditions in the anaerobic digestion, or at least that of pig waste. It appears, that the coli or streptococci, which form a large proportion of the flora of the digester inputs, are only kept up by this continued inoculation and it seems difficult to allot to them any fundamental role in the digestion, although they play some part, if they proliferate at all, in the general fermentation of simple sugars and conversion of input nitrogen to bacterial cells. They also play probably a major part in reducing the environment of the digester to an E_h low enough for growth of the methanogenic and other strictly anaerobic bacteria. Both the flora and the digestion process were similar in established pig waste digesters whether the digester was initially seeded or not.

(iv) *The activity of anaerobic bacteria in the digesters*

A detailed description and classification of the bacteria is not needed. Rather, we shall consider a few properties helping to assign to the bacteria a function in the digestion process. All the bacteria have general fermentative properties which allow them to play a part in degradation of the low M.W. compounds in the waste, but only comparatively few isolates were capable of hydrolysing more complex substances. This is rather akin to the situation in the rumen, for instance, where, although the ruminant's diet is largely cellulose, only a small proportion of the flora is actively cellulolytic, and only certain strains of species such as *Butyrivibrio* are cellulolytic.

The ability to hydrolyse starch was the most common hydrolytic activity amongst the bacteria isolated and did not seem to be confined to any one group. Amongst the bacteria isolated from the digesters on the nonselective anaerobic medium were a group of Gram negative, short to medium length, straight rods with rounded ends, provisionally grouped as bacteroides. The majority were amylolytic and some were proteolytic, and products of sugar fermentations were a mixture of acetic and butyric acids, with in some cases formic. Another group which could be active in starch hydrolysis, but which were generally nonproteolytic, were Gram negative, pleomorphic coccobacilli isolated from the domestic digester. These could also ferment cellobiose, produced by cellulolytic bacteria and glycerol from lipid hydrolysis. Three major groups of clostridia were isolated. The majority of isolates (group I) were amylolytic, but not proteolytic or cellulolytic. Although the isolates varied in their ability to use ammonia as sole nitrogen source, they were in general very similar to *Cl. butyricum*, with acetic and butyric acids as fermentation products.

The ability to hydrolyse hemicellulose (xylan) appeared to be confined principally to one type, which was present in numbers $> 4 \times 10^4$/ml. Of 10 isolates from hemicellulose medium studied in detail 9 were classified with the rumen bacterium, *Bacteroides ruminicola:* they were Gram negative coccobacilli. The sugars fermented varied somewhat, but no strains were cellulolytic and only a minority were amylolytic. The fermentation products were largely acetic, propionic and succinic acids. The one isolated not classified as *B. ruminicola* has

MAP–5

not been identified, but was a long, slender, Gram negative rod, again producing acetic and propionic acids.

The cellulolytic bacteria were a heterogeneous group, which have not yet been classified. Generally $10^4 - 10^5$/ml were present. Only one isolate was Gram positive and this was a rod, generally curved and often in short chains. It produced mainly propionic acid, with a trace of formic and succinic acids and occasionally acetic acid, from cellulose. This isolate remained actively cellulolytic in stock cultures, almost all the other strains lost activity. The remainder of the isolates were Gram negative coccobacilli or rods of various morphologies, forming a variety of VFA as fermentation products. None of the bacteria isolated resembled the ruminococci, which are amongst the most actively cellulolytic bacteria of the rumen.

Group II clostridia were proteolytic, but not amylolytic or cellulolytic, and fermentation products from sugar media were acetic and *iso*valeric acids. Group III were proteolytic but did not ferment sugars, and presumably obtained energy from amino acid fermentations. Clostridia, so far, appear to be amongst the most actively proteolytic bacteria in the digesters. *Clostridium perfringens* was found in the domestic anaerobic digester.

Other types of bacteria were also isolated in small numbers, for instance, lactobacilli and staphylococci, but their contribution to the digestion process is uncertain.

The lipolytic bacteria were the major group not specifically sought, and these may be important, especially in domestic anaerobic digesters, although even in the pig waste digestion some "crude fat" disappeared (Table 3), presumably by hydrolysis and incorporation into bacteria. Several of the bacteria, besides these specifically mentioned, fermented glycerol produced by lipid hydrolysis.

Deamination was another activity not specifically tested for but, by analogy with the rumen, *B. ruminicola* might be expected to provide a major part of this activity, although the clostridia would undoubtedly contribute as well. The increase in $NH_3 - N$ during the digestion of pig waste (Table 3) is in accord with deaminative activity in the bacteria and also with the fact that of the bacteria so far tested only a minority were capable of growth on ammonia as sole N source. The majority required a complex medium, although some may use NH_3, with amino acids as N source. From the point of view of the purity of the effluent from the digesters, it seems desirable to increase NH_3 utilization. Unless energy sources are the limiting factor this might be brought about by inoculation of NH_3-utilizing bacteria.

The end products of the hydrolyses and fermentations carried out by the bacteria described here contained a mixture of lower VFA. A net loss of acetic acid from the cultures was sometimes found, the acetic generally being incorporated into higher acids. Hydrogen, as a fermentation product, was not checked, and formation of CO_2 could not be tested because of the bicarbonate buffer and CO_2 atmosphere used in cultures. Some isolates produced lactic or succinic acids, and, although this was another group not specifically tested for, one might expect lactate- and succinate-fermenting bacteria to be present in the digester, converting these acids to VFA, because the former acids were not found in the digesters.

(b) *Methanogenic bacteria*

The medium of Siebert & Hattingh (1967), with 20% of clarified digester fluid added, was used to count the number of bacteria producing methane from formate, acetate, propionate or butyrate in a CO_2 atmosphere. These liquid media were inoculated with diluted digester fluid in the same way as the nonmethanogenic cultures and 2 solid media were used to isolate the methanogenic bacteria. One was the medium G of Siebert *et al.* (1968): this is the synthetic one containing as substrates the gaseous phase of $H_2:CO_2$ in the ratio 80:20 and all the lower VFA. The other medium was that of Smith (1965): this again has the substrate gas phase and contained 30% of digester fluid. In one series of tests sodium acetate was also added to try to enrich for acetate-utilizing bacteria. All cultures were incubated for 4 weeks at 38°.

In the experiments on methane bacteria attention was focused on the digestions developed from the pig waste without a seed of domestic sludge. One series of tests on the domestic sludge used to seed the digesters showed, whichever medium was used, methanogenic bacteria to be present in lower numbers than in the fully developed pig waste digesters. Methane was produced from the Siebert & Hattingh (1967) medium only when formate was the substrate.

As already mentioned (p. 113) methanogenic bacteria developed over a period of weeks as the digester was established. The numbers varied somewhat with loading rate, but were 10^5-10^6/ml. Of the acids tested as substrates under CO_2 only formic or butyric acid gave rise to methane and the numbers of bacteria using these substrates were about equal. A large number of examinations was made of isolates from cultures using the multisubstrate media of Siebert *et al.* (1968) or Smith (1965), or the modified Smith medium, but all the bacteria were the same and were classified as *Methanobacterium formicicum*. This was a Gram negative rod of variable length. It produced methane from CO_2 + H_2 or formate. It did not produce methane from acetate, propionate, butyrate, *iso*butyrate, valerate, *iso*valerate, succinate, pyruvate, glucose, ethanol, propanol, butanol or *iso*propanol.

5. General Discussion

The overall reactions occurring during the anaerobic digestion of waste were outlined in the first part of this paper. The bacteriological results show that the initial hydrolysis of the waste is carried out mainly, if not entirely, by a mixed population of strictly anaerobic bacteria. The further fermentation of the hydrolysis products to acidic or gaseous endproducts is brought about by the metabolism of anaerobic and facultatively anaerobic bacteria, the latter probably also being responsible for keeping the E_h value of the digestion at a low level.

The acidic fermentation products of the carbohydrate-fermenting bacteria in pure cultures were varying amounts of the lower VFA with lactic and succinic acids. The latter would be fermented to volatile acids in a mixed culture. Small amounts of branched chain and higher acids in fermentation products probably came from metabolism of amino acids, and fermentation of amino acids by bacteria such as the clostridia leads to VFA production. The pig waste input to the digesters, which contain the unabsorbed products of bacterial fermentations in the intestines and the products of fermentations occurring while the material stood under the pig house, contained a mixture of acids. The general composition c. 65% of acetic, 20% of propionic and 8% of butyric along with small amounts of valeric and branched-chain acids: formic acid was not found. During the course of building up the digestion from pig waste the proportions of the acids changed until acetic acid formed 87—100% of the total with only, at the most, 5—6% of propionic acid and 1—2% of the higher acids. The total acid concentration in the pig waste was 7000—17,000 p/m (as acetic) and that in the digesters varied 120—2000 p/m.

The difference between the fermentation products of the bacteria in pure cultures and the measured fermentation products in the digesters could be due to a number of causes. Fermentation products can vary with the pH value of the culture. The pH of the pure cultures are initially rather lower than the pH of the digesters and fall during growth of the bacteria, so that much of the fermentation products is formed at a pH near 6. The waste standing under the pig house was also generally lower in pH than the digesters, and is more akin to the pure culture conditions. Fermentation products may also change with growth rate of bacteria and overall one would expect very different growth rates of the bacteria in pure culture and in the digesters. It has also been found, for instance with rumen bacteria, that fermentation products vary with the composition of the medium. There are, thus, many difficulties in the way of equating fermentation products *in vitro* and in the environment of the digester.

Apart from these considerations we have the major one that the acids assayed in the digesters are only the residual products of growth of a mixed culture in which end products of one bacterium may be utilized by another. In a number of the pure cultures a net loss of acetate was observed, and this could have been converted into higher acids, as is well known for the butyric acid bacteria. Some lower VFA also act as growth factors for anaerobic bacteria and are converted into cellular substance. But the greatest utilization of volatile acid fermentation products in the digestion process is generally accepted to be by methanogenesis, and it has been observed by us and by others that a reduction in acid concentration in digesters coincides with development of a methanogenic fermentation.

The results of our experiments on the isolation of methanogenic bacteria are, not in accord with the suggestion of general utilization of VFA in methanogenesis. Although the results of the gas analyses during the development of digestion suggests $H_2 + CO_2$ as a source of methane, acetate is generally considered to be the main precursor of methane. But in the liquid media with different substrates only formate (or the equivalent $H_2 + CO_2$) and butyrate gave evidence of methanogenesis, and all the bacteria tested in isolation belonged to one species of formate, or $H_2 + CO_2$, utilizing bacterium. This bacterium has previously been isolated in high numbers from domestic anaerobic sludge (Mylroie & Hungate, 1964; Smith, 1966) and it appears to be generally found in anaerobic digesters. Schnellen (1947) and Smith (1966) also isolated *Methanosarcina barkerii* which utilizes $H_2 + CO_2$ or acetate. Smith also isolated a number of other methanogenic bacteria which utilized $H_2 + CO_2$ or formic acid, but utilization of other acids appeared not to have been specifically tested Burazewski (1964), while isolating *Methanobacterium formicicum* also claimed to· have isolated bacteria producing methane from a wide variety of acids and alcohols, but this was done only by enrichment methods and the relative numbers of the different types could not be determined. It would thus seem that formate or $H_2 + CO_2$ utilizing bacteria are present in large numbers in anaerobic digesters, but the contribution of bacteria using other acid substrates is difficult to determine. Although a number of experiments have been shown that addition of acids other than formic to digester contents, or incubation of diluted contents in acid-containing media (as we have done) apparently produces methanogenesis, there remains some discrepancy between the bacteria isolated in pure culture and the supposed substrates for methanogenesis.

Several explanations are possible. The most obvious is that the media used are inadequate for isolating a variety of methanogenic bacteria but, on the other hand, bacteria apparently using each of the VFA have been isolated at various times by others using these or very similar media. However, the media for methanogenic bacteria need more investigation. If the bacteriological results are true there could be in the digesters a fermentation pattern biased towards the lower acids with a possible increase in formic acid or $H_2 + CO_2$ production by a "mass action" effect of its rapid removal by methane production and with acetic acid accumulating as a nonutilized product. The amount of acetic acid in the digesters makes this unlikely. There could be a conversion of higher VFA to formic acids which is then utilized, although our experimental results suggest that only butyric acid could be so converted. This is brought about by unknown bacteria present in the liquid, mixed cultures which produced methane from butyrate (p. 117), but it seems rather unlikely on present knowledge of bacterial metabolism. Or the *Methanobacterium formicicum* isolated can ferment higher VFA but only in the presence of unknown factors produced in the mixed culture of the digester and not present in the pure cultures. This seems unlikely. A final, again unlikely, alternative is that if the higher VFA are produced in quantity they are utilized by bacteria for purposes other than methanogenesis. The overall process of anaerobic digestion of wastes and the role of various bacteria are thus explained to some extent, but there remain many problems, some of them common to investigations of complex microbial habitats, which are in need of further study.

This work was carried out with the help of a grant from the Agricultural Research Council.

6. References

Blackburn, T. H. & Hobson, P. N. (1962). Further studies on the isolation of proteolytic bacteria from the sheep rumen. *J. gen. Microbiol.* 29, 69.

Bryant, M. P., Wolin, E. A., Wolin, M. J. & Wolfe, R. S. (1967). *Methanobacillus omelianskii*, a symbiotic association of two species of bacteria. *Arch. Mikrobiol.* 59, 20.

Buraczewski, G. (1964). Methane fermentation of sewage sludge. I. The influence of physical and chemical factors on the development of methane bacteria and the course of fermentation. *Acta. microbiol. Polon.* 13, 321.

Butterworth, J. P., Bell, S. E. & Garvock, M. G. (1960). Isolation and properties of the xylan-fermenting bacterium 11. *Biochem. J.* 74, 180.

Henderson, C., Hobson, P. N. & Summers, R. (1969). The production of amylase, protease and lipolytic enzymes by two species of anaerobic rumen bacteria. In, *Continuous Cultivation of Micro-organisms*. Eds. I. Málek, K. Beran, Z. Fencl, V. Munk, J. Ricica and H. Smrcková *et al.* London: Academic Press.

Hobson, P. N. (1969). Growth of mixed cultures and their biological control. In *Microbial Growth*. Eds A. B. Meadow & C. D. Pirt. Cambridge: University Press.

Hobson, P. N. (1969a). Rumen bacteria. In *Methods in Microbiology* 3b. Eds. J. R. Norris & D. W. Ribbons. London: Academic Press.

Hobson, P. N. (1971). Rumen micro-organisms. *Prog. industrial Microbiol.* 9, 41.

Hobson, P. N. & Summers, R. (1967). The continuous culture of anaerobic bacteria. *J. gen. Microbiol.* 17, 63.

Kirsch, E. J. & Sykes, R. M. (1971). Anaerobic digestion in biological waste treatment. *Progr. ind. Microbiol.* 9, 155.

Loesche, W. J. (1969). Oxygen sensitivity of various anaerobic bacteria. *Appl. Microbiol.* 18, 723.

Maki, L. R. (1954). Experiments on the microbiology of cellulose decomposition in a municipal sewage plant. *Antonie van Leeuwenhoek* 20, 185.

Mann, S. O. (1968). An improved method for determining cellulolytic activity in anaerobic bacteria. *J. appl. Bact.* 31, 241.

Mylroie, R. L. & Hungate, R. E. (1954). Experiments on the methane bacteria in sludge. *Can. J. Microbiol.* 1, 55.

Nottingham, P. M. & Hungate, R. E. (1969). Methanogenic fermentation of benzoate. *J. Bact.* 98, 1170.

Schnellen, C. G. T. P. (1947). Onderzoekingen over de Methaangisting. Dissertation Technische Hoogeschool, Delft.

Siebert, M. L. & Hattingh, W. H. J. (1967). Estimation of methane-producing bacteria by the most probable number (MPN) technique. *Wat. Res.* **1**, 13.

Siebert, M. L., Toerien, D. F. & Hattingh, W. H.J. (1968). Enumeration studies on methanogenic bacteria. *Wat. Res.* **2**, 545.

Smith, P. H. (1965). Pure culture studies of methanogenic bacteria. *Proc. 20th Purdue Waste Conf.* p. 583.

Smith, P. H. (1966). The microbial ecology of sludge methanogenesis. *Devs ind. Microbiol.* **7**, 156.

Tempest, D. W., Herbert, D. & Phipps, P. J. (1967). Studies on the growth of *Aerobacter aerogenes* at low dilution rates in a chemostat. In *Microbial Physiology and Continuous Culture.* Eds. Powell London: H.M.S.O.

Toerien, D. F. & Hattingh, W. H. J. (1969). Anaerobic digestion. 1. The microbiology of anaerobic digestion. *Wat. Res.*, **3**, 385.

The Microbial Ecology of the Activated Sludge Process*

E. B. Pike and C. R. Curds

*Department of the Environment, Water Pollution Research Laboratory,
Elder Way, Stevenage, Herts, England*

CONTENTS

1. Introduction

THE ACTIVATED SLUDGE process is, in general, a continuous culture process. In contrast to the simple chemostat, its dominant features are the continuous inoculation of the aerated culture with micro-organisms present in the incoming sewage and the feedback of most of the cell yield, as activated sludge, to the culture vessel. These serve to encourage rapid adsorption, uptake and oxidation of pollutants, and to maintain stable operation over a wide range of dilution rates and incoming nutrient concentrations, caused by fluctuations in the strength and flow of sewage. Because satisfactory operation of a plant requires the concentrations of nutrients and cells in the final effluent to be low and because the cell yield, drawn off from the plant as waste sludge, presents disposal problems, dilution rates in the aeration tank are typically low, so that specific growth rates of the sludge organisms and nutrient levels are also low. It is also essential that the operating conditions in the aeration tank should give rise to compact, flocculent growth, able to settle rapidly. Difficulties in operation arise when growth of filamentous micro-organisms produce a "bulking sludge" condition, in which case settling is retarded and cells escape in the effluent, thus giving it a high content of suspended solids. Because of these practical interests in the process, there has been much research into the kinetics of nutrient uptake and into the organisms and conditions giving rise to flocculation and sludge bulking. These apart, the microbial ecology of the process has not been explored to any great extent; this paper attempts to survey what is known and to provide suggestions for future study.

2. Typical Plants and Their Operation

The reader is referred to the works of McCabe & Eckenfelder (1956), Bolton & Klein (1961), and Hawkes (1963), and to papers by Ainsworth (1966), Abson & Todhunter (1967), and Coackley (1969) for full details. The mechanisms for aeration, which may be by supplying diffused air or by vigorous mechanical agitation, also serve to mix the activated sludge liquor. Mixing is not normally complete, except possibly in small aeration tanks, and conditions usually diverge from the ideal of a homogeneous, completely mixed reactor with feedback (Herbert, 1961) to those of a tubular, pipe flow reactor, to an extent depending on the design, particularly where flow is through long, deep aeration channels, or through a series of individual homogeneous aeration pockets. Oxygen uptake is most rapid in the initial stages of aeration; this is sometimes provided for by installing a greater degree of aeration at the influent end of the tank (tapered aeration), or by dividing the influent flow and injecting it at intervals along the aeration tank (step aeration).

Activated sludge treatment is usually a secondary process, treating an influent which has already been subjected to primary settling to remove grosser solid material. Its principal advantage over other treatments — a direct result of the feedback mechanism — is its compactness for a given rate of substrate removal. In conventional treatment of wastes with a 5 days biochemical oxygen demand (BOD) of c. 250 mg/l, the mixed liquor suspended solids content, a rough indication of the micro-organism concentration, is limited to 6000 mg/l or less by the ability of the settling stage to concentrate and return sludge reliably Were more efficient methods of cell concentration and return to be evolved, it would no doubt be possible to increase the efficiency of the process, or to handle wastes much richer in nutrients, as from the food and fermentation industries, or from farms (Hardt, Clesceri, Nemerow & Washington, 1970; Pirt & Kurowski, 1970). Such wastes can be given initial aerobic treatment by so-called "dispersed growth" systems, in which the sludge is retained as far as possible within the aeration system, as in the Pasveer oxidation ditch, and is not fed back.

3. The Activated Sludge Community and Selective Mechanisms

The activated sludge community is specialized and shows a lower diversity of species than does a biological filter. The community is dominated by heterotrophic bacteria, both aggregated in the sludge flocs and freely dispersed in the liquor. The bacteria, together with saprobic protozoa, form the basic trophic level, followed by holozoic protozoa feeding on bacteria. Fungi are poorly represented, except in certain bulking conditions where they may predominate. Small numbers of rotifers and nematode worms may be present, but algae are normally absent. The implications of ecology in activated sludge treatment are discussed by Hawkes (1963) and Pipes (1966).

Considerable selective pressures operate within the process and arise from the intrinsic nature of the reactor, its modes of operation and the nature of the waste.

Although it is possible to apply basic kinetic theory to the activated sludge plant, this might be thought to present considerable difficulties, since one is dealing with a mixed population and a complex substrate which varies in concentration and flow rate. Nevertheless, if simplifying assumptions are made, ecological behaviour can be predicted from theory, and these predictions agree reasonably with observations. This is discussed specifically in relation to nitrification by Downing, Painter & Knowles (1964), and later in this paper in connection with mixed systems of bacteria and protozoa. Under steady conditions the growth rate of the sludge organisms is equivalent to the specific sludge wastage rate (mass of organisms removed in unit time/total mass of organisms in the aeration tank), which is independent of the dilution rate in the reactor (influent flow rate/aerator volume). Thus, short mean aeration times and high sludge wastage rates encourage high growth rates of sludge organisms, suppressing higher trophic levels and selecting fast-growing bacteria, although removal of nutrients is less complete. Downing et al. (1964) show that for slow growing bacteria, such as *Nitrosomonas* spp, to become established a minimum aeration period must be given, and this is roughly equal to the fractional increase in the concentration of activated sludge during aeration divided by the growth rate constant of the organism in the mixed liquor.

By applying the same argument, it can be seen that the settling stage selects flocculating organisms; as freely suspended bacteria are removed constantly in the effluent and by ciliated protozoa, they are required to grow faster than flocculating organisms to become established.

Selective pressures will also be exerted by the presence of toxic materials in the influent, excess or deficiency of certain nutrients, pH value and the concentration of dissolved oxygen in the aeration tank.

4. Bacteria in the Activated Sludge Process

(a) Study of populations

Although bacteria predominate in activated sludge, the study of their populations presents technical difficulties, which are discussed by Pike, Carrington & Ashburner (1972) and are only mentioned briefly here. Before total or viable counts can be made homogenization must be used to release individual bacteria from the flocs with minimum cell damage. Various combinations of homogenization and suspending medium have been advocated (Allen, 1944; Gayford & Richards, 1970; Williams, Stafford, Callely & Hughes, 1970; Williams, Forster & Hughes, 1971) and these generally result in increased viable counts of roughly one or two orders of magnitude over counts obtained before homogenization.

Work at this laboratory (Table 1) and by Sladká & Zahrádka (1970) shows that there is a great discrepancy between total microscopic counts of bacteria from sewage treatment processes and viable aerobic counts. This can be attributed in part to the difficulty of distinguishing microscopically between bacteria and suspended debris and to inadequacies of cultural and physiological conditions provided in the plate count procedure. The data (Table 1) suggest that particles of bacterial size comprise much of the suspended material in activated sludge liquor and indeed in other stages of sewage treatment: a total cell count of $1.4 \times 10^{12}/g$ of dry suspended biomass could be given by rod-shaped particles ($1.5 \times 0.7\mu m$) of density 1.1 g/cm^3. It is reasonable to suppose that many of the bacteria in activated sludge may be moribund or dead, since the process is typically poised at the stationary or decline phases of growth under low nutrient concentrations. Greatly decreased viability has been shown to occur in continuous cultures of *Aerobacter aerogenes* grown at very low dilution rates (Tempest, Herbert & Phipps, 1967). Unz & Dondero (1970) succeeded in cultivating only 203 of 1498 bacterial cells micromanipulated from activated sludge. It is also probable that the physiological conditions for substrate accelerated death (Postgate & Hunter, 1963) may be present, because the viable counting procedure involves suspending cells grown under near-starvation conditions in a nonnutrient diluent and transferring them to a rich growth medium.

Although some taxonomic or population studies have been made in which a single nonselective medium, such as nutrient agar (Allen, 1944), nutrient agar with skim milk (Jasewicz & Porges, 1956), tryptone-glucose-extract-skim milk agar (Adamse, 1968) or sewage agar (Dias & Bhat, 1964), was used for isolating or enumerating strains, the experiments of Prakasam & Dondero (1967a, b) and Lighthart & Oglesby (1969) show that no single medium can be relied on to support growth of all the nutritional types of activated sludge. Various workers have attempted to group isolates by their nutritional requirements, following the

Table 1

Numbers of total and viable bacteria in samples from different stages of sewage treatment and in the suspended biomass

| Source (and no.) of samples § | Bacterial count* | | | | % of bacteria viable |
| | In samples (no./ml) | | In biomass† (no./g) | | |
	Total	Viable	Total	Viable	
Settled sewage (22)	$6 \cdot 8 \times 10^8$	$1 \cdot 4 \times 10^7$	$3 \cdot 2 \times 10^{12}$	$6 \cdot 6 \times 10^{10}$	$2 \cdot 0$
Activated sludge mixed liquor (20)	$6 \cdot 6 \times 10^9$	$5 \cdot 6 \times 10^7$	$1 \cdot 4 \times 10^{12}$	$1 \cdot 2 \times 10^{10}$	$0 \cdot 85$
Filter slimes (18)	$6 \cdot 2 \times 10^{10}$	$1 \cdot 5 \times 10^9$	$1 \cdot 3 \times 10^{12}$	$3 \cdot 2 \times 10^{10}$	$2 \cdot 5$
Secondary effluents (10)	$5 \cdot 2 \times 10^7$	$5 \cdot 7 \times 10^5$	$4 \cdot 3 \times 10^{12}$	$4 \cdot 7 \times 10^{10}$	$1 \cdot 1$
Tertiary effluents (10)	$3 \cdot 4 \times 10^7$	$4 \cdot 1 \times 10^4$	$3 \cdot 4 \times 10^{12}$	$4 \cdot 1 \times 10^9$	$0 \cdot 12$

*Counts are geometric means. Total counts using Helber counting chamber, viable counts by plate dilution-frequency method upon CGY agar, incubated at 22° for 6 days (Pike, Carrington & Ashburner, 1972).
†dry wt of suspended solids retained by Whatman GF/C paper.
§from sewage works and laboratory plants treating mainly domestic sewage.

procedure of Lochhead & Chase (1943) for soil bacteria. Dias & Bhat (1964) found that only 24% of 110 raw sewage isolates and 8% of 150 activated sludge isolates required neither vitamins nor amino acids for growth in a basal medium containing glycerol, succinate and ammonium nitrate. Of 71 *Arthrobacter* strains isolated by Adamse (1970) from a dairy waste plant, 55% needed added vitamins for growth with inorganic nitrogen (N) and only 17% for growth with amino acids. Prakasam & Dondero (1967*a, b*) found that agar media containing only activated sludge extract as a nutrient gave counts generally higher than the other media tested, but that the efficacy of an extract varied with source of the extract and of the sample plated. Approximately half of the 127 strains isolated on activated sludge extract agar failed to grow on any of the defined media containing glucose, amino acids, vitamins, yeast extract and mineral salts.

The classical approach to description of bacterial populations in ecosystems by taxonomic study and identification of.isolates has drawbacks, in that (a) many isolates have to be examined; (b) and even then, important, but minority, classes of bacteria may not be isolated; and (c) the results can only be qualitative or partly quantitative. The labour of screening many isolates can be reduced by replica plating (Lighthart & Oglesby, 1959; Prakasam & Dondero, 1967*b, c,* 1970), or by using simultaneously a variety of diagnostic or selective media to count different populations directly (Harris & Sommers, 1968; Department of the Environment, 1971). The data obtained by these methods of study is suitable for interpretation by the methods of numerical taxonomy (Sokal & Sneath, 1963) to enable either samples or bacteria (isolates or populations) to be grouped on the basis of mutual association or correlation, as in the studies of activated sludge populations by Lighthart & Oglesby (1969) and Prakasam & Dondero (1970). Fully quantitative data are suitable for factor analysis by the method of principal components, which enables the multivariate behaviour of the ecosystem to be represented by a small number of factors, as has been attempted for populations of soil bacteria (Sundman & Gyllenberg, 1967; Sundman, 1970) and for expressing changes in chemical, bacterial and biochemical characters in a nitrogen-limited sludge digester (Toerien *et al* 1969).

(b) *Bacterial genera in activated sludge*

The majority of the genera which have been described are of Gram negative bacteria (Table 2). Coli-aerogenes bacteria account for only a small fraction of the viable population in sewage (Harkness, 1966) and in activated sludge (Allen, 1944). The proportion of *Escherichia coli* in the total coli-aerogenes population was shown to decline from 75% in raw sewage to 25 and 30%, respectively, in activated sludge and effluent (Dias & Bhat, 1965). Although *Nitrosomonas* spp. can be readily isolated from activated sludge by enrichment (Loveless & Painter, 1968), the presence and activity of this organism and of *Nitrobacter* spp. is most readily gauged from chemical analysis of liquors.

Table 2

The principal genera which have been recorded in
taxonomic studies of activated sludge bacteria

Genus*	References
Pseudomonas	Allen (1944), Jasewicz & Porges (1956), van Gils (1964), Adamse (1968), Lighthart & Oglesby (1969), Tezuka (1969), Unz & Dondero (1970)
Comamonas	Dias & Bhat (1964)
Lophomonas	van Gils (1964)
Nitrosomonas	Loveless & Painter (1968)
Zoogloea	Butterfield (1935), Dias & Bhat (1964), Tezuka (1969)
Sphaerotilus	Harkness (1966), Austin & Forster (1969)
Large, filamentous micro-organisms	Cyrus & Sladká (1970), Sladká & Zahrádka (1970)
Azotobacter	Dias & Bhat (1965)
Chromobacterium	Allen (1944)
Achromobacter	Allen (1944), van Gils (1964), Austin & Forster (1969), Lighthart & Oglesby (1969)
Flavobacterium	Adamse (1968), Austin & Forster (1969), van Gils (1964), Jasewicz & Porges (1956), Lighthart & Oglesby (1969), Tezuka (1969), Unz & Dondero (1970)
Coli-aerogenes bacteria	Dias & Bhat (1965), Austin and Forster (1969)
Micrococcus, Staphylococcus	Allen (1944), Jasewicz & Porges (1956)
Bacillus	Jasewicz & Porges (1956)
Arthrobacter, coryneforms	Dias & Bhat (1965), van Gils (1964), Adamse (1968, 1970)
Nocardia, Mycobacterium	Anderson (1968)
Bacteriophage, Bdellovibrio	Dias & Bhat (1965)

*The authors' terminologies, generally.

5. Protozoa in the Activated Sludge Process

Anyone who has looked at activated sludge through a microscope will know that the first organisms noticed after the bacterial sludge flocs are the protozoa. For many years these animals have been known to be present in both aerobic and anaerobic stages of sewage treatment, but it is only during the last decade that a clear picture of their role and importance in aerobic sewage treatment processes has began to emerge. In activated sludge, many species of protozoa have been identified and total numbers of protozoa of the order of 50,000 cells/ml of mixed liquor are frequently recorded. Populations of this size have been estimated (Ministry of Technology, 1968) to constitute c. 5% of the suspended solids in the mixed liquor. About 228 species of protozoa have been reported to occur in activated sludge plants, and this includes 17 species of phytoflagellates, 16 of zooflagellates, 25 of amoebae (both naked and thecate), 6 actinopods and 160 species of ciliates. In addition to being the class represented by the greatest variety of species, the ciliates are usually the dominant protozoa both numerically and from biomass estimations. But this is not always the case, because Brown (1965), Sydenham (1968) and Schofield (1971) have reported that amoebae of various types can be the dominant forms in activated sludge plants. It is not clear, however, how frequently this situation arises. Curds & Cockburn (1970a) examined the protozoan populations of 56 activated sludge plants in England, Scotland and Wales and concluded that ciliated protozoa were by far the most common dominant forms, although on occasion both amoebae and flagellates were seen in small numbers. It seems likely, from the literature, that although flagellates are occasionally present in activated sludge they are not as common in that process as they are in percolating filters, and the evidence available suggests that flagellates are associated with organically overloaded activated sludges.

Although many species of ciliated protozoa have been identified, 34% (54 species) of the ciliate species recorded belong to a single order, the Peritrichida, examples of which are *Vorticella* spp., *Opercularia* spp., and *Epistylis* spp. All the peritrichous species found in activated sludge are sessile types and attach themselves, usually by means of a stalk, to the sludge floc; furthermore, these are generally numerically the dominant ciliates found in this process. Two other orders are represented by significant numbers of species, these are the orders Gymnostomatida and Hypotrichida, and species of the latter order are usually present in greater numbers than those of the former. Curds & Cockburn (1970a) suggested that the species shown in Fig. 1 were the most important protozoa in activated sludge, because they were found most frequently and in the largest numbers. From this the following points arise: (i) all of the most important protozoa are ciliates; (ii) all but one (*Trachelophyllum pusillum*) are known to feed on bacteria; and (iii) most of the important protozoa are either sessile or crawl over the surface of sludge flocs.

(a) *Protozoa as indicator organisms*

Several authors (Agersborg & Hatfield, 1929; Cramer, 1931; Tomlinson, 1939; Reynoldson, 1942; Barker, 1946; Baines, Hawkes, Hewitt & Jenkins, 1953; and Curds & Cockburn, 1970b) have suggested the use of protozoa as indicator organisms in the assessment of the condition of the sludge, and there is a correlation between the condition of the sludge develops (Curds, 1966).

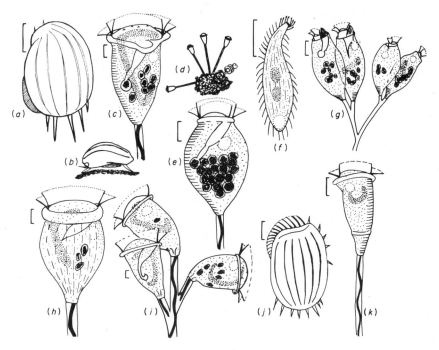

Fig. 1. The most commonly found protozoa in activated sludge. *(a) Aspidisca costata* (dorsal view), *(b) A. costata* (lateral view to show crawling habit), *(c) Vorticella convallaria, (d) V. convallaria* attached to sludge floc, *(e) Vorticella microstoma, (f) Trachelophyllum pusillum, (g) Opercularia coarctata, (h) Vorticella alba, (i) Carchesium polypinum, (j) Euplotes moebiusi, (k) Vorticella fromenteli.* Adjacent scales represent 10 μm.

Generally, sludge is suggested to be in a poor condition when there are few ciliates and many flagellates present; the latter diminish as the sludge improves until, in a good sludge, ciliates predominate. There are scattered reports which indicate that the effluent quality is at its best when the ciliate population consists predominantly of peritrichs (Reynoldson, 1942; Baines *et al.* 1953). Many authors have suggested that certain species, or small groups of species, indicate particular sludge conditions, whereas in the scheme proposed by Curds & Cockburn (1970*b*) all the species of protozoa present are taken into account. They have suggested a method whereby the approximate range of the BOD of the effluent may be predicted from a knowledge of the species structure of the protozoan populations of the sludge. A field trial of 36 plants showed that 80% of the predictions were correct and that those incorrectly predicted were mainly borderline cases. Curds & Cockburn (1970*b*), however, stress that all indicator species systems should be applied with caution since they tend to oversimplify extremely complex ecological situations and there is not really sufficient fundamental knowledge to permit positive predictions to be made.

(b) Role of protozoa in the activated sludge process

Protozoa were originally thought to be harmful to the activated sludge process and their removal was suggested by Fairbrother & Renshaw (1922). Most other authors, however, have since proposed that protozoa are of some benefit to the process for a variety of reasons and to a variable extent. Pillai & Subrahmanyan (1942, 1944), for example, expressed the view that bacteria were of secondary importance in the purification processes claiming that the peritrich *Epistylis*, even in the absence of bacteria, could account for a 70% reduction in the permanganate value, as well as the albuminoid and ammonia N concentrations in sewage, and that without protozoa sludge would not grow, nor was there any clarification. Furthermore, they claimed that the *Epistylis* sp. was also responsible for 80% of the nitrification in activated sludge. Only a few of these claims have been substantiated and now most authors agree that protozoa play a secondary but important role in aerobic waste treatment processes.

Curds, Cockburn & Vandyke (1968) were able to grow protozoa-free activated sludges from mixed bacterial populations which they had isolated from a sludge obtained from a full-scale plant. The bacterial sludge was grown initially as a batch culture and then inoculated into 6 replicate bench scale, activated sludge plants which were completely enclosed; the sludges were kept free from protozoa by dosing them with cool, heat treated sewage. This made it possible for the first time to study the effect of the subsequent addition of protozoa to the plants, and therefore to assess the role of protozoa and to quantify the magnitude of their effect upon effluent quality.

Under protozoa-free condition all 6 plants produced highly turbid effluents of inferior quality, and the turbidity was significantly related to the presence of very large numbers of bacteria suspended in the effluent. Without protozoa the mean 5 days BOD of the effluents was high, as were the contents of organic carbon and suspended solids and other effluent parameters (Table 3). Cultures of ciliated protozoa were then added to 3 of the plants whilst the other 3 were ~~used as controls without protozoa. After a few days,~~ during which time the protozoan population became properly established, there was a dramatic improvement in the quality of the effluents from the plants containing protozoa. Their clarity was greatly improved and this was associated with a significant decrease in the concentration of viable bacteria. Furthermore, the effluent BOD and concentration of suspended solids also decreased significantly. The 3 units still operating without protozoa continued to deliver turbid, low quality effluents. The ranges of effluent qualities obtained from bench-scale pilot plants operating in the presence and absence of protozoa are summarized in Table 3, and from these experimental data it was concluded that ciliated protozoa have a highly significant beneficial effect on activated sludge effluents.

It has been suggested that protozoa play a direct part in the formation of the

Table 3

Effect of ciliated protozoa on the effluent quality of bench scale, activated sludge plants

Effluent analysis	Value	
	Without ciliates	With ciliates
BOD (mg/1.)	53–70	7–24
COD (mg/1.)	198–250	124–142
Permanganate value (mg/l)	83–106	52–70
Organic nitrogen (mg/l)	14–21	7–10
Suspended solids (mg/l)	86–118	26–34
Optical density at 620 nm	0·95–1·42	0·23–0·34
Viable bacteria count (millions/ml)	106–160	1–9

sludge by causing or enhancing some of the flocculation (Viehl, 1937; Barritt, 1940; Jenkins, 1942; Pillai & Subrahmanyan, 1942; Hardin, 1943; Watson, 1945; Barker, 1943, 1946; Sugden & Lloyd, 1950; Curds, 1963). This has been shown to be due to mucus secreted from the peristome in the case of *Balantiophorus minutus* (Watson, 1945) and due to a mucus and ·a polysaccharide in the case of *Paramecium caudatum* (Curds, 1963). However, it is well known that many bacteria are able to flocculate or grow in flocculant forms, e.g. *Zoogloea ramigera,* without the aid of protozoa. Furthermore, studies on the feeding rates of protozoa using both batch (Curds & Cockburn, 1968) and continuous culture (Curds & Cockburn, 1971) techniques have shown that if the protozoan populations of activated sludge ingest bacteria at rates similar to those of the ciliate *Tetrahymena pyriformis* in pure culture the removal of dispersed bacteria in the effluents noted in the bench scale, activated sludge experiments could be explained simply by predation alone, without the necessity of considering the possibilities of protozoan induced flocculation.

6. The Occurrence of Fungi

Although mould fungi and, to a less extent, yeast can be isolated apparently quite regularly from activated sludge (Cooke, 1956; Cooke & Pipes, 1969) they do not normally form an appreciable part of the flora, except under certain bulking conditions. Consequently the ecology of fungi has been extensively studied by only a few authors. Cooke (1963) has published a comprehensive laboratory manual for isolating and identifying fungi in waste waters. Cooke & Pipes (1969) isolated 20 species of fungi and yeasts from 35 samples taken from 18 activated sludge plants; the widest distribution in the samples was of *Geotrichum candidum, Trichosporon* sp. and *Penicillium* spp. but the highest proportions of the total colonies observed were of *Cephalosporium* spp. (38%), *Cladosporium cladosporiodes* (22%) and *Penicillium* sp. (19%). Only 1% of the total colonies were yeasts. The fungal types found could be considered to belong to the ecological group of sugar fungi, characteristically, although not exclusively, using simple carbohydrates. The authors found no evidence for antibiotic activity in activated sludge tanks, although this might have been supposed from the presence of *Trichoderma viride* and *Penicillium* spp. The smallest numbers and the fewest species of fungi were found in the winter; numbers increased to a peak in the autumn.

The predacious fungi, *Zoophagus* sp. and *Arthrobotrys* sp., have been reported in pilot plants by Cooke & Ludzack (1958), Pipes & Jenkins (1965) and Pipes (1965), where they were capturing rotifers and nematodes. The last two authors gave evidence for a succession, in which large developing populations of rotifers were invaded and controlled by *Zoophagus* spp., which then became dominant and caused a transitory bulking condition in the sludge. Sladká & Zahrádka (1970), recorded the predacious fungus *Dactylaria* in an experimental plant.

Sludge bulking may be caused by *Geotrichum* spp., which can regularly dominate the flora in a plant (Hawkes, 1963; Schofield, 1971). This organism has been mistakenly identified for *Sphaerotilus* in bulking sludge (Pipes & Jones, 1963) and appears to have a competitive advantage over the microbial population, as a whole, when nitrogen or phosphorus is limiting (Jones, 1965).

7. Flocculation

Flocculent growth is not necessary for efficient substrate removal during aeration, but is essential for ensuring a clear effluent and sufficient concentration of the sludge returned from the gravitational settling tank. Several theories have been proposed to account for flocculation, and they have been reviewed by Painter & Hopwood (1967, 1969), Coackley (1969) and Sladká & Zahrádka (1970). The early view, that a specific bacterium, *Zoogloea ramigera,* was responsible for flocculation, by secreting a gelatinous matrix to which other organisms adhered and became enveloped (Butterfield, 1935), is not tenable, as a variety of bacteria can be isolated from activated sludge flocs (Table 2) and those able to flocculate in aerated pure cultures belong to a variety of genera

(McKinney & Horwood, 1952; McKinney & Weichlein, 1953; McKinney, 1956; Dias & Bhat, 1964; Anderson, 1968). There seems to be general agreement that flocculation is explicable by the principles of colloid science, but there is disagreement over the actual mechanisms involved. It is recognized that operation of plants at high dilution rates and high substrate concentrations tends to give a more dispersed or filamentous growth, and that flocculation is encouraged by the opposite conditions (Tenney & Stumm, 1965; Sladká & Zahrádka, 1970). McKinney (1952, 1956) suggested that the normal dispersed condition of bacterial cultures was caused by mutual repulsion between negatively charged cell surfaces and that flocculation in older cultures resulted from their lack of energy. Peter & Wuhrmann (1971) present evidence suggesting that bacterial suspensions resemble protected dispersoids and that, since flocculation can be brought about experimentally by adding optimum amounts, in relation to the bacterial surface, of cationic polyelectrolytes, flocculation in treatment processes could be caused by the presence of naturally occurring polyelectrolytes, such as humic acids, in the liquor. Similarly, Tenney & Stumm (1965) suggest that flocculation is caused by other natural polyelectrolytes, such as complex polysaccharides and polyamino acids, excreted at the cell surface during the decline and endogenous growth phases.

Although the accumulation of intracellular poly-β-hydroxybutyric acid (PHB) has been associated with flocculation of pure cultures of activated sludge isolates (Crabtree, McCoy, Boyle & Rohlich, 1965), later work indicates that this may not be the sole mechanism involved (Painter & Hopwood, 1969), or that there may be little causal connection with flocculation in activated sludge plants. Dias & Bhat (1964) found that 51% of their isolates and 95% of *Zoogloea* isolates were capable of forming sudanophilic inclusions. Although Crabtree *et al.* (1965) reported that their activated sludge solids contained 12% (dry wt) of PHB much smaller amounts, or none, were reported by van Gils (1964) and Painter, Denton & Quarmby (1968) in activated sludge. Experiments with *Zoogloea* spp. (Angelbeck & Kirsch, 1969) and *Flavobacterium* spp. (Tezuka, 1969) from sludge did not relate PHB levels to the ability to flocculate.

Flocculation in pure cultures of zoogloea and other bacteria was shown to be independent of C:N ratio in the medium (van Gils, 1964; Tezuka, 1967) and to be induced by low pH values, whereas the effect of calcium and magnesium ions was conflicting, either bringing about flocculation (van Gils, 1964; Tezuka, 1967) or reversing aggregation (Angelbeck & Kirsch, 1969; Tezuka, 1969).

The evidence given suggests that the flocculation mechanism is still incompletely understood, and much of the evidence relates to pure cultures of zoogloeia and other bacteria producing flocculent growth in laboratory media. There would seem much to be gained from future studies of the activated sludge biocoenosis under different conditions of growth and of the chemical nature of the surface of sludge bacteria and from the applications of the principles of colloid science.

8. Bulking of Activated Sludge

Bulking is a growth condition in which the sludge has poor settling qualities (high "sludge volume index") because of loose, flocculent, cotton wool-like growth of filamentous micro-organisms. Uncritical observers in the past have ascribed the condittition exclusively to *Sphaerotilus* spp., although it is now realized that bulking sludge represents a community, dominated by *Sphaerotilus* spp. or other filamentous micro-organisms such as *Bacillus, Nocardia, Beggiatoa, Thiothrix* (Pipes, 1969), Vitreoscillaceae, *Leucothrix, Lineola* (Cyrus & Sladká, 1970), *Geotrichum* (Pipes & Jones, 1963; Schofield, 1971) and *Zoophagus* (Cooke & Ludzack, 1958; Pipes, 1965).

Because bulking seriously interferes with plant operation and effluent quality and because of the nuisance caused by the closely related "sewage fungus" communities growing in polluted water, much research has been made into the physiological conditions encouraging these communities and, in particular, the growth of *Sphaerotilus* spp. (Mulder, 1964; Phaup, 1968; Curtis, 1969). *Sphaerotilus* spp. can utilize a variety of carbohydrates and either amino or inorganic N, the last also requiring vitamin B_{12}. Bulking appears to be encouraged by low dissolved oxygen concentrations, the combined effect of high C:N and C:P ratios (Hattingh, 1963), or primarily by N deficiency (Dias, Dondero & Finstein, 1968). Bulking activated sludge and pure cultures of *Sphaerotilus* spp. have a low DNA content compared with normally settling sludge and *Zoogloea ramigera* (Genetilli, 1967).

It also appears that a low "sludge age" (the reciprocal of the specific sludge wastage rate defined above) can result in bulking (Sladká & Zahrádka, 1970); if so, it is easily deduced that bulking may be a self stable condition, as poor settling will result in further loss of sludge organisms in the effluent and a further decrease in sludge age. This is equivalent to a selection for fast growing organisms.

9. Removal of Bacteria and Pathogenic Organisms

Pathogenic micro-organisms, including enteric viruses, can be isolated regularly from sewage; their presence is indicative of the general state of health of the community, although some may originate from other than human sources (Kabler, 1959; Clarke & Kabler, 1964; Malherbe, 1964; Berg, 1966*a*; Grabow, 1968; Leclerc *et al.*, 1971).

Among sewage treatments, the activated sludge process is particularly effective in achieving a reduction in numbers of bacteria, including pathogens, and of enteric viruses in effluents, compared with numbers in incoming sewage, although its effect on cysts of parasitic protozoa and worm ova is usually not great (Kabler, 1959). Reductions of 90–99% have been quoted in the numbers of total bacteria, *E. coli, Salmonella* spp., *Shigella* spp., and *Vibrio cholerae* entering treatment (Kabler, 1959; Keller, 1959; Kampelmacher & Noorle Jantzen, 1970). Although some of these bacteria may be removed from the ecosystem with the waste sludge, which tends to show an increase in numbers compared with the incoming sewage (Kabler, 1959), there is evidence that the prime agency is the grazing of ciliated protozoa, since in the laboratory experiments of Curds & Fey (1969) the half life of *E. coli* in a plant containing protozoa-free activated sludge was 16 h, compared with 1·8 h in the presence of protozoa. It appears that, although bacteriophages and *Bdellovibrio* sp., specific for pseudomonads and enterobacteria, are present in activated sludge liquor, they are not effective at removing bacteria (Dias & Bhat, 1965).

Activated sludge treatment achieves a significant, but not complete, removal of enteroviruses and reductions of 79–99·98% are quoted for full-scale and experimental plants (Clarke & Kabler, 1964; Clarke, Stevenson, Chang & Kabler, 1961; Malherbe, 1964; Berg, 1966b; Lund, Hedström & Jantzen, 1969). The removal mechanisms appear to be by adsorption of virus particles on the flocs, obeying a Freundlich adsorption isotherm (Clarke, Stevenson, Chang & Kabler, 1961).

10. Mathematical Approach to the Ecology of Activated Sludge

Many of the observations made on protozoa in activated sludge can now be explained on the basis of the growth kinetics and settling properties of the organisms, and on the operational characteristics of the plant. Many workers have derived mathematical models for the activated sludge process (Herbert, 1961; McKinney, 1962; Grieves, Milbury & Pipes, 1964; Schulze, 1964, 1965; Reynolds & Yang, 1966); all of which have considered sludge to consist of a single flocculating micro-organism: a few (Downing & Wheatland, 1962; Downing & Knowles, 1966) have considered the dynamics of different organisms and the effects they might have upon plant performance. Recently, Curds (1971a) used mathematical models and computer simulation techniques to study the population dynamics of bacteria and protozoa growing together in a single-stage continuous culture. He also considered (1971b) the dynamics and fate of a variety of micro-organisms growing together in a completely mixed, activated sludge reactor fitted with feedback. In this, he dealt with a hypothetical activated sludge plant which receives and contains a number of different micro-organisms defined by the following characteristics:

(a) *dispersed sewage bacteria,* which are those bacteria borne in sewage in considerable quantities but which on entering the reactor do not flocculate:

they remain in suspension in the sedimentation tank and are evenly dispersed throughout the effluent and recycle flows;

(b) *sludge bacteria,* which form the bulk of the sludge mass and are assumed not to be present in the sewage flow and always to flocculate completely;

(c) *soluble substrate,* which for the purposes of the model is considered to be the limiting substrate contained in the sewage flow, all other necessary elements being regarded as present in excess; this substrate is utilized by both type of bacteria but not ciliated protozoa; and

(d) *ciliated protozoa,* which are assumed to feed entirely on dispersed bacteria since it is argued that the sludge bacteria are present only as flocculated masses too large for ingestion. It is assumed that these organisms are not present in sewage and that there are two types. First, free swimming types which because of their habit never flocculate and remain evenly dispersed in the sedimentation tank; second, sedentary protozoa which always flocculate since they are assumed to be attached directly to the masses of flocculant sludge bacteria.

Of the above populations only the dispersed bacteria, free swimming ciliates and soluble substrate are able to leave the plant with the effluent; sludge bacteria and sedentary protozoa never leave with the effluent but some are continually pumped away at a fixed rate after sedimentation. One further important assumption was also made; that all organisms obey Michaelis-Menten growth kinetics, the specific growth rates of the two bacteria being related to the concentration of soluble substrate and the specific growth rate of the ciliate population to the concentration of dispersed bacteria present in the reactor. All simulations were carried out on an IBM S/360 digital computer using the highly sophisticated language Continuous System Modeling Program (CSMP/S/360). The values of the various kinetic constants and the substrate and bacterial concentrations of the sewage used for these studies were thought to be feasible, and in some instances were based on the results of experimental work. Arbitrary starting values were assigned to the populations and the computer was programmed to integrate these populations against time using the appropriate mass balance and kinetic equations.

Initially two separate simulations were carried out to ascertain whether the activated sludge populations were dynamically stable; both simulations included dispersed sewage and sludge bacteria, but in one the ciliate population consisted only of free swimming forms, and in the other the ciliates were only attached sedentary forms. The two simulations were therefore carried out as a direct comparison of the effect of ciliate habit on the dynamics and concentrations of the various microbial populations. In each case the initial starting values for the population sizes were assumed to be low and equal; as time proceeded the concentrations of sludge bacteria increased, whereas the connections of dispersed sewage bacteria and substrate decreased. The attached ciliate population increased with time but the free swimming one decreased. All populations, however, asymptotically approached steady-state conditions without signs of oscillation, so it was concluded that the model was dynamically stable. Under these circumstances it is valid to solve, manually, the mass balance equations for simple steady-state solutions at various rates. This has been done over a wide range of sludge specific wastage rates and sewage dilution rates (keeping the wastage rate at 1/10th that of the dilution rate) under 3 situations: (a) when the protozoa are attached forms; (b) when the protozoa are free swimming ciliates, and (c) when ciliated protozoa are not present. The results are illustrated in Fig. 2.

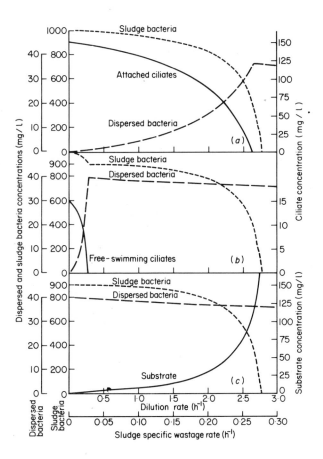

Fig. 2. Steady state populations of micro-organisms at various dilution and sludge wastage rates. *(a)* Attached ciliates present; *(b)* free swimming ciliates present; *(c)* ciliates absent. Substrate concentration curve in *(c)* was also obtained in *(a)* and *(b)*.

It can be shown by simple algebra that under steady state conditions the specific growth rate of settling organism (sludge bacteria and attached protozoa) is equal to the specific wastage rate of the sludge. This explains why in all 3 situations the concentration of substrate is precisely the same at any given wastage rate. The microbial populations, however, were different for each situation. From the mathematics it seems that substrate concentration is independent of ciliates; however, it is clear that the turbidity of the effluent, caused by dispersed sewage bacteria, is completely dependent on protozoa. Large numbers of dispersed bacteria were present when protozoa were absent (no predation), fewer when free swimming ciliates were present, and least when the ciliate population was composed of attached forms. The last two cases are easily explained; the growth rate of the attached forms is very low and equals the sludge specific wastage rate whereas the specific growth rate of the free swimming forms is always much higher and equals the sewage dilution rate; it follows, therefore, from Michaelis-Menten kinetics that the higher the growth rate then the higher is the concentration of substrate (which in the case of ciliates is dispersed bacteria) present. Furthermore free swimming ciliates were more easily washed out from the system than were the attached forms. Thus, from these theoretical observations an overall high-quality effluent would be expected when attached ciliates were dominant, an effluent of slightly worse quality when only free swimming ciliates were present and a poor effluent when no ciliated protozoa were present. These ideas are in fact frequently expressed in the literature, being based on practical experience, but now mathematical modelling is supplying quantitative explanations of these phenomena. For example, free swimming ciliates and flagellates were considered by McKinney & Gram (1956) to be indicative of a low efficiency plant and the peritrichous ciliates (attached forms) a high one. The indicator species work of Curds & Cockburn (1970b) lists the free swimming ciliates as being those more frequently found in plants producing inferior quality effluents (BOD $\geqslant$ 21 mg/l) whereas the attached species of *Vorticella*, *Epistylis* and *Opercularia* were commonly found in plants producing good quality effluents (DOD < 20 mg/l).

The presence of protozoa and the habit of the dominant species, therefore, influence directly the concentration of dispersed bacteria present and so indirectly the concentration of sludge bacteria present, because these two types of bacteria are in competition with each other for the soluble substrate. Thus, the greater the concentration of dispersed bacteria, the smaller the concentration of sludge bacteria which will be present.

Successions of protozoan types in a variety of cultures are well documented (Woodruff, 1912; Barritt, 1940; Bick, 1958, 1960), and Curds (1966) showed that when an activated sludge plant is started without the addition of a sludge inoculum, a succession of protozoa appears with time. Initially flagellated protozoa are dominant, and this peak is followed by a succession of peaks of free swimming ciliates, crawling ciliates and finally attached sedentary ciliates. At the same time the sludge accumulates and the effluent quality improves. The usual explanation given for these successions may be summarized as their being the result of changes in the environment which successively favour the growth of new types. Evidence and experience of computer simulations, however, suggested that successions of this nature might equally well be explained simply on a basis of growth kinetics and upon the settling properties of the organisms. A computer programme was written (Curds, 1971*b*) to describe the situation in a hypothetical activated sludge reactor which included sludge bacteria, dispersed

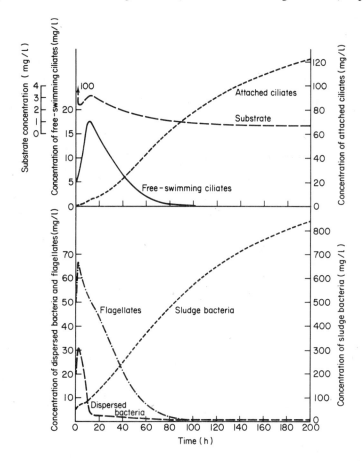

Fig. 3. Computer simulation of successions of micro-organisms during establishment of an activated sludge.

sewage bacteria, flagellated protozoa, free swimming ciliates and attached ciliates. Only the flagellates have not been mentioned above, but for the purposes of this simulation they were assumed to utilize the soluble substrate, they would not settle and that they were not present in the inflowing sewage.

The results of this simulation (Fig. 3) demonstrate tha the use of the above criteria gave a succession of organisms which was qualitatively similar to that found in practice. Flagellated protozoa and dispersed bacteria were the first dominant organisms, but as time progressed both declined. The flagellates declined slowly owing to competition for substrate with the two bacteria, while the dispersed bacteria declined rapidly owing to competition for substrate and to the predatory activities of the ciliates present. The free swimming ciliates reached a small peak after the dispersed bacteria but were soon displaced by the attached ciliate population owing to competition for bacteria. It is clear that any organism which settles and is hence returned back to the aeration tank has a distinct advantage over an organism which does not settle, and is therefore continually washed out in the effluent. If crawling ciliates were assumed to settle, but less readily than attached ciliates, then a peak of crawling forms might be expected to occur between the other 2 ciliate peaks as is observed in practice.

It would appear from the above mathematical considerations that many of the observations made in the past concerning protozoa can be explained, but this does not necessarily mean that they are correct. The activated sludge community like any other ecological society is highly complex and it is possible that the results of such a simple model are fortuitous. More complex models which more closely approximate the activated sludge process are being devised and it is hoped that in the future mathematical models and computer simulations will explain many more of the observations made on all forms of microbial life in the activated-sludge process.

11. References

Abson, J. W. & Todhunter, K. H. (1967). Effluent disposal. In *Biochemical and Biological Engineering Science,* Vol. 1. Ed. N. Blakeborough. London: Academic Press.

Adamse, A. D. (1968). Formation and final composition of the bacterial flora of a diary waste activated sludge. *Wat. Res.* **2**, 665.

Adamse, A. D. (1970). Some characteristics of arthrobacters from a dairy waste activated sludge. *Wat. Res.* **4**, 797.

Agersborg, H. P. K. & Hatfield, W. D. (1929). The biology of a sewage-treatment plant—a preliminary survey—Decatur, Illinois. *Sewage Wks J.* **1**, 411.

Ainsworth, G. (1966). An introduction to the activated sludge process. *Process Biochem.* **1**, 15.

Allen, L. A. (1944). The bacteriology of activated sludge. *J. Hyg., Camb.* **43**, 424.

Anderson, R. E. (1968). Isolation and generic identification of the bacteria from activated-sludge flocs, with studies of floc formation. Thesis, University of Wisconsin. Abstract in *Wat. Pollut. Abstr.* **42**, 263.

Angelbeck, D. I. & Kirsch, E. J. (1969). Influence of pH and metal cations on aggregative growth of non-slime-forming strains of *Zoogloea ramigera. Appl. Microbiol.* **17**, 435.

Austin, B. L. & Forster, C. F. (1969). The microbial ecology of a Lübeck activated sludge plant. *Wat. Waste Treat. J.* **12**, 208.

Baines, S., Hawkes, H. A., Hewitt, C. H. & Jenkins, S. H. (1953). Protozoa as indicators in activated-sludge treatment. *Sewage ind. Wastes* **25**, 1023.

Barker, A. N. (1943). The protozoan fauna of sewage disposal plants. *Naturalist, Hull* p. 65.

Barker, A. N. (1946). The ecology and function of protozoa in sewage purification. *Annls appl. Biol.* **33**, 314.

Barritt, N. W. (1940). The ecology of activated sludge in relation to its properties and the isolation of a specific soluble substance from the purified effluent. *Annls appl. Biol.* **27**, 151.

Berg, G. (1966a). Virus transmission by the water vehicle, I. Viruses. *Hlth Lab. Sci.* **3**, 86.

Berg, G. (1966b). Virus transmission by the water vehicle. II. Virus removal by sewage treatment processes. *Hlth Lab. Sci.* **3**, 90.

Bick, H. (1958). Ökologische Untersuchungen an Ciliaten fallaubreicher Kleingewasser. *Arch. Hydrobiol.* **54**, 506.

Bick, H. (1960). Ökologische Untersuchungen an Ciliaten und andern Organismen aus verunreinigten Gewassern. *Arch. Hydrobiol.* **56**, 378.

Bolton, R. L. & Klein, L. (1961). *Sewage Treatment: Basic Principles and Trends.* London: Butterworths.

Brown, T. J. (1965). A study of the protozoa in a diffused-air activated-sludge plant. *Wat. Pollut. Control* **64**, 375.

Butterfield, C. T. (1935). Studies of sewage purification. II. A zoogloea-forming bacterium isolated from activated sludge. *Publ. Hlth Rep., Wash.* **50**, 671.

Clarke, N. A. & Kabler, P. W. (1964). Human enteric viruses in sewage. *Hlth Lab. Sci.* **1**, 44.

Clarke, N. A., Stevenson, R. E., Chang, S. L. & Kabler, P. W. (1961). Removal of enteric viruses from sewage by activated sludge treatment. *Am. J. publ. Hlth* **51**, 1118.

Coackley, P. (1969). Some aspects of activated sludge. *Process Biochem.* **4**, 27.

Cooke, W. B. (1956). Check list of fungi isolated from polluted water and sewage. *Sydowia* **9**, 146.

Cooke, W. B. (1963). *A Laboratory Guide to Fungi in Polluted Waters, Sewage and Sewage Treatment Systems. Their Identification and Culture.* Cincinnati, Ohio: U.S. Dept. of Health, Education and Welfare, Public Health Service.

Cooke, W. B. & Ludzack, F. J. (1958). Predacious fungi behaviour in activated sludge systems. *Sewage ind. Wastes* **30**, 1490.

Cooke, W. B. & Pipes, W. O. (1969). The occurrence of fungi in activated sludge. *Proc. 23rd ind. Waste Conf. Purdue Univ., Engng Extn Ser.* No. 132, p. 170.

Crabtree, K., McCoy, E., Boyle, W. C. & Rohlich, G. A. (1965) Isolation, identification and metabolic role of the sudanophilic granules of *Zoogloea ramigera. Appl. Microbiol.* **13**, 218.

Cramer, R. (1931). The role of protozoa in activated sludge. *Ind. Engng Chem. ind. Edn.* **23**, 309.

Curds, C. R. (1963). The flocculation of suspended matter by *Paramecium caudatum. J. gen. Microbial.* **33**, 357.

Curds, C. R. (1966). An ecological study of the ciliated protozoa in activated sludge. *Oikos* **15**, 282.

Curds, C. R. (1971a). A computer-simulation study of predator-prey relationships in a single-stage continuous-culture system. *Wat. Res,* in press.

Curds, C. R. (1971b). Computer simulations of microbial population dynamics in the activated-sludge process. *Wat. Res,* in press.

Curds, C. R. & Cockburn, A. (1968). Studies on the growth and feeding of *Tetrahymena pyriformis* in axenic and monoxenic culture. *J. gen. Microbiol.* **54**, 343.

Curds, C. R. & Cockburn, A. (1970a). Protozoa in biological sewage-treatment processes – I. A survey of the protozoan fauna of British percolating filters and activated-sludge plants. *Wat. Res.* **4**, 225.

Curds, C. R. & Cockburn, A. (1970b). Protozoa in biological sewage-treatment processes – II. Protozoa as indicators in the activated-sludge process. *Wat. Res.* **4**, 237.

Curds, C. R. & Cockburn, A. (1971). Continuous monoxenic culture of *Tetrahymena pyriformis. J. gen. Microbiol.* **66** 95.

Curds, C. R., Cockburn, A., & Vandyke, J. M. (1968). An experimental study of the role of the ciliated protozoa in the activated-sludge process. *Wat. Pollut. Control* **67**, 213.

Curds, C. R. & Fey, G. J. (1969). The effect of ciliated protozoa on the fate of *Escherichia coli* in the activated-sludge process. *Wat. Res.* **3**, 853.

Curtis, E. J. C. (1969). Sewage fungus: its nature and effects. *Wat. Res.* **3**, 289.

Cyrus, Z. & Sladká, A. (1970). Several interesting organisms present in activated sludge. *Hydrobiologia* **35**, 383.

Department of the Environment (1971). *Water Pollution Research, 1970.* London: H.M. Stationery Office.

Dias, F. F. & Bhat, J. V. (1964). Microbial ecology of activated sludge. I. Dominant bacteria. *Appl. Microbiol.* **12**, 412.

Dias, F. F. & Bhat, J. V. (1965). Microbiol ecology of activated sludge. II. Bacteriophages, *Bdellovibrio*, coliforms and other organisms. *Appl. Microbiol.* **13**, 257.

Dias, F. F., Dondero, N. C. & Finstein, M. S. (1968). Attached growth of *Sphaerotilus* and mixed populations in a continuous-flow apparatus. *Appl. Microbiol.* **16**, 1191.

Downing, A. L. & Knowles, G. (1966). Population dynamics in biological treatment plants. *Proc. 3rd Int. Conf. Wat. Pollut. Res.* **2**, 117.

Downing, A. L., Painter, H. A. & Knowles, G. (1964). Nitrification in the activated-sludge process. *J. Proc. Inst. Sew. Purif.* p. 130.

Downing, A. L. & Wheatland, A. B. (1962). Fundamental considerations in biological treatment of effluents. *Trans. Instn chem. Engrs* **40**, 91.

Fairbrother, T. H. & Renshaw, A. (1922). The relation between chemical constitution and antiseptic action in the coal tar dyestuffs. *J. Soc. chem. Ind. Lond.* **41**, 134.

Gayford, C. G. & Richards, J. P. (1970). Isolation and enumeration of aerobic, heterotrophic bacteria in activated sludge. *J. appl. Bact.* **33**, 342.

Genetelli, E. J. (1967). DNA and nitrogen relationships in bulking activated sludge. *J. Wat. Pollut. Control Fed.* **39**, R32.

Van Gils, H. W. (1964). *Bacteriology of Activated Sludge.* IG-TNO Report No. 32, The Hague: Research Institute for Public Health Engineering.

Grabow, W. O. K. (1968). The virology of waste water treatment. *Wat. Res.* **2**, 675.

Grieves, R. B., Milbury, W. F., & Pipes, W. O. (1964). A mixing model for activated sludge. *J. Wat. Pollut. Control Fed.* **36**, 619.

Hardin, G. (1943). Flocculation of bacteria by protozoa. *Nature, Lond.* **151**, 642.

Hardt, F. W., Clesceri, L. S., Nemerow, N. L. & Washington, D. R. (1970). Solids separation by ultrafiltration for concentrated activated sludge. *J. Wat. Pollut. Control Fed.* **42**, 2135.

Harris, R. F. & Sommers, L. E. (1968). Plate-dilution frequency technique for assay of microbial ecology. *Appl. Microbiol.* **16**, 330.

Hattingh, W. H. J. (1963) Activated sludge studies. 3 – Influence of nutrition on bulking. *Wat. Waste Treat.* **9**, 476.

Hawkes, H. A. (1963). *The Ecology of Waste Water Treatment.* Oxford: Pergamon Press.

Herbert, D. (1961). A theoretical analysis of continuous culture systems. In *Continuous Culture of Micro-organisms,* Monograph No. 12, p. 21. London: Society of Chemical Industry.

Jazewicz, L. & Porges, N. (1956). Biochemical oxidation of dairy wastes. III. Isolation and study of sludge micro-organisms. *Sewage ind. Wastes* **28**, 1130.

Jenkins, S. H. (1942). The role of protozoa in the activated-sludge process. *Nature, Lond.* **150**, 607.

Jones, P. H. (1965). The effect of nitrogen and phosphorus compounds on one of the micro-organisms responsible for sludge bulking. *Proc. 20th ind. Waste Conf. Purdue Univ., Engng Extn. Ser.* No. 118, p. 297.

Kabler, P. (1959). Removal of pathogenic micro-organisms by sewage treatment processes. *Sewage ind. Wastes* **31**, 1373.

Kampelmacher, E. H. & Noorle Janszen, L. M. van (1970). Salmonella—its presence and removal from a wastewater system. *J. Wat. Pollut. Control Fed.* **42**, 2069.

Keller, P. (1959). Bacteriological aspects of sewage pollution and river pollution. *J. Hyg., Camb.* **57**, 410.

Leclerc, H., Perchet, A., Savage, C., Andrieu, S. & Nguematcha, R. (1971). Microbiological aspects of sewage treatment. *Proc. 5th. int. Conf. Wat. Poll. Res., San Francisco, 1970.* Oxford: Pergamon Press.

Lighthart, B. & Oglesby, R. T. (1969). Bacteriology of an activated sludge wastewater treatment plant – a guide to methodology. *J. Wat. Pollut. Control Fed.* **41**, R267.

Lochhead, A. G. & Chase, F. E. (1943). Qualitiative studies of soil micro-organisms. V. Nutritional requirements of the predominant bacterial flora. *Soil Sci.* **55**, 185.

Loveless, J. E. & Painter, H. A. (1968). The influence of metal ion concentration and pH value on the growth of a Nitrosomonas strain isolated from activated sludge. *J. gen. Microbiol.* **52**, 1.

Lund, E. Hedström, C. E. & Jantzen, N. (1969). Occurrence of enteric viruses in wastewater after activated sludge treatment. *J. Wat. Pollut. Control Fed.* **41**, 169.

Malherbe, H. H. (1964). The recovery of viruses from sewage. *J. Proc. Inst. Sew. Purif.* p. 210.

McCabe, J. & Eckenfelder, W. S., Eds (1956). *Biological Treatment of Sewage and Industrial Wastes,* Vol. I. *Aerobic Oxidation.* Eds. J. McCabe & W. W. Eckenfelder. New York: Reinhold Publishing Corpn.

McKinney, R. E. (1952). A fundamental approach to the activated sludge process. II. A proposed theory of floc-formation. *Sewage ind. Wastes* **24**, 280.

McKinney, R. E. (1956). Biological flocculation. In *Biological Treatment of Sewage and Industrial Wastes,* Vol. I. *Aerobic oxidation.* Eds J. McCabe & W. W. Eckenfelder. New York: Reinhold Publishing Corpn.

McKinney, R. E. (1962). Mathematics of a completely mixed activated sludge. *J. sanit. Engng Div. Am. Soc. civ. Engrs* **88**, SA3, 87.

McKinney, R. E. & Gram, A. (1956). Protozoa and activated sludge. *Sewage ind. Wastes* **28**, 1219.

McKinney, R. E. & Horwood, M. P. (1952). Fundamental approach to the activated sludge process. I. Floc-forming bacteria. *Sewage ind. Wastes* **24**, 117.

McKinney, R. E. & Weichlein, R. G. (1953). Isolation of floc-producing bacteria form activated sludge. *Appl. Microbiol.* **1**, 259.

Ministry of Technology (1968). Protozoa in sewage-treatment processes. *Notes on Water Pollution* No. 43. London: H.M.S.O.

Mulder, E. G. (1964). Iron bacteria, particularly those of the *Sphaerotilus-Leptothrix* group, and industrial problems. *J. appl. Bact.* **27**, 151.

Painter, H. A. Denton, R. S. & Quarmby, C. (1968). Removal of sugars by activated sludge. *Wat. Res.* **2**, 427.

Painter, H. A. & Hopwood, A. P. (1967). Microbiology of waste treatment processes. *Rep. Prog. appl. Chem.* **52**, 491.

Painter, H. A. & Hopwood, A. P. (1969). Microbiology of waste treatment processes. *Rep. Prog. appl. Chem.* **54**, 405.

Peter, G. & Wuhrmann, K. (1971). Contribution to the problem of bioflocculation in the activated sludge process. *Proc. 5th int. Conf. Wat. Poll. Res., San Francisco, 1970.* Oxford: Pergamon Press.

Phaup, J. D. (1968). The biology of *Sphaerotilus* species. *Wat. Res.* **2**, 597.

Pike, E. B., Carrington, E. G. & Ashburner, P. A. (1972). An evaluation of procedures for enumerating bacteria in activated sludge. *J. appl. Bact.* In press.

Pillai, S. C., & Subrahmanyan, V. (1942). Role of protozoa in activated sludge. *Nature, Lond.* **150**, 525.

Pillai, S. C., & Subrahmanyan, V. (1944). Role of protozoa in aerobic purification of sewage. *Nature, Lond.* **154**, 179.

Pipes, W. O. (1965). Carnivorous plants in activated sludge. *Proc. 20th ind. Waste Conf. Purdue Univ., Engng Extn. Ser.* No. 118, 647.

Pipes, W. O. (1966). The ecological approach to the study of activated sludge. *Adv. appl. Microbiol,* 8, 77.

Pipes, W. O. (1969). Types of activated sludge which settle poorly. *J. Wat. Pollut. Control Fed.* 41, 714.

Pipes, W. O. & Jenkins, D. (1965). *Zoophagus* in activated sludge – a second observation. *Int. J. Air Wat. Pollut.* 9, 495.

Pipes, W. O. & Jones, P. H. (1963). Decomposition of organic wastes by *Sphaerotilus. Biotechnol. Bioengng* 5, 287.

Pirt, S. J. & Kurowski, W. M. (1970). An extension of the theory of the chemostat with feedback of organisms. Its experimental realization with a yeast culture. *J. gen. Microbiol.* 63, 357.

Postgate, J. R. & Hunter, J. R. (1963). Acceleration of bacterial death by growth substrates. *Nature, Lond.* 198, 273.

Prakasam, T. B. S. & Dondero, N. C. (1967b). Aerobic heterotrophic bacterial populations of of sewage and activated sludge. I. Enumeration. *Appl. Microbiol.* 15, 461.

Pakasam, T. B. S. & Dondero, N. C. (1967b). Aerobic heterotrophic bacterial populations of sewage and activated sludge. II. Method of characterization of activated sludge bacteria. *Appl. Microbiol.* 15, 1122.

Prakasam, T. B. S. & Dondero, N. C. (1967c). Aerobic heterotrophic bacterial populations of sewage and activated sludge. III. Adaption in a synthetic waste. *Appl. Microbiol.* 15, 1128.

Prakasam, T. B. S. & Dondero, N. C. (1970). Aerobic heterotrophic bacterial populations of sewage and activated sludge. V. Analysis of population structure and activity. *Appl. Microbiol.* 19, 671.

Reynolds, I. D. & Yang, J. T. (1966). Model of the completely-mixed activated-sludge process. *Proc. 21st Industr. Waste Conf. Purdue Univ., Eng. Extn. Ser.* No. 121, p. 696.

Reynoldson, T. B. (1942). Vorticella as an indicator organism for activated sludge. *Nature, Lond.* 149, 608.

Schofield, T. (1971). Some biological aspects of the activated-sludge plant at Leicester. *Wat. Pollut. Control* 70, 32.

Schulze, K. L. (1964). The activated-sludge process as a continuous flow culture. Part. I. *Wat. Sewage Wks* 111, 526.

Schulze, K. L. (1965). The activated-sludge process as a continuous flow culture. Part II. *Wat. Sewage Wks* 112, 11.

Sladká A. & Zahrádka, V. (1970). *Morphology of Activated Sludge.* Prague-Podbaba: Water Research Institute, Technical Paper No. 126.

Sokal, R. R. & Sneath, P. H. A. (1963). *Principles of Numerical Taxonomy.* San Francisco: W. H. Freeman.

Sugden, B. & Lloyd, L. (1950). The cleaning of turbid waters by means of the ciliate *Carchesium:* a demonstration. *J. Proc. Inst. Sew. Purif.* p. 16.

Sundman, V. (1970). Four bacterial soil populations characterized and compared by a factor analytical method. *Can. J. Microbiol.* 16, 455.

Sundman, V. & Gyllenberg, H. (1967). Application of factor analysis in microbiology. *Suomal. Tiedeakat. Toim.,* Sarja IV, No. 112.

Sydenham, D. H. J. (1968). The ecology of protozoa and other micro-organisms in activated sludge. PhD. thesis, University of London.

Tempest, D. W., Herbert, D. & Phipps, P. J. (1967). Studies on the growth of *Aerobacter aerogenes* at low dilution rates in a chemostat. In *Microbial Physiology and Continuous Culture.* (Eds E. O. Powell, C. G. T. Evans, R. E. Strange & D. W. Tempest). London: H.M.S.O.

Tenney, M. W. & Stumm, W. (1965). Chemical flocculation of micro-organisms in biological waste. *J. Wat. Pollut. Control Fed.* 37, 1370.

Tezuka, Y. (1967). Magnesium ion as a factor governing bacterial flocculation. *Appl. Microbiol.* 15, 1256.

Tezuka, Y. (1969). Cation–dependent flocculation in a *Flavobacterium* species predominant in activated sludge. *Appl. Microbiol.* **17**, 222.

Tomlinson, T. G. (1939). The biology of sewage purification. *J. Proc. Inst. Sew. Purif.* 225.

Toerien, D. F., Hattingh, W. H. J., Kotzé, J. P., Thiel, P. G. & Siebert, M. L. (1969). Factor analysis as an aid in an ecological study of anaerobic digestion. *Wat. Res.* **3**, 129.

Unz, R. F. & Dondero, N. C. (1970). Nonzoogloeal bacteria in wastewater zoogloeas. *Wat. Res.* **4**, 575.

Viehl, K. (1937). Investigations concerning the nature of self purification and artificial biological purification. *Z. Hyg. InfektKrank.* **119**, 383.

Watson, J. M. (1945). Mechanism of bacterial flocculation caused by protozoa. *Nature, Lond.* **155**, 171.

Williams, A. R., Forster, C. F. & Hughes, D. E. (1971). Using an ultrasonic technique in the enumeration of activated sludge bacteria. *Effluent Wat. Treat. J.* **11**, 83.

Williams, A. R., Stafford, D. A., Callely, A. G. & Hughes, D. E. (1970). Ultrasonic dispersal of activated sludge flocs. *J. appl. Bact.* **33**, 656.

Woodruff, L. L. (1912). Observations on the origin and sequence of the protozoan fauna of hay infusions. *J. exp. Zool.* **12**, 205.

MAP–6

Disposal by Dilution? —An Ecologist's Viewpoint

H. A. HAWKES

*Applied Hydrobiology Section, University of Aston in
Birmingham, England*

CONTENTS

1. Introduction

UNTIL THE Industrial Revolution the population in Britain remained small and
was largely rural. In such an agrarian based society the practice of waste disposal
by burrowing it in the soil where it was broken down by natural processes of
putrefaction was adequate for such scattered communities. In the cities,
however, the problem was more acute and in the 18th century shocking sanitary
conditions prevailed. During the late 18th and early 19th centuries the growth of
industry and the resultant concentration of populations in towns and cities
aggravated the situation. Man's activities like those of other animals inevitably
affect his environment. With increases in the density of populations this effect

tends to be more evident. Dilution of these wastes, therefore, into the environment appears at first to be an attractive method of disposal. Many factories were sited on the banks of rivers resulting in industrial wastes and some domestic wastes from associated dwellings being discharged directly to the rivers.

The introduction early in the 19th century of the water carriage system enabled the wastes to be carried from the dense centres of population into the rivers. Although this reduced the filth in the streets its discharge to water courses converted many of them to open sewers and by the middle of the 19th century public conscience revolted against the conditions of the rivers in the industrial areas of Yorkshire, Lancashire, the Midlands and in London. The concern on public health grounds was justified by the subsequent cholera outbreaks of 1866 and 1872 in London. Two Royal Commissions of 1865 and 1868 were set up to study the problem of river pollution. They recommended the treatment of sewage on land before discharge to rivers and they initiated scientific investigations into the methods of sewage treatment.

2. The Role of Dilution in the Disposal of Wastes

In waste disposal generally, to reduce the harmful effects on the environment several methods based on different principles are practised:

(i) the waste may be concentrated and then isolated from the environment. This is practised with relatively small quantities of highly noxious wastes which are disposed either underground or in the depths of the ocean. The danger here is the effectiveness of the method of isolation to prevent their ultimate escape into the environment. Some such materials disposed of underground have found their way into aquifers and have polluted water resources. The disposal of biogenic wastes by this method is not only uneconomical because of the amounts involved, but constitutes a loss of materials to the natural cycles in ecosystems;

(ii) the waste is treated either mechanically, chemically or biologically to render it harmless;

(iii) the waste is diluted in the environment to a concentration at which it is not harmful.

Of the three possible diluting media the terrestrial environment is the least suitable and the atmosphere the most suitable physically, water being intermediate in this respect. Although we are here considering the aquatic environment, the possibility of passage of pollutants between the different media needs to be borne in mind. Run-off from soils into surface waters and the contamination of aquifers by noxious wastes by underground disposal methods are dangers that are fully recognized. The disposal of gaseous effluents and the volatilization of materials such as pesticides into the atmosphere from the earth's surface may result in these being precipitated onto land or water surfaces via rain. The global distribution of some pollutants in waters may be due to this path of dispersal in addition to that by oceanic currents.

Dilution as a method of disposal by itself is only possible when a relatively large volume of water in relation to the effluent is available to give the necessary degree of dilution.

In many cases of disposal two or all of the above listed methods are used in combination to prevent pollution. The treatment processes on a typical sewage works illustrates this. Sewage, the waterborne wastes of man's domestic and industrial activities, is already considerably diluted, containing, as it does, 99·9% of water. In the initial stages of treatment particulate solids are removed by mechanical and physical processes of screening and settlement in tanks. The removed solids, known as primary sludge, is further concentrated by physical means of pressing or vacuum filtration or by biological treatment under anaerobic conditions known as "sludge digestion". The digested sludge is then de-watered and dried in beds. The concentrated solids may then be dispersed in the terrestrial environment by application to agricultural land but, in the case of sewage sludges derived from sewages containing industrial toxic wastes such as metals, they are best isolated by tipping. Alternatively, further treatment and concentration by incineration is now practised.

The nonseparable material in solution and suspension is subjected to biological treatment either by the activated sludge process or by percolating filters. In both of these processes part of the organic matter is oxidized and part concentrated by biosynthesis into microbial growths which are subsequently removed by sedimentation as secondary sludge. This is treated along with the primary sludge before disposal. The effluent containing the end products of respiration – carbon dioxide, phosphates and nitrates and any unoxidized organic matter, together with a high bacterial content – is discharged to the receiving water. To remove these materials and organisms from the effluent would require expensive tertiary treatment processes. Thus most sewage works' effluents, even from efficient works, require dilution in the receiving water. This need for dilution as a final process in the disposal of sewage was early recognized by the Royal Commission on sewage disposal, who based their standards for BOD (biochemical oxygen demand) on the assumption that the effluent discharging to inland streams would be diluted with a minimum of 8 vol of river water.

The process of dilution is also used in some purification processes. The efficiency of percolating filters has been increased by diluting the feed liquor applied with a proportion of the returned effluent, a process known as recirculation. Among other advantages the dilution of the feed reduces the growth rate of the microbial film, especially when fungi are present, thus reducing the accumulation of film in the upper layers of the filter and thereby preventing the choking of the filter which is a common limitation to the rates at which sewage can be treated by this process. In the activated sludge method the immediate dilution of the waste in the complete mixing process, under certain circumstances, has proved beneficial over the plug flow system in which there is no immediate dilution of the waste. Dilution, therefore, is involved in many aspects of waste disposal in the aquatic environment.

3. The Nature and Types of Pollution

Before assessing the effectiveness of the process of dilution in preventing pollution it is first necessary to define what we mean by pollution; although it is difficult to define objectively, failure to do so leads to much fruitless discussion. Many biologists have defined river pollution as any discharge which affects the biota of the river. This, however, is not an acceptable definition in practical terms of water management. A more useful definition is the discharge of anything to the water which directly or indirectly changes the nature of the water to such an extent that its suitability for man's legitimate uses is impaired. Such uses would normally include public and industrial supply, agriculture and recreational uses and general aesthetic considerations. Although ecological changes induced by discharges are significant in indicating changes in environmental conditions, these in themselves may or may not constitute pollution. Pollution can only be defined subjectively in relation to water use. This is well illustrated by reference to *Sphaerotilus natans* which forms macroscopic fungoid growths known as "sewage fungus" in organically enriched waters. Under many circumstances such growths are aesthetically objectionable; they cause nuisance to anglers and give rise to pollution. In other circumstances, however, it was found in experimental trout farming in upland streams that the addition of carbohydrates induced *Sphaerotilus* growths which supported a chironomid larval population upon which the trout fed and grew more rapidly. In these circumstances the presence of *Sphaerotilus* represented increased productivity.

Although effluents rarely affect the receiving water in a single simple manner it is convenient for our purpose to consider the effectiveness of dilution in relation to 3 groups of pollutants.

    (i) harmful or undesirable organisms introduced into a water body via an effluent;

   (ii) substances not naturally occurring in water, e.g. pesticides, or of substances in unnatural concentrations so as to be directly harmful, e.g. heavy metals;

(iii) natural biogenic materials present in unnaturally high concentrations so as to change adversely the natural balance of populations within the ecosystem, e.g. organic pollution, eutrophication.

The role and effectiveness of dilution as a method of disposal in the aquatic environment will be discussed in relation to the above three types of pollutant. Although micro-organisms and their activities will receive special attention, to appreciate the full effect of pollutants one needs to consider the whole ecosystem affected, which includes other groups of organisms.

4. Physical and Hydrological Considerations in the Dispersion and Mixing of Effluents in the Aquatic Environment

The effectiveness of dilution as a method of disposal of effluents containing any type of pollutant depends on the rate of dispersion of the effluent in the receiving waters. Tracer techniques have been developed using tracers such as rhodamine B or radioactive ones such as KBr^{82} and such organisms as *Serratia indica* and spores of *Bacillus subtilis* var. *niger* (Pile, Bufton & Gould, 1969). Neither organism occurs naturally in rivers or sea. *Serratia indica* was found to be dispersed similarly to coli-aerogenes bacteria; it disappeared after a few days and was rapidly killed in sea water exposed to sunlight. However, the continuous release of *Serr. indica* into the sea was a suitable method of measuring the dispersion field around a discharge. Single doses could be used to provide evidence of dispersion processes and transit time to the shore. Although *B. subtilis* was also killed after 3 days in seawater exposed to light, it was more resistant and did not die or germinate in the dark in sea water over a 9 days test period. It was considered that the spores of *B. subtilis* var. *niger* would not undergo change during the period required in tracer experiments and, because they were found to disperse similarly to KBr^{82} and *Serr. indica*, they could be used to measure physical dilution independently of mortality.

Although with bacteria it may well be that they are dispersed in a similar pattern to substances, other organisms with greater powers of locomotion may not. Based on the results of tracer experiments and basic physical diffusion coefficients, mathematical models have been developed in an attempt to predict the dispersion of effluents under different hydrological conditions.

(a) *Dispersion in rivers*

Most of the models for rivers relate to the mixing of an effluent in a channel of flowing water. Although these are useful in relation to the body of water in predicting water quality, the effect on the river ecosystem as a whole is more complex. Many, but not all, of the benthic and periphytic organisms live out of the water flow of the stream, either in the river bed or towards the banks. The water in which these organisms live is different from that in the main stream which is that predicted by most models. There is a direct exchange of material between the flowing water and the "dead water", thus the quality of the latter is related to that of the main flow even though it is not the same. Although many benthic micro-organisms live in the dead water others such as *Sphaer. natans* and *Cladophora* spp. by adopting plumose or filamentous growth forms, live in the main flow from which they derive their nutrient requirements. Upon death and decay the materials assimilated are transferred to the benthic community thus providing another path of transfer. Hays & Krenkel (1968) observed that there were usually significant discrepancies between observed field results on the dispersion of effluents in rivers and mathematical models, although the latter had been successfully tested in laboratory flumes. Whereas in models the distribution is normal with respect to distance, the observed results on rivers are usually skewed, having truncated leading edges and extended tails on the trailing edges. To explain this discrepancy they proposed a new model based on a 2-zone system. A river cross section was considered to consist of the main stream, where the dominant mass transport mechanism is longitudinal, and a stagnant or dead water zone where the longitudinal velocity is comparatively negligible and lateral mass transport dominates. The model accounts for part of the material introduced into the main flow being transferred to the dead water zone to be subsequently released to the main flow producing the observed tails.

The rate of dispersion of an effluent is related to the relative volumes of effluent and diluting water. In a river, however, this is complicated by the changes in flow rate with volume of diluting water. At high flows, in larger rivers, although the degree of dispersion of the dilution water is affected, the absolute concentration of the effluent was still lower than at low flow (Kisiel *et al.*, 1964). The direction of the discharge relative to the river flow is another factor influencing dispersion. Some effluents are discharged across the width of the river from pipes across the river bed to assist dispersion. The effect of any pollutional curtain so formed, however, must be considered in respect of migratory fish.

The relative densities of the effluent and the receiving water is a further factor influencing the speed of dispersion and mixing. Density differences may be due to differences in salinity and temperature. In Yugoslavia the mixing of the tributary Spreca, which was polluted with industrial waste waters, and the River Bosna was incomplete 11 km downstream of the confluence (Preka & Lipold, 1965). Observations on the polluted Buffalo River which empties sluggishly into Lake Erie has shown that during the summer due to thermal stratification of the river caused by thermal discharges, the cooler lake water moves upstream under the river water and mixes with it, thus improving the quality of the river water and allowing benthic organisms to live in the lower stretches of the river (Blum, 1965). In this case stratification is advantageous in bringing diluting water upstream into the river. The degree of turbulence and hydrographical features such as the shape of the river bed, types of bank and the depth-width ratio also affect the dispersion. The Susquehanna River in Pennsylvania receives tributaries on its west side containing high concentrations of calcium and bicarbonate ions whilst those on the east side draining anthracite coal fields have high concentrations of calcium and sulphate ions and low pH values. Incomplete mixing is evident several miles downstream. This is attributed to the very small depth-width ratio and the width of the river (Anderson, 1963).

(b) *Dispersion in estuaries*

The situation in estuaries is more complicated with the longer retention time during which tidal mixing occurs. A classical study has been carried out on the Thames estuary by the Department of Scientific and Industrial Research (1964). A mathematical model was developed to satisfy the conditions in the Thames where only slight stratification occurs. Water in a particular cross section of the estuary is carried upstream and downsteam during each tidal cycle, the average tidal excursion being 9 miles. An effluent entering the estuary at a steady rate is dispersed during one tidal cycle throughout the volume of water passing the discharge point flowing upsteam or downstream. Because of the inflow of fresh water it is displaced seaward to be further dispersed by tidal mixing. The model assumed that during each tidal cycle a proportion of the water in the cross section under consideration was dispersed over a distance of 6 miles downstream and another proportion similarly dispersed upstream, the remainder returning at the end of the cycle to that at the beginning. The overall retention time in the estuary is related to the freshwater flow entering; at low flows the average retention time for discharges into the upper reaches is *c.* 3 months.

(c) *Dispersion in lakes*

The dispersion of effluents discharged directly to lakes or polluted rivers entering lakes is difficult to generalize. It will be affected by the type of stratification and the relative positions of the discharge and any outflow from the lake. Blum (1965) found that influence of the polluted Buffalo River in Lake Erie was determined by the main flow of the river water towards Black Rock Canal and Niagara River. Csanady (1966) attempted to define the dispersion of effluents in the Great Lakes in terms of currents and eddies. Work on Lake Huron indicated that horizontal dispersal is produced by currents whilst vertical dispersal depends on eddies. He concluded that the concentration of an effluent at any point in a lake at a given time could only be predicted as a probability.

(d) *Dispersion in the sea*

Most of the work on sea dispersion has been concerned with the pollution of beaches. This is dealt with specifically in a subsequent section on the coastal dispersion of bacteria.

5. Disposal by Dilution of Effluents containing Harmful or Undesirable Organisms

(a) *Organisms in effluents*

Large numbers of organisms of enteric origin, including many pathogens, are discharged via sewage effluents into receiving waters such as rivers, lakes, estuaries and the sea. In the latter half of the 19th century the association of the outbreaks of cholera and typhoid fevers with the polluted condition of the rivers drew attention to the need for adequate sewage treatment, even though the causative organisms were then unknown. Although the early stages of development of sewage treatment were concerned with the public health aspects, later developments were more biochemical being concerned with the oxidation of organic matter and the removal of BOD rather than the removal of pathogens, although these may incidentally be reduced considerably in the process. A sewage effluent satisfactory in regards to its BOD may contain large numbers of pathogens, especially at times of epidemic (Harold, 1935). Allen, Tomlinson & Norton (1944) found that although single percolating filtration removed > 90% of the bacteria in the summer the percentage removed in the winter fluctuated considerably and was sometimes quite low. Modern improved methods of treatment although producing more oxidized effluents at higher loadings are often less efficient at removing bacteria. Tertiary treatment processes may however further reduce the number of bacteria. Allen *et al.* (1949) found that although sand filtration of sewage effluent further reduced the numbers of bacteria this process was not considered reliable to produce consistently an effluent of good bacterial quality. Harkness (1966) reported that by irrigating effluent over grass plots the numbers of *Escherichia coli* type I were appreciably reduced to average counts of <100/ml from counts of thousands/ml in the effluent from the oxidation process. Similar reductions can be effected by lagooning. Data on the efficiency of tertiary sewage treatment processes in removing bacteria, reported by the Department of Scientific and Industrial Research (1963), showed that grass plots were most effective (87–97% reduction), lagoons next (67%), followed by slow sand filters (52–62%) and rapid gravity filters (32%); microstrainers were least effective (6%). Kabler (1959), reviewing the literature on this subject, concluded that although there was a marked reduction in the numbers of pathogenic micro-organisms during sewage treatment they were still present in the final effluent in significant numbers.

Where untreated or partially treated sewage is discharged, as into some estuaries and coastal waters, the numbers of organisms are many times greater but so is the dilution factor. At times of high flow, 6 x the designed "dry weather flow", the sewers overflow and large numbers of organisms enter the rivers. Besides domestic sewage some industrial effluents may also contain pathogenic organisms, e.g. slaughter houses, farming, tanneries and animal laboratories. Of the list of the commoner waterborne pathogens of man (Table 1) the enteroviruses and salmonellae are probably the most significant in British waters. Burman (1966) reported that it was possible to isolate salmonellae from most sewage effluents and the receiving rivers at any time of the year. Although there was strong circumstantial evidence of the presence of shigellae it was rarely isolated from sewage effluents or river water; this, Burman considered, was probably due to the inadequacy of the techniques.

Helminth parasites are usually associated with warmer countries but eggs of cestodes and nematodes are more common in sewage effluents in Europe than is commonly realized (Shephard, 1971).

(b) *Principles of disposal by dilution*

The presence of such organisms in water reduces its value as raw water for public supply and presents a potential health hazard to man and his cattle in its several uses. Although adequate treatment is given to such waters by water treatment processes before it is put into public supply, direct use of the water for bathing and other forms of aquatic recreation, crop irrigation, cattle watering, cooling water and shellfish and water cress culture, all present potential health hazards. The principle of disposal of such effluents by dilution requires that the concentration of organisms in the effluent is reduced by the physical process of dispersion in the diluting water, and by other means, so that at the point of use their concentration does not present a health hazard in relation to the respective use or uses. In practice the difficulty is in determining what these acceptable concentrations should be in relation to the several uses. For bathing, for example, in some American states the standard for a good water is that it should not carry > 50 coli-aerogenes bacteria/100 ml. In New Zealand a maximum coliform count of 100/100 ml is recommended. In Britain a Medical Research Council Memorandum considered bacteriological standards impracticable for bathing waters on beaches in this country. They considered that except for a few "aesthetically revolting beaches" the risk to health of bathing in sewage contaminated sea water was negligible (Klein, 1962). As pointed out by Burman (1966), although a single *Salm. typhi* cell, if drunk in water, might cause infection, other salmonellae would not. If the water was used in food processing, however, they could multiply rapidly in certain foods to give rise to food poisoning. The effectiveness of dilution as a method of disposal depends initially on the physical processes of dispersion and eventually on the death rate of the organisms in

Table 1

Pathogenic organisms in sewage and polluted waters. (After Wilson, 1949)

Organism	Disease	Remarks
Virus	Poliomyelitis	Exact mode of transmission not yet known. Found in effluents from biological sewage purification plants
Vibrio cholerae	Cholera	Transmitted by sewage and polluted waters
Salmonella typhi	Typhoid fever	Common in sewage and effluents in times of epidemics
Salmonella paratyphi	Paratyphoid fever	Common in sewage and effluents in times of epidemics
Salmonella spp.	Food poisoning	
Shigella spp.	Bacillary dysentery	Polluted waters main source of infection
B. anthracis	Anthrax	Found in sewage. Spores resistant to treatment
Brucella spp.	Brucellosis – Malta fever in man. Contagious abortion in sheep, goats and cattle	Normally transmitted by infected milk or by contact. Sewage also suspected
Mycobacterium tuberculosis	Tuberculosis	Isolated from sewage and polluted streams. Possible mode of transmission. Care with sewage and sludge from sanatoria
Leptospira icterohaemorrhagiae	Leptospirosis (Weil's disease)	Carried by sewer rats
Entamoeba histolytica	Dysentery	Spread by contaminated waters and sludge used as fertilizer. Common in warmer countries
Schistosoma spp.	Bilharzia	Probably killed by efficient sewage purification
Taenia spp.	Tape worms	Eggs very resistant, present in sewage sludge and sewage effluents. Danger to cattle on sewage-irrigated land or land manured with sludge
Ascaris spp. *Enterobius* spp.	Nematode worms	Danger to man from sewage effluents and dried sludge used as fertilizer

relation to the rate at which they are discharged. In a closed system such as a lake, a reservoir or the sea, a death rate is necessary to prevent an increase in the concentration of organisms after dilution.

Thus the principle of disposal by dilution depends on other agencies causing the necessary control of the population of the introduced pathogens. The factors influencing the physical dispersion of the organisms in the receiving water have been outlined; agencies controlling the increase in the population of the dispersed organisms will now be considered.

(c) *Factors affecting the survival of enteric pathogens in the aquatic environment*

Although many of the enteric pathogens of man are regarded as waterborne diseases, when dispersed in natural waters most of them find the environment inhospitable and die off over a period, thus failing to establish a self-perpetuating population. In rivers, following a rapid increase in numbers of enteric bacteria associated with the discharge of a sewage effluent there is a rapid decline in the numbers with increasing distance downstream. Hannay (1945) reported that the presumptive counts of both coliform bacteria and faecal streptococci followed this pattern. Below one effluent, for example, the count of faecal streptococci decreased by 97% in < 3 miles and in another by the same amount in < 2 miles. The decrease in both cases was not due to dilution. In lentic waters similar reductions take place in respect of time. Storage of water in reservoirs is known to improve its bacterial quality.

Physical, chemical and biotic agencies contribute to bring about this decline in the numbers of enteric micro-organisms.

(i) *Physicochemical agencies*

Flocculation and subsequent settlement remove some micro-organisms from the water. Carlson *et al.* (1968) have demonstrated in laboratory experiments the inactivation of viruses, T2 bacteriophage and poliovirus 1 by adsorption on clay particles. They found that the inactivation depended on the concentration of cations in the clay producing a necessary charge distribution inducing a clay-cation-virus bridge so that the virus was removed from the water along with the clay. The presence of extraneous organic matter inhibited this inactivation of the virus by competing for adsorption sites on the clay. In the presence of sufficient organic matter there was evidence of desorption of virus occurring.

Light is considered to have an adverse effect on many enteric bacteria when dispersed in the aquatic environment. Investigating the seasonal fluctuation of *E. coli* in a coastal lagoon which receive fresh water and sewage effluent, Lagarde & Castellvi (1965) concluded that the decrease in pathogenic bacteria in the summer when the pollution load was greatest was not due to the occurrence of an increase in salinity. In the discussion in the above paper the results of experiments with bottles in the dark and in the light were quoted which suggested the importance of light in suppressing the bacterial population. Reynolds (1965) investigated the death rate of *E. coli* in a mixture of sewage and sea water in large experimental concrete tanks. He observed the death rate was not uniform throughout the tank and whereas there was a logarithmic reduction in numbers during daylight there was little significant change at night. This effect was only slightly affected by cloud and also occurred in glass bottles; he therefore concluded that the effective radiation was not ultraviolet.

Differences in the natural chemistry of the water may also be significant. Bukovsky (1958) working on streams in Czechoslavakia showed that the survival of coliform bacteria in surface waters was assisted by increased concentrations of calcium. His field results were confirmed by laboratory experiments.

(ii) *Biotic agencies*

Predatory, parasitic and ectocrinal activities probably play a major role in aquatic ecosystems in reducing the population of enteric micro-organisms. Ciliate protozoa are bacteria feeders; Gray (1952) associated the ciliate fauna of a chalk stream with the bacterial flora, *Paramoecium* and *Colpoda* being associated with Gram negative rods. The dense population of the ciliate *Carchesium*, which assumes sewage fungus-like growths on the beds of organically polluted streams, must affect the bacterial population of the water flowing over them. Bacteriophages have been commonly isolated from polluted waters and it has been suggested that their presence could be used as an indication of faecal pollution. More recently other parasitic activity has been demonstrated. Guelin & Lamblin (1966) found that when a suspension of *Salm. typhi* was added to a sample of sewage or polluted water it was clarified in a few days. The lytic agent was thermolabile and filterable; although it could be subcultivated indefinitely in suspensions of *Salm. typhi,* it did not grow on ordinary nutrient culture media. As it also multiplied on dead as well as on living suspensions of *Salm. typhi* it was not a bacteriophage. They suggested the lytic agent may be a minute parasitic flagellated vibrio, *Bdellovibrio bacteriovorus*. In further work (1969) these workers showed that this organism also used *E. coli, Pseudomonas, Erwinia, Salmonella* and *Shigella* spp. as host organisms but not Gram positive bacteria. Suspension of these were however cleared rapidly when added to sewage and river water indicating the presence of some other active bactericidal agent. A survey indicated that *Bdellovibrio* only occurred in polluted waters, up to 100,000/ml being present in sewage effluents. They concluded that *Bd. bacteriovorus* alone or in conjunction with other agents plays an important role in the self-purification of water (Guelin, Lepin & Lamblin, 1968). Mitchell, Yankofsky & Jannasch (1967) found that in samples of sea water taken from different parts of the Atlantic Ocean the reduction in the numbers of *E. coli* was strongly affected by the marine bacterial population. There was evidence that a specific lytic microbial population developed after introduction of *E. coli* into fresh sea water. Bacteriological examination of sea water off the Swedish coast into which sewage discharged showed that the numbers of *E. coli* declined sharply with depth and distance from the outfalls. *Clostridium perfringens,* which was found in all polluted waters, declined in a similar manner. This decline, it was concluded, was due to other factors as well as dilution. Antibiotic activity was demonstrated in shallow quieter waters. Bacilli which produced antibodies inhibitory to *E. coli* were isolated and identified as *B. licheniformis* and *B. coagulans,* both planktonic species (Bonde, 1968). Aubert, Lebout & Aubert (1965), investigating the bactericidal activity of sea water, demonstrated the importance of the indigenous marine plankton, mostly phytoplankton, as the agents.

Duff, Bruce & Antia (1966) confirmed the bactericidal activity of *Scenedesmus obliquus* against *Salm. typhimurium.* They found an inverse relationship between the growth of the algae and the density of the Gram negative rods. Sieburth (1965) observed that the antibacterial activity of estuarine waters varied seasonally associated with changes in the phytoplankton. Although there was no close correlation between maximum anticoliform activity and the incidence of a particular species of phytoplankton, the closest relationship was with the termination of the blooms of the predominant diatom of the water being studied *Skelotonema costatum.* Extracts of *Skel. costatum,* but not of other diatoms from the bay, were found to inhibit *Vibrio* and *Pseudomonas* spp.

The antibacterial activity was increased by illumination and by storage in dark, under which conditions the autolysis of senescent cells occurred. This indicated that a light activated autolytic product of *Skel. costatum* was an important agent in the natural control of the bacterial population in estuarine waters.

(d) *Concentration of pathogenic micro-organisms*

After the successful dispersion of micro-organisms in the diluting water they may be re-concentrated under certain circumstances. Deposition of bacteria from water may result in their accumulation in the surface layers of benthic muds (Allen, Grindley & Brooks, 1953). More active concentration results from the feeding activity of benthic invertebrates which appear to select predominantly Gram negative coli-aerogenes organisms. The filter-feeding forms would be expected to concentrate large numbers of bacteria from the large volumes of water they filter. Of public health significance in this respect are the Pelecypoda; enteric viruses and bacteria are commonly found in shellfish living in polluted waters. They may be present after the organisms have disappeared from the water, the retention being favoured by temperatures of $< 10°$. By remaining in nonpolluted water the shellfish cleanse themselves. Viruses are reported to survive for longer periods than bacteria on leaves of plants irrigated with polluted water (Grigorieva *et al.*, 1965).

The helminth pathogens, in contrast to enteric micro-organisms, do not find the aquatic environment so inhospitable. The life cycle of many of the flukes involves a stage in an aquatic snail. Although the miracidia larvae must find a molluscan host within a relatively short space of time, their dispersion in sewage effluents diluted with water ensures their widespread dissemination. Within the host they rapidly multiply to produce a high population of the infective stage. Silverman & Griffiths (1955) produced evidence to suggest that the increase in bovine cysticercosis caused by the bladder-worm stage of *Taenia saginata* in cattle, after World War II, was probably due to overloaded sewage treatment plants. The mild nature of the infections suggested a widespread dissemination of eggs which would result from their dispersal in sewage effluents diluted with the river water.

Factors influencing the concentration of enteric bacteria discharged from nearby sewage outfalls on bathing beaches have been the subject of many studies. In one such study, Gameson, Bufton & Gould (1967) investigated the distribution of coliform bacteria in coastal waters in relation to a sewer outfall situated well beyond low-water mark. Although the mean count near the beach was relatively low, 73-100 organisms/ml, there were very wide variations from sample to sample and from day to day. They concluded that the most important factors causing these variations were the roughness of the sea, the nearby discharge from a river, and sunlight.

Gameson, Pike & Barrett (1968) compared the bacterial pollution of 2 beaches, one with a short outfall (20 m) from which crude sewage discharged and the second with a long outfall (430 m), the sewage discharge of which had been comminuted. From the results obtained they attempted to develop a mathematical model to predict the amount of pollution reaching the beaches, but they concluded that this was not entirely satisfactory. In more recent work (Cameron, Munro & Pike, 1970) a study was made for a length situation where the outfall extended only c.20 m beyond the average low-water mark. The results from this survey showed a significant relationship between coliform count and both tidal state and wind velocity. The log of the coliform count at each station varied approximately sinusoidally with tidal state, there being roughly a 5-fold variation in coliform count. This variation was not greatly influenced however by the tidal range. There was also a pronounced relation between coliform count with wind velocity component, the highest counts usually being associated with wind induced currents parallel to the shore and in the direction from the outfall to the sampling point. Roughness of the sea, although difficult to distinguish from wind effects, was considered to be an important factor, the highest counts being associated with rougher seas. Although the results from such a short outfall are unlikely to be directly applicable to future outfalls which will be much longer, they provided data for the construction of a model to predict the variations in bacterial numbers associated with tidal state. This was based on the simple Fickian diffusion equation to give the rate of change of coliform concentration, C, in the absence of bacterial mortality or any wind effects:

$$\frac{\partial C}{\partial t} = D_x \frac{\partial^2 C}{\partial x^2} + D_y \frac{\partial^2 C}{\partial y^2} + D_z \frac{\partial^2 C}{\partial z^2} - V_x \frac{\partial C}{\partial x}$$

where x is the longshore distance from the outfall; y is the distance offshore; z is the depth below water surface; D_x is the longshore diffusion coefficient; D_y is the offshore diffusion coefficient; D_z is the vertical diffusion coefficient; V_x is the longshore velocity.

After testing the model, using coefficients based on the most relative available data against observed results, D_x and D_y were adjusted. There was then a satisfactory agreement between the observed and calculated tidal states at which the maximum and minimum coliform counts occurred at each station. The calculated maximum at each station were within a factor of 2 of the observed maxima. At the stations most distant in the outfall, however, the minima were appreciably too low.

6. Disposal by Dilution of Effluents Containing Toxic or Other Directly Harmful Substances

The principle involved in this method of disposal is to so dilute the toxic or harmful substance in the receiving water that it is then present at subharmful concentrations. Difficulties again arise however in determining at what concentrations such substances are harmless. Other problems are the progressive accumulation of such materials in the water to harmful levels and the possibility of the re-concentration of the pollutant in organisms and food chains.

(a) *Permissible concentrations after dilution*

Most of the published results are of laboratory work designed to determine the short term tolerance of different species, mostly adult fish, to different toxicants. This however does not necessarily predict all harmful effects if used for public consumption or of undesirable tastes and odours in the water itself or in fish used as food by man. The most popular and convenient way of expressing the acute toxicity of a substance is the LC_{50}, i.e. that concentration of the substance in water which kills 50% of the test animals in a specified time, e.g. 24 or 48 h. Such tests which determine the lowest concentration of a poison which kills fish in a relative short time, 1-2 days, are useful in predicting the effects of

acute toxicity, as produced, for example, by a plug of toxic substance passing
down a river. A series of recent papers has shown that with many of the poisons
commonly associated with industrial sewage effluents, the joint toxicity of 2 or
more of these is simply additive (Lloyd, 1961; Herbert, 1962; Herbert &
Shurben, 1964; Herbert & Vandyke, 1964; Brown, Mitrovic & Start, 1968;
Brown, Jordan & Tiller, 1969). By expressing the concentration of each poison
present as a proportion of the LC_{50} value the combined resultant toxicity can
be predicted by addition; if greater than unity the water is likely to be lethal
(Brown, 1968). Using this technique Lloyd & Jordon (1963, 1964) compared
the predicted toxicity of 24 sewage effluents on the basis of the concentrations
of cyanide, ammonia, copper, zinc and phenol and as determined
experimentally. In 13 of the 18 toxic effluents, the toxicity was predicted
within ± 30% of the observed values. Six effluents were correctly predicted to be
nontoxic (acutely) and in only 2 cases was the effluent more toxic than
predicted. Similarly the toxicity of fishless rivers in the industrial Midlands were
largely accounted for by the additive effect of toxicities of ammonia, copper,
zinc and phenol (Herbert, Jordon & Lloyd, 1965).

Although the advances reviewed above make possible the prediction of acute
toxicity where the nature and concentrations of the poisons are known the
prediction of chronic or subacute toxicity is more difficult. As pointed out by
Edwards & Brown (1967), although it is now possible to state for many common
poisons what limiting concentration will kill fish within a few days, it is still
rarely possible to know the highest concentration which does not have a
significant adverse effect on a natural population exposed to a poison more or
less continuously. Some workers have derived a permissible concentration by
applying a factor to the LC_{50} value; Jones (1964) quotes a factor 0·1 but also
quotes Beak (1958) who considers this factor to be little more than an
intelligent guess. Later work, in which fish cell cultures were used to assess
toxicity (Rachlin & Perlmutter, 1968), showed that the concentration of zinc
which exerted no observable toxic effect on the cell cultures was, in fact, 1/10
of the LC_{50} value. Edwards & Brown (1967), on the basis of field data,
suggest a factor of 0·3—0·4. Holden (1964) considered a factor of 0·01 was
appropriate for pesticides. Sprague (1969) recommends the incipient LC_{50} after
long exposure as the most useful single criterion of toxicity: if impracticable, the
LC_{50} 4 days was considered a useful substitute and often the equivalent, the
acute lethal process in most cases, ceasing within 4 days. Ball (1967) considers
that the slope of the % kill-concentration response curve is as important a
character as the LC_{50} value in determining a permissible level.

Some poisons are now known to exert delayed lethal effects on fish e.g.
pesticides (Alabaster & Abram, 1965), cadmium and acrylonitrile (Ministry of
Technology, 1969), and these must be taken into consideration when
determining permissible levels. Herbert (1961), discussing the relevance of

toxicity tests to fish populations in rivers, suggested that a practicable permissible level in rivers receiving sewage effluents and industrial discharges, would be that concentration which, under low flow conditions would not kill more than a small, fixed proportion, 1–5%, of fish in a 3 months period. With these concentrations some kills could be expected under hot dry summer conditions. This value could be determined by extrapolation from data obtained in laboratory tests.

Attempts have been made to determine experimentally the concentrations of poisons which produce a definite physiological effect on organisms. The effects on the respiration rate, serum protein content, histological and histochemical changes and changes in the red blood count have been used as criteria. However, as Mount (1967) states, although "death is obviously adverse" the ecological significance of many of the physiological changes is less evident.

There is little published work on the susceptibility of different developmental stages to toxicity and most of these refer to salmonids or tropical fish. Skidmore (1965) found that the resistance of zebra fish eggs to zinc was greatest when newly laid and then declined and were most sensitive 4-13 days after hatching. Penaz (1965) also found that the eggs of brown trout were more tolerant to ammonia during the early stages of development. With zinc however the eggs of brown trout were most sensitive in the early "eyed" stage; as embryonic development proceeded sensitivity decreased and the young fish showed maximum resistance immediately after hatching. Thereafter resistance gradually decreased until the end of the alevin stage when it was similar to that of the adult fish (Ministry of Technology, 1969).

The presence of substances although nontoxic even over a long period, may be harmful to the species and eventually affect the population. In many rivers detergents are present at sublethal concentrations. Mann & Schmid (1965) have shown that at sublethal concentrations of the detergent tetrapropylenebenzene sulphonate the growth rate of guppies was impaired in a test period of 6 weeks. Lemke & Mount (1963) found that the growth rate of bluegills was reduced in concentrations of ABS detergent $c.1/3$ of the LC_{50} value in a 30 days test. Bardach, Fujiya & Holl (1965) found that concentrations of detergent as low as 0·5 mg/l, much lower than the lethal level, damaged the chemoreceptors of the yellow bullhead and affected its swimming and feeding. Histological examination revealed the erosion of the taste buds thus impairing their receptor functions. It was considered that this could affect the ability of fish to detect and avoid poisons in the water. Dugan (1967) found that long term exposure of goldfish to sublethal concentrations of detergents rendered them more susceptible to the toxic effects of dieldrin and DDT. Detergents have been reported as enhancing the undesirable flavours in fish flesh due to phenols, tar derivatives and mineral oils (Mann, 1964). The overall effect of such substances on the ecosystem is probably best reflected by changes in the overall productivity.

(b) *Persistence in the aquatic environment*

The rate of accumulation of such substances depends on their susceptibility to degradation. Although some of the more natural compounds such as phenols are readily degraded, many of the modern synthetic organic compounds, such as the organophosphorus pesticides, are less degradable. Heavy metals may be inactivated by chelation. The biodegradation of some synthetic organic compounds under lentic stratified conditions was studied by De Marco, Symons & Robeck (1967). The detergents sodium lauryl suphate and linear alkylate sulphonate and the herbicide 2,4-D were all biodegraded both aerobically and anaerobically. The rate of degradation was reduced however by low temperatures and to a greater extent in the case of aerobic breakdown by low oxygen concentrations. Thus under stratified conditions in impoundments such synthetic organic compounds could accumulate in the hypolimnion.

Pesticides differ considerably in their resistance to degradation. The marked differences in resistance to biodegradation shown by different pesticides is probably due to their molecular structure. 2,4-D is readily decomposed but 2,4,5-T, which differs by only a single chlorine on the aromatic ring, is very resistant (Alexander, 1964). Okey & Bogan (1965) compared the relative degradation of synthetic organic pesticides and found that the presence of 3 or more chlorine atoms on an aromatic ring restricted microbial metabolism severely. This, they considered, was due to electronic effects of chlorine on the substrate. The resistance of lindane, BHC and dieldrin compounds and 2,4,5-T to microbial attack, they considered was due to the electron-poor nature of the unsubstituted carbon atoms.

Many workers have reported the accumulation of radionuclides in the bottom muds of rivers. These are significant in themselves but they also indicate the possible accumulation of less readily detectable substances in bottom muds. In the Clinch and Tennessee rivers the sediments were found to contain 21% of the Cs_{137} that had been discharged to the river over a 20 years period (Pickering *et al.*, 1966). Although no accumulation of pesticides is likely to take place in the river water itself there is evidence that it accumulates in the coastal waters around Britain (Ministry of Technology, 1965-6). Workers have reported that the concentration of pesticides in bottom deposits are commonly at least 100 times those in the overlying water (Warnick, Gaufin & Gaufin, 1966).

(c) *Re-concentration by organisms*

Work with radioactive tracers have demonstrated the ability of aquatic organisms to concentrate nonessential elements from dilute concentrations in their surrounding environment. This uptake may be either direct or via food chains. There is evidence that when they are obtained via a food chain they are retained to a greater extent than when absorbed directly; they are presumably chemically bound in the food organisms (Williams & Pickering, 1961).

Aquatic organisms live in a more intimate relationship with their medium than do nonaquatic ones. Whereas in nonaquatic animals most of the materials, except oxygen, is taken in by feeding and drinking, aquatic organisms are also able to take in ions by absorption from the aquatic medium. In many, special organs have been developed for this purpose. By the same pathway however nonessential ions also gain entry. The minnow *(Phoxinus phoxinus)* when exposed to subacute concentrations of metals for prolonged periods was found to accumulate different metals in the internal organs. Although histopathological studies revealed no serious gill damage, as in acute toxicity, the internal organs were affected, especially after copper exposure. Histochemical and analytical studies revealed a continuous uptake and storage of copper in the internal organs. In copper concentrations of 0·1 mg/l in the water the copper content of the liver rose from a normal level of 55 mg/l in control fish to 960 mg/l in fish exposed for 90 days: in the kidneys, the copper content rose from a control average of 48 mg/l to 138 mg/l. Throughout the period the copper content of all organs except the gills continued to increase. Lead and nickel were accumulated to a less extent and, although zinc was detected, there was no continued accumulation (Preston, 1971). Metals such as zinc, lead, mercury and cadmium are known to be concentrated in aquatic organisms.

A now classical example of mercury poisoning is that of Minamata Bay, Japan. In 1953 inhabitants in the vicinity of Minamata Bay were found to be suffering from a severe neurological disorder which was attributed to the mercury content of the fish food they consumed taken from the bay. A factory manufacturing acetaldehyde from acetylene, using mercuric oxide in sulphuric acid as a catalyst, discharged its waste water, containing mercury, to the bay. A wet sludge produced in the process contained methyl mercury chloride. This substance was isolated from shellfish in the bay and it produced experimentally symptoms of the disorder which has become known as Minamata disease (Irukayama, 1967). In discussion of this paper the danger of the highly toxic alkyl metal compounds generally was stressed. Tetraethyl lead, for example, is not removed by chemical coagulation which is effective as a waste water treatment process for removing inorganic lead compounds. It is probably that inorganic metal compounds discharged to the aquatic environment are transformed into organic metal compounds in passing through food chains. In experiments in aquaria and a lake, mercuric chloride was converted to methyl mercury in this way (Jensen & Jernelöv, 1969). The algae, *Microcystis aeruginosa, Anabaena cylindrica, Scendesmus quadvicanda* and *Oedogonium* have been found in laboratory culture experiments to concentrate the pesticides aldrin, dieldrin, endrin and DDT >100-fold from dilute solutions (Vance & Drummond, 1969). Relatively high concentrations or residual DDT and dieldrin have been found in the flesh and eggs of fish-eating ducks from the River Tweed; it was considered that the pesticides were concentrated through the food chains (Ministry of Technology, 1964-5).

Besides the dangers involved in the concentration of toxic material in organisms which are the food of man, the activity and death of such organisms results in the transfer of the concentrated pollutant to the benthos. Differences in the feeding habits of benthic animals at different depths in York River, U.S.A. affected the degree of removal of materials from the surrounding water and the subsequent incorporation of the removed material as biodeposits in the subsurface sediments (Haven, 1966).

In investigations in an Italian lake the concentration of each radionuclide differed in the different granulometric fractions of the sediment. The radioactivity of the smallest particles in the sediment played an important role in the uptake of radioactivity by the snail *Viviparus ater.* This appeared to concentrate radioactive strontium into its shell and embryo and to concentrate radioactive ruthenium from the water into its faeces and thence into the bottom deposits (Ravera, 1964).

7. Disposal by Dilution of Effluents Containing Biogenic Materials

Sewage effluents are by far the commonest form of discharge to the aquatic environment. These contain natural organic materials and their degradation products such as nutrient salts. The discharge of such biogenic materials may be regarded as augmenting the natural transfer of materials between ecosystems within the biosphere. This may result however in adverse conditions being induced in the aquatic environment due to the imbalance created in the ecosystem. An ecosystem may be regarded as a system in which there is a cyclic interchange of material between the biotic and abiotic. Most ecosystems have 3 components, producer, consumer and decomposer populations. Thus, introduced biogenic material is rapidly assimilated in the ecosystem and results in an imbalance depending on the nature of the introduced material. Organic matter tends to encourage the development of the heterotrophic decomposer component (organic pollution) whilst nutrient salts favour the producers (eutrophication). In either case the materials dispersed in the water are trapped and concentrated in the ecosystem.

(a) *Organic pollution*

The most serious adverse effect of discharging organic matter to a river is the reduction in the dissolved O_2 concentration of the water resulting from the increased respiratory demand of the saprobic micro-organims. This reduction may adversely affect the other aquatic organisms, including fish.

The resultant O_2 profile in a river downstream of an organic effluent is the resultant of two opposing forces each having sets of contributory processes. One set tends to deoxygenate the stream water to satisfy the O_2 demand. These demands include that of the soluble and suspended organic matter in the water, the demand of the benthic deposits and the natural respiratory demand of other organisms present in the water and in the benthos. The opposing set of forces tends to reoxygenate the stream water up to saturation level. These include physical surface re-aeration and photosynthetic activity of phytoplankton and benthic algae and plants. Both processes proceed as a time function in accordance with definite laws. Deoxygenation in each unit of time is proportional to the remaining concentration of oxidizable organic matter (BOD). Reoxygenation in each unit of time is proportional to the remaining degree of unsaturation of dissolved O_2 (the oxygen saturation deficiency). On a time basis these opposing forces result in the typical O_2 curve, which may be expressed mathematically as

$$\frac{\partial D}{\partial t} = K_1 L - K_2 D$$

where D represents the oxygen saturation and L the momentary oxygen demand at the same time, K_1 and K_2 being coefficients defining the rates of deoxygenation and reoxygenation respectively (Streeter, 1958). Dilution by reducing the concentration of organic load reduces L and thus limits the intensity and extent of the O_2 sag. The diluting water also contributes a supply of oxygen to the budget.

Gameson & Barrett (1958) caculated that water entering the Thames estuary over Teddington weir contributed 77 tons of O_2/day during 1950-53. This however, even with a further 7 tons supplied by tributaries, represented only a small proportion of the O_2 demand most of which was satisfied by surface re-aeration in the estuary. By minimizing the depletion of O_2 in waters receiving organic wastes, dilution plays a most important role, and the use of dilution as a method of disposal of such effluents is probably the most acceptable use of this method of disposal.

Apart from the deoxygenating effect, organic matter affects benthic communities directly. Nutritional changes favours heterotrophic growths such as the sewage fungus. *Sphaerotilus natans* is associated with effluents from dairies, beetsugar processing, breweries, textile mills, flour mills and spent sulphite liquors from pulping mills besides sewage treatment works. A feature of such growths is their presence in water having very low concentrations of the waste. Infestations have been recorded from rivers in which sulphite liquors were present as low as 1 mg/l corresponding to a BOD of 0·2 mg/l. It would appear that *Sphaerotilus* is physiologically well adapted to utilize efficiently organic nutrients from dilute solutions which its benthic mode of life enables it to concentrate. This dilution of organic effluents although reducing the degree of deoxygenation does not necessarily control the development of sewage fungus.

Although dilution may decrease the intensity of the growth/unit area, the length of the infestation would be extended. In contrast to disposal by dilution, the intermittent discharge of the wastes has been practised to control sewage fungus growths in rivers receiving pulp mill effluents (Amberg & Cormack, 1960). By discharging the effluent during 1 day in 6 and storing it for the intervening 5 days the amount of sewage fungus growth was considerably less than when the waste was discharged continuously to provide maximum dilution.

Thus the disposal of organic wastes by dilution although preventing deoxygenation is less useful in preventing the development of objectionable benthic conditions. When the sewage fungus dies and decomposes it creates a secondary oxygen demand.

(b) Eutrophication

The discharge of effluents such as oxidized sewage effluents containing mineral salts such as nitrates and phosphates may cause marked changes in the ecology of the receiving water. Some of these changes such as blanket-weed (Cladophora) growths in rivers and algal blooms in lakes and reservoirs are detrimental to man's interests. In lakes eutrophication is a natural process in the life history of the lake involving a progressive increase in fertility and productivity, associated with which there are changes in the dominant species. The process is accelerated by the discharge of nutrient minerals in effluents. The lake ecosystem effectively traps such materials to increase the fertility. The continuous discharge of such effluents into lakes inevitably results in the accumulation of nutrients and accelerates eutrophication whatever the initial dilution factor. For such effluents disposal by dilution in lakes is likely to give rise to problems eventually. In lakes nutrients are rapidly cycled within the ecosystem and are taken up from quite dilute solutions. It would appear that phosphorus (P) could be a key to determining the degree of eutrophy, yet there is no evidence of a direct relationship between the degree of eutrophy and P concentration. Asterionella is capable of taking up P from natural lake water containing as little as 0·001 mg/l, and there is evidence that it is made available for resynthesis by lysis of dead algal cells as they sink through the euphotic zone. Barlow & Bishop (1965) found that under stratified conditions in the late summer, 80% of the P was liberated in the epilimnion of Cayuga Lake, N.Y., mostly by activity of cladocerans. Beers (1962-4) also considers the importance of zooplankton excretion of phosphates and ammonia in relation to marine phytoplankton nutrients.

Phosphorus may also be stored on benthic muds. At low concentration P was found to be tightly bonded, probably as aluminium phosphate, but at higher concentrations it was more loosely bound independent of the aluminium content of the sediment (Harter, 1968).

Although simple nutrients such as N and P are important considerations, the degree of eutrophy is not readily defined in such simple terms. The critical levels of nutrients needed to promote algal growths under natural conditions has not been determined except for the silica requirement for the diatom *Asterionella formosa*. Observations on Lake Windermere have shown over a number of years that the population of this diatom declined each season when the silica concentration fell below 0·5 mg/l (Lund, 1950).

Culture experiments using artificial media do not provide results on limiting nutrient concentrations applicable to field conditions. Apart from the dynamic balance of essential nutrients in natural waters, which is probably not simulated in culture media, it is now being realized that many algae are partially heterotrophic in requiring organic growth factors, such as vitamin B_{12}, thiamine and biotin (Pravasoli, 1964); Fogg (1965) estimates that 70% of planktonic algae require vitamin B_{12}. It is also significant that these and trace elements would be supplied in domestic sewage effluents. Fogg quotes Rodhe's findings that in artificial culture media at least 0·04 mg/l of P was needed to sustain maximum growth of *Asterionella formosa*, whereas in natural lakes 0·002 mg/l sufficed. He suggests the presence of some factor other than nutrients, trace elements, chelating materials, vitamin B_{12} and thiamine in the natural water which enables the algae to use P at these low concentrations. Natural waters, he concludes, cannot be looked upon as simple solutions of mineral salts plus certain definite organic substances having chelating or growth promoting properties, and he draws attention to the dissolved organic matter in water about which we know very little.

Fogg & Nalewajko (1964) found that phytoplankton under natural conditions release glycolic acid into the water as an extracellular product of photosynthesis. The fact that this substance can be re-absorbed by algae from the water suggests that there is an equilibrium between intra- and extracellular concentrations. This, it is suggested, would affect the growth rate of algae; in waters low in glycollic acid content, as in a lake at the end of winter, photosynthetic activity of the algae would be devoted to the production of extracellular glycollic acid so that cell growth and multiplication would not occur until an equilibrium concentration was established in the water. Under other conditions, net absorption of glycolic acid by the algal cells could occur which would support heterotrophic growth of the phytoplankton. Beside algae, several freshwater bacteria can use glycolic acid as their sole carbon source.

All this evidence points to the fact that we are not likely to define clearly the parameters of eutrophy in simple nutritional terms alone. Indeed, as with many ecological phenomena, it may not be defined in autoecological terms but necessitates our considering in more detail the synecological aspects, especially the competition between species. In other fields it has been found that factors affect populations of species more often by determining the outcome of interspecific competion rather than by direct effects. Over a range of any factor, one species dominates and eliminates other species and beyond the limit of this range a second species replaces the first. Thus at the limit of these ranges a small change in the factor results in a marked change in the biota, whereas the same degree of change in the middle of the range would have little effect. In this connection the sudden changes in species of the phytoplankton in the process of eutrophication not associated with marked changes in observable nutritional conditions may be significant. Although algal studies have naturally been most common in relation to eutrophication the role of bacteria in cycling material in eutrophic waters is important. The better growth of some algae in culture in the presence of bacteria may be attributed to a symbiotic effect such as the production of the vitamin B_{12} or CO_2 (Lange, 1967). They play an essential role in the cycling of macronutrients such as P. Golterman (1964) found in laboratory studies that 50% of the P in killed *Scenedesmus* cells was liberated by lysis into the water in a few hours. Phosphates in the nucleic acids were not liberated by autolysis but were probably attacked by bacteria, while N was liberated only slowly by autolysis. He concluded that autolysis of the dying bloom induced bacterial breakdown. Hayes (1963) however considered that bacteria compete for the inorganic P and convert it into organic P, in which form it is unavailable to plants.

Golterman (1970) considered that eutrophication may contribute to the increased survival time of *E. coli* as a result of shading by phytoplankton.

It would appear therefore that eutrophication is better defined by overall nutrient balances and turnover rates rather than by concentrations. In guidelines given by OECD predicting eutrophication in lakes (Vollenweider, 1969) it was suggested that if waters contain 10 mg of P and 200–300 mg of N/m^3 in the spring, they were in danger of becoming eutrophic if the incoming nutrients exceeded 0·2–0·5 g of P and 5–10 g of N/m^2 year. Golterman (1970) stresses the importance of the phosphate turnover period in assessing eutrophication rather than P concentration. On these considerations the disposal by dilution in lakes of effluents rich in plant nutrients is not likely to prevent eutrophication, in some waters also benthic microorganisms remove and accumulate nutrients from dilute solutions. Growths of *Cladophora*, although associated with sewage effluents, are found in profuse amounts over a wide range of nutrient concentrations. Nevertheless it has been found that rivers containing a high proportion of oxidized sewage effluent are quite stable in terms of both water quality and the restricted benthic community they support. This is in contrast to the succession of changes in benthic communities which become established in relation to organic pollution. When such eutrophic rivers are diluted with good quality river water a natural rich benthic fauna is rapidly established even when the dilution factor is low. Such conditions exist in the River Ray, a tributary of the upper Thames where a restricted fauna dominated by *Asellus aquaticus* exists along its whole length until it joins the Thames below the confluence with which a typical rich fauna exists although the dilution afforded by the Thames is only small (Hawkes & Davies, 1971). Thus in eutrophic rivers dilution could be beneficial.

8. Practice of Disposal by Dilution

In discharging effluents into lakes or the sea one must necessarily accept the fixed dilution factor available and reduce the concentration of materials and density of organisms by treatment methods to such levels that they do not accumulate to cause adverse conditions, taking into account the rate of decline and the possibilities of reconcentration by agencies discussed above. In practice the only degree of control of dilution is in the method and positioning of the discharge and in the regulation of the flow of the effluent.

In rivers a greater degree of control of dilution is afforded by controlling the flow in the river. River flows naturally fluctuate widely and in many circumstances although the normal flow in a river provides adequate dilution for the effluents it receives, at times of low flow the dilution may not be sufficient to prevent adverse conditions arising in the river below the effluents. By regulating the flow in the river by storing water in impoundments at times of high flow it can be released in calculated amounts as required at times of low flow. Storage is usually provided by regulating impoundments which change the natural flow and the distribution of the water mass spatially. Besides providing for flow regulation for diluting effluents such schemes provide other amenities.

These schemes are, however, not without problems; the evaluation of benefit and detriment of stream flow regulation by impoundments has been fully considered at a symposium on this subject in Cincinnati (U.S. Dept. of Health, Education and Welfare, 1965). The main adverse effect was considered to be the deterioration in the quality of the hypolimnion water under stratified conditions in impoundments, so that if this water were discharged it could have serious effects downstream. One type of pollution, however, that is most effectively reduced by dilution is that of thermal effluents. At times of low flow the augmentation with cold hypolimnion water would be especially beneficial in this respect. Under some climatic conditions serious loss of water by evaporation could occur and this would increase the concentration of dissolved solids.

A further difficulty is that by augmenting the flow at times of low flow, although an effluent may initially be diluted, because of the increased rate of flow, the time taken to reach a given abstraction point downstream is reduced; this could affect the degree of self purification achieved so that at the point of abstraction the water quality could be inferior to that in the river without flow augmentation. Thus the flow regulation programme should be planned for each

individual river system taking into account such considerations as:
  (i) low flows and their frequency;
  (ii) the quantity, quality and position of the different effluents discharging to the river;
  (iii) the position and quantity of abstractions and minimal quality of water required;
  (iv) waste assimilative capacity of the river and of the diluting water;
  (v) the water quality standard required in different stretches of the stream in relation to water use.

Hull (1967) discusses the relative merits and cost analysis of advanced waste treatment and low flow augmentation as alternative methods of pollution control. Except for the removal of suspended solids and pathogens he recommended low flow augmentation as an economic substitute for advanced treatment, especially to maintain the necesary oxygen level in relation to fisheries.

Although flow regulation for water quality control is compatible with the amenity requirements, abstraction from the upper stretches of rivers and from ground water in the vicinity reduces the amount of water available for dilution. In this connection the Water Resources Board policy of supplying raw water for public supply *via* regulated river flows rather than directly from upland impoundments, is beneficial in maintaining more diluting water available in the rivers.

9. Conclusions

Having surveyed the literature related to disposal by dilution we may now consider the two opposing views expressed in their extremes as "the solution to pollution is dilution" and "dilution is no solution to pollution". Obviously both these views are exaggerated generalizations and need to be considered more critically in relation to different types of pollution. Generally however, even when effluents have received a high degree of treatment, dilution is always beneficial.

Today, it is technologically feasible to give full treatment to most wastes before discharging them, but often such treatments are economically prohibitive. In such cases it would appear reasonable to utilize the homeostatic selfpurifying capacities of natural freshwater ecosystems providing no serious adverse effects resulted. With the most common effluent, sewage effluent, the serious effects of deoxygenation can be reduced by dilution. At the same time the enteric organisms present also die off rapidly, so that dilution would appear to be a practicable method of disposal of normal sewage effluent. Because of the concentrating forces associated with benthic growths in lotic waters and phytoplankton in lentic waters, dilution of biogenic materials may not prevent problems. For the same reason dilution is not suitable for the disposal of nonbiodegradable, toxic or undesirable materials.

Dilution is beneficial with heated effluents but where suspended matter is involved, although it may be beneficial in reducing the concentration of suspended matter, the adverse effects on benthic communities when this settles is not alleviated by dilution.

Thus, although dilution is not a solution to all pollution, in conjunction with other pollution control measures it can be a most practicable and economic final stage of disposal of many effluents. As such the degree of treatment and degree of dilution are inter-related and should be planned in relation to each discharge.

Dilution should be practised scientifically rather than hopefully.

10. References

Alabaster, J. S. & Abrams, F. S. H. (1965). Development and use of a direct method of evaluating toxicity to fish. *Adv. Wat. Pollut. Res. 2nd Int. Conf., Tokyo 1964.* **1,** 44.

Alexander, M. (1964). Microbiology of pesticides and related hydrocarbons. Principles and applications in aquatic microbiology. *Proc. Rudolfs Res. Conf., Rutgers State Univ.* p.15.

Allen, L. A., Brookes, Eileen & Williams, Irene L. (1949). Mechanical filtration of sewage effluents. 11. Effects on the bacterial population of treatment of sewage effluents in sand or anthracite filters. *J. Proc. Inst. Sew. Purif.* **4,** 411.

Allen, L. A., Tomlinson, T. G. & Norton, I. L. (1944). The effect of treatment in percolating filters on bacterial counts. *J. Proc. Inst. Sew. Purif.* p. 115.

Allen, L. A., Grindley, J. & Brooks, E. (1953). Some chemical and bacterial characteristics of bottom deposits from lakes and estuaries. *J. Hyg., Camb.* **51,** 185.

Amberg, J. R. & Cormack, J. F. (1960). Factors affecting slime growth in the Lower Columbia River and evaluation of some possible control measures. *Pulp. Paper Can.* Feb. p.1.

Anderson, P. W. (1963). Variations in the chemical character of the Susquehanna river at Harrisburg, Pennsylvania. U.S. Geol. Surv. Wat. Supply Pap. 1779 – B. U.S. Govt. Printing Office, Washington, D.C. (*via Wat. Pollut. Abstr.* (1966) **39.** Abstr. no. 1451).

Aubert, M., Lebout, H. & Aubert, J. (1965). Effect of marine plankton in destruction of enteric bacteria. *Proc. 2nd int. Conf. Wat. Pollut. Res., Tokyo.* **3,** 303.

Ball, I. R. (1967). The relative susceptibilities of some species of freshwater fish to poisons – I Ammonia. *Wat. Res.* **1,** 767.

Bardach, J. E., Fujiya, M. & Holl, A. (1965). Detergents: effects on the chemical senses of the fish *Ictalurus natalis* (le Suceur). *Science, N.Y.* **148,** 1605.

Barlow, J. P. & Bishop, J. W. (1965) Phosphate regeneration by zooplankton in Cayuga Lake. *Limnol. Oceanogr.* **10,** 15.

Beak, T. W. (1958). Toleration of fish to toxic pollution. *J. Fish. Res. Bd Can.* **15,** 559.

Beers, J. R. (1962–4), Ammonia and inorganic phosphorus excretion of some marine zooplankton. 4th Meeting Ass. Isl. Mar. Labs., Curacao, 11, 848. (*via Wat. Pollut. Abstr.,* 1965, **38,** no. 934).

Blum, J. L. (1965). Interaction between Buffalo river and lake Erie. Great Lakes Res. Div., Publ. No. 13, 25 (*via Wat. Pollut. Abstr.,* 1967, **40,** abstr. no. 1227).

Bonde, G. P. (1968). Studies on the dispersion and disappearance of enteric bacteria in the marine environment. *Rev. int. Oceanogr. med., Nice* **9,** 17.

Briscou, J., Tysset, C. & Roy, Y.d. L. (1965). Study on the microbiology of the benthos. In *Symp. Commn. int. Explor. Scient. Mer. Medit., Monaco* 1964.

Brown, V. M. (1968). The calculation of the acute toxicity of mixtures of poisons to rainbow trout. *Wat. Res.* **2**, 723.

Brown, V. M., Mitrovic, V. V. & Stark, G. T. C. (1968). Effects of chronic exposure to zinc on toxicity of a mixture of detergent and zinc. *Wat. Res.* **2**, 255.

Brown, V. M., Jordon, D. H. M. & Tiller, B. A. (1969). The acute toxicity to rainbow trout of fluctuating concentrations of mixtures of ammonia, phenol and zinc. *J. Fish Biol.* **1**, 1.

Bukovsky, L. (1958). The influence of calcium on the survival of coliform microbes in surface waters. *Cslka Hyg.* **3**, 228 (*via Wat. Pollut. Abstr.*, 1964, **37**, abstr. no. 1517).

Burman, N. P. (1966). Discussion on Harkness, N. Bacteria in sewage treatment processes. *J. Proc. Inst. Sew. Purif.* **6**, 5.

Carlson, G. F., Woodward, F. E., Wentworth, D. F. & Sproul, J. (1968). Virus inactivation on clay particles in natural waters. *J. Wat. Pollut. Control Fed.* **40**, No. 2 Res. Suppl.

Csanady, G. T. (1966). Dispersal of foreign matter by currents and eddies of the Great Lakes. *Great Lakes Res. Div. Un. Mich.* Publ. No. **15**, 283.

Davies, A. W. (1969). Hydrological and chemical aspects of river pollution. *J. Soc. Wat. Treat. Exam.* **18**, 161.

DeMarco, J., Symons, J. M. and Robeck, G. G. (1967). Behaviour of synthetic organics in stratified impoundments. *J. Am. Wat. Wks Ass.* **59**, 965.

Department of Scientific and Industrial Research (1964). Effects of polluting discharges on the Thames Estuary. The reports of the Thames Committee and of the Water Pollution Research Laboratory. *Wat. Pollut. Res.* **Tech. Pap.** No. 11.

Department of Scientific and Industrial Research (1963). "Polishing" of sewage works effluent. *Notes Wat. Pollut.* no. 22, p. 4.

Duff, D. C. B., Bruce, D. L. and Antia, N. J. (1966). The antibacterial activity of marine planktonic algae. *Can. J. Microbiol.* **12**, 877.

Dugan, P. R. (1967). Influence of chronic exposure to anionic detergents on toxicity of pesticides to goldfish. *J. Wat. Pollut. Control Fed.* **30**, 63.

Edwards, R. W. & Brown, V. M. (1967). Pollution and Fisheries: A progress report. *Wat. Pollut. Contr.* **66**, 63.

Fogg, G. E. (1965). *Algal Cultures and Phytoplankton: Ecology.* University of London: Athlone Press.

Fogg, G. E. & Nalewajko, C. (1964). Glycollic acid as an extra-cellular product of phytoplankton. *Verh. int. Ver. theor. angew. Limnol.* **15**, 806.

Gameson, A. L. H. & Barrett, M. J. (1958). Oxidation, reaeration and mixing in the Thames estuary. Oxygen relationships in streams. *Robert A. Taft Sanit. Engr. Center* Tech. Rep. W 58–2, p.63.

Gameson, A. L. H., Bufton, A. W. J. & Gould, D. J. (1967). Studies of the coastal distribution of coliform bacteria in the vicinity of a sea outfall. *J. Wat. Pollut. Control,*

Gameson, A. L. H., Munro, D. & Pike, E. B. (1970). Effects of certain parameters on bacterial pollution at a coastal site. *Inst. Wat. Pollut. Control. Symp. Bournemouth.*

Gameson, A. L. H., Pike, E. B. & Barrett, M. J. (1968). Some factors influencing the concentration of coliform bacteria on beaches. *Rev. int. Oceanogr. med., Nice* **9**, 255.

Golterman, H. L. (1964). Mineralisation of algae under sterile conditions or by bacterial breakdown. *Verh. int. Ver. theor. angew. Limnol.* **15**, 544.

Golterman, H. L. (1970). Consequences of phosphate eutrophication of freshwater. *Ingenieur Grav.,* **82**, G 99–G 106. (*via Wat. Pollut. Abstr.* 1970, **43**, abstr. no. 2451).

Gray, E. (1952). The ecology of the ciliate fauna of Hobson's Brook, a Cambridgeshire chalk stream. *J. gen, Microbiol,* **6**, 108.

Grigorieva, L. V., Gorodetsky, A. S., Omelyanets, T. G. & Bogdanenko, L. A. (1965). Survival of bacteria and viruses on vegetables irrigated with infected water. *Gig. Sanit.* **30**, no. 12, 28 (*via Wat. Pollut. Abstr.* 1967, **41**, abstr. no. 14).

Guelin, A. & Lamblin, D. (1966). The bactericidal power of water. *Bull. Acad. natn. Med.* **150**, 526. (*via Wat. Pollut. Abstr.* 1968, **41**, abstr. no. 937).

Guelin, A., Lepine, P. & Lamblin, D. (1967). Bactericidal activity of water and the part played by *Bdellovibrio bacteriovorus*. *Annls Inst. Pasteur, Paris* 113, 666.

Guelin, A., Lepine, P. & Lamblin, D. (1968). On the self purification of water. *Rev. int. Oceanogr. med.* 10, 221.

Hannay, C. L. (1945). Some problems in the bacteriology of rivers. *Proc. Soc. appl. Bact.*

Harter, R. D. (1968) Adsorption of phosphorus by lake sediment. *Proc. Soil. Sci. Soc. Am*, 31, 514. (*via Chem. Abstr.* 1968, 69, 6513).

Harkness, N. (1966). Bacteria in sewage treatment processes. *J. Proc. Inst. Sew. Purif.* 6, 542.

Harold, C. H. H. (1935). *Thirtieth Annual Report Metropolitan Water Board.*

Haven, D. S. (1966). Concentration of suspended radioactive wastes into bottom deposits. Progress report, September 1, 1965–August 31, 1966. *U.S. Atom. Enger. Commn.* ORO – 2789 – 17. (*via Wat. Pollut. Abstr.* 1967, 40 abstr. no. 1558).

Hawkes, H. A. & Davies, L. J. (1971). Some effects of organic enrichment on benthic invertebrate communities in stream riffles. In *The Scientific Management of Animal and Plant Communities for Conservation.* British Ecological Society Symposium. Int. Symposium. Norwich, 1970.

Hayes, F. R. (1963). The role of bacteria in the mineralisation of phosphorus in lakes. *Symp. Mar. Microbiol. Springfield, Ill.* p. 654 (*via Biol. Abstr.* 1964, 45, 393).

Hays, J. R. & Krenkel, P. A. (1968). Mathematical modelling of mixing phenomena in rivers. *Un. Tex. Wat. Resour. Symp.* no. 1, p.111.

Herbert, D. W. M. (1961). Freshwater fisheries and pollution control. *Proc. Soc. Wat. Treat. J.* 10, 135.

Herbert, D. W. M. (1962). The toxicity to rainbow trout of spent still liquors from the distillation of coal. *Ann. appl. Biol.* 50, 755.

Herbert, D. W. M., Jordan, D. H. M. & Lloyd, R. (1965). A study of some fishless rivers in the industrial Midlands. *J. Proc. Inst. Sew. Purif.* 6, 569.

Herbert, D. W. M. & Shurben, D. S. (1964) The toxicity to fish of mixtures of poisons. 1. Salts of ammonia and zinc. *Ann. appl. Biol.* 53, 33.

Herbert, D. W. M. & Vandyk, J. M. (1964). The toxicity to fish of mixtures of poisons. 11. Copper-ammonia and zinc-phenol mixtures. *Ann. appl. Biol.* 53, 415.

Holden, A. V. (1964). The possible effects on fish of chemicals used in agriculture. *J. Proc. Inst. Sew. Purif.* 361.

Hull, C. H. J. (1967). *River Regulation, River Management.* (Ed. P. C. C. Isaac). London: Maclaren and Sons.

Irukayam, K. (1967). The pollution of Minamata bay and Minamata disease. *Proc. 3rd. int. Conf. Wat. Pollut. Res., Munich.* 1966 3, 153.

Jensen, S. & Jernelöv, A. (1969). Biological methylation of mercury in aquatic organisms. *Nature, Lond.* 223, 753.

Jones, J. R. E. (1964). *Fish and River Pollution.* London: Butterworth.

Kabler, P. (1959). Removal of pathogenic micro-organisms by sewage treatment processes. *Sewage ind. Wastes* 31, 1373.

Kisiel, C. C., Shapiro, M. A., Ficke, J. F., Morgan, P. V. & Spear, R. D. (1964). Tracer studies with rhodamine-B in a 3.2 Km reach of the upper Ohio river. *Verh. int. Ver. theor. angew. Limnol.* 1962. 15, 265.

Klein, L. (1962). *River Pollution.* 11. *Causes and Effects.* London: Butterworth.

Lagarde, E. & Castellvi, J. (1965). Aspects of bacterial pollution in a brackish region on the Mediterranean coast. *Symp. Commn. int. Explor. scient. Mer. Medit., Monaco, 1964.* (*via Wat. Pollut. Abstr.* 1967, 40, abstr. no. 1015).

Lange, W. (1967). Effects of carbohydrates on the symbiotic growth of planktonic blue-green algae with bacteria. *Nature, Lond.* 215, 1277.

Lemke, A. E. & Mount, D. I. (1963). Some effects of alkylbenesulphonate on the bluegill, *Sepomis macrochirus. Trans. Am. Fish. Soc.* 92, 372.

Levina, R. I. (1966). Experimental bacteriacidal activity of protococcal algae against the bacillus of murine typhus. *Zdravooknr. Belorussii,* 12. (9), 46. (*via Wat. Pollut. Abstr.* 1968, 71, abstr. no. 552).

Lloyd, R. (1961). The toxicity of mixtures of zinc and copper sulphates to rainbow trout (*Salm. gairdnerii* Richardson). *Ann. app. Biol.* **49**, 535.

Lloyd, R. & Jordon, D. H. M. (1963). Predicted and observed toxicities of several sewage effluents to rainbow trout. *J. Proc. Inst. Sew. Purif.* **2**, 167.

Lloyd, R. & Jordon, D. H. M. (1964). Predicted and observed toxicities of several sewage effluents to rainbow trout: a further study. *J. Proc. Inst. Sew. Purif.* **2**, 183.

Lund, J. W. G. (1950). Studies on *Asterionella formosa* Hass II. Nutrient depletion and spring maximum. *J. Ecol.* **38**, 1.

Mann, H. (1964). Effects on the flavour of fishes by oils and phenols. *Symp. Commn. Explor. Scient. Mer. Medit. Monaco, 1964,* p. 371. (*via Wat. Pollut. Abstr.* 1967, **40**, abstr. no. 1030).

Mann, H. & Schmid, O. J. (1965). Effects of sub-lethal concentrations of detergents (tetrapropylenebenzene-sulphonate) on the growth of *Lebistes reticulatus. Arch. Fisch. Wiss.* **16**, 16. (*via Wat. Pollut. Abstr.* 1966, **39** abstr. no. 1000).

Ministry of Technology, (1965–66) Reports of the Government Chemist, 1964 and 1965. *Wat. Pollut. Abstr.* 1968, **41**, *abstr. no.* 550.

Ministry of Technology (1969). Some effects of pollution on fish. II Notes on Water Pollution, no. 45, p.3.

Mitchell, R., Yankofsky, S. & Jannasch, H. W. (1967). Lysis of *Escherichia coli* by marine micro-organisms. *Nature, Lond.* **215**, 891.

Mount, D. I. (1967). Considerations for acceptable concentrations of pesticides for fish production. *Am. Fish. Soc. Spec. Publ.* No. 4.

Okey, R. W. and Bogan, R. H. (1965). Apparent involvement of electronic mechanisms in limiting microbial metabolism of pesticides. *J. Wat. Pollut. Control Fed.* **37**, 692.

Penaz, M. (1965). The influence of ammonia on eggs and young *Salmo trutta var. fario. Zool. Listy,* **14**, No. 1, p. 47. (*via Wat. Pollut. Abstr.* 1967, **40**, abstr. no. 1936).

Pickering, R. J., Carrigan, P. H., Tamura, T., Abee, H. H., Beverage, J. W. & Andrew, R. W. (1966). Radioactivity in bottom sediments of the Clinch and Tennessee rivers. *Proc. Symp. int. Atom. Energ. Agency* p.57.

Pike, E. B., Bufton, A. W. J. & Gould, D. J. (1969). The use of *Serratia indica* and *Bacillus subtilis var. niger* spores for tracing sewage dispersion on the sea. *J. appl. Bact.* **32**, 206.

Parvasoli, L. (1964). Effect of external metabolities on algae. *Verh. int. Ver. theor. angew. Limnol.* **15**, 804.

Preka, N. A. & Lipold, N. A. (1965). Discussion on Harlemen, D. R. F. *Proc. 2nd int. Conf. Wat. Pollut. Res., Tokyo.,* 1964. **1**, 279.

Preston, J. R. (1971). Histopathologic, histochemical and dissolved oxygen tolerance studies on the minnow *(Phoxinus phoxinus)* in acute and subacute heavy metal toxicity. Ph.D. Thesis, Univ. of Aston in Birmingham.

Rachlin, J. W. & Perlmutter, A. (1968). Fish cells in culture for study of aquatic toxicants. *Wat. Res.* 2.

Ravera, O. (1964). Symposium in Karlsruhe (German Federal Republic 26–28th June 1963. Europ. Fed. for Protection of Waters, Inform. Bull. No. 10, p.61 *via (Wat. Pollut. Abstr.* 1967, **40**, abstr. no. 1738).

Reynolds, N. (1965). The effect of light on the mortality of E. coli in sea water. *Symp. Commn. int. Explor. Scient. Mer Medit., Monaco,* 1964, p. 241. (*via Wat. Pollut. Abstr.* 1967, **40**, abstr. no. 899).

Shephard, M. R. N. (1971). The role of sewage treatment in the control of human helminthiases. *Helminth. Abstr. Series A. Anim. Hum. Heminth.* **40**, 1.

Sieburth, J. M. (1965). Role of algae in controlling bacterial populations in estuarine waters. *Symp. Commn. int. Explor. scient. Mer Medit., Monaco,* 1964, p.217. (*via Wat. Pollut. Abstr.* 1967, **40**, abstr. no. 902).

Silverman, P. H. & Griffiths, R. B. (1955). A review of methods of sewage disposal in Great Britain, with special reference to the epizootiology of *Cysticercus bovis. Ann trop. Med. Parasit.* **49**, 436.

Skidmore, J. F. (1965). Resistance to zinc sulphate of zebra-fish *(Brachydanio rerio* Hamilton – Buchanan) at different phases of its life history. *Ann. appl. Biol.* **56**, 47.

Sprague, J. B. (1969). Measurement of pollutant toxicity to fish. 1 – Bioassay methods for acute toxicity. *Wat. Res.* **3.**

Streeter, H.W. (1958). The oxygen sag and dissolved oxygen relationships in streams. Oxygen relationships in streams. *Robert A. Taft, Sanit. Engr. Center,* Tech. Rep. W 58 – 2, p.25.

U.S. Department of Health, Education, and Welfare, Public Health Service. (1965). Streamflow Regulation for Quality Control. *Publ.* no. 999, WP 30. Washington, D.C.: U.S. Dept of Health.

Vance, B. D. and Drummond, W. (1969). Biological concentration of pesticides by algae. *J. Am. Wat. Wks Ass.* **61,** 360.

Vollenweider, A. (1969). Eutrophication of waters. *Mitt. Verein. dt. Gewass Schutz,* no. 1/2, p. 4. (*via Wat. Pollut. Abstr.* 1970, **43,** abstr. no. 1216).

Warnick, S. L., Gaufin, R. F. & Gaufin, A. R. (1966) Concentration and effects of pesticides in aquatic environments. *J. Am. Wat. Wks Ass.* **58,** 601.

Williams, L. G. & Pickering, Q. H. (1961). Direct and food-chain uptake of caesium [137] and strontium [85] in bluegill fingerlings. *Ecology* **42,** 205.

Wilson, H. (1949–50). Diseases from sewage, Publ. Hlth. Johannesburg, 1949, **13,** 332; 363; (1950), **14,** 13, 45, 77, 209, 140.

MAP–7

Microbiological Aspects of the Pollution of Fresh Water with Inorganic Nutrients

D. R. KEENEY

Department of Soil Science,
University of Wisconsin,
Madison, Wisconsin, 53706, U.S.A.

R. A. HERBERT* AND A. J. HOLDING

Department of Microbiology, University of Edinburgh,
School of Agriculture, West Mains Road,
Edinburgh EH9 3JG, Scotland

CONTENTS

1. Introduction

IN FRESHWATER ecosystems an increase in the input of the major inorganic nutrients, in particular nitrogen (N) and phosphorus (P) compounds can accelerate the complex series of biological, chemical and physical interactions which govern the growth of plant, animal and microbial populations. Any increase in the concentration of available nutrients in an aquatic system whether promoted by naturally occurring or manmade processes is referred to as eutrophication. The gross changes most commonly observed as a result of eutrophication include increased growth of the littoral vegetation, development of 'nuisance' algal blooms and deoxygenation of hypolimnic waters. Ultimately, the amenity value of the aquatic system is reduced and water abstraction for industrial and public water supplies becomes more difficult and

* Present address: Department of Bacteriology, The University, Dundee DD1 4HN, Scotland.

eventually uneconomic. The general aspects of eutrophication have been reviewed recently (Lund, 1965; Nursall 1966; Vollenweider, 1968; Brezonik & Lee, 1968). Although there is qualitative agreement on the effects of eutrophication it is difficult to quantify the parameters involved. Little distinction is made between the term 'eutrophication' and 'pollution' yet the terms are not synonymous because pollution also includes the addition to waters of a wide range of contaminating materials including solid wastes, organic and inorganic nutrients, heated wastes and toxic substances.

The availability of both N and P has an important influence on the general ecology of aquatic systems; however in this paper greater emphasis is placed on the microbial processes governing N availablity, since micro-organisms appear to have only an indirect, but nonetheless important, effect on P availability. As an introduction some data are given on the sources and concentration of these nutrient substances entering fresh waters, but no consideration is given to the important processes involved in their release and/or immobilization prior to their entry into fresh water ecosystems.

In addition to these two major groups of nutrient substances brief consideration is given to certain microbial processes, particularly relating to the oxidation reduction potential (E_h) of the environment, which affect the availability of iron (Fe), manganese (Mn), carbon (C) and sulphur (S) compounds.

Although brief reference is made to the predominant types of bacteria occurring in fresh waters no attempt has been made to discuss in detail the more general ecological aspects of fresh water microbiology. This review is concerned primarily with lake systems, but many of the principles involved can be assumed to be applicable to river systems when comparable environmental conditions prevail.

2. Nutrients Limiting Microbial Growth

(a) Algae

Numerous reviews of the nutrients controlling biological productivity in aquatic systems suggest that N and P are the limiting factors (Lund, 1965; Fruh, 1967; Mackenthun, 1969). Vollenweider (1968) analysed the data of Thomas (1953, 1963) and concluded that algal productivity is largely governed by the concentration of P compounds. Several recent reports however suggest that N might be the limiting factor under certain conditions. Lund (1965) states that the low summer levels of nitrate (NO_3-N) commonly observed in surface waters of eutrophic lakes might suggest N limitation at this time. Nitrogen has also been suggested as the important algal growth limiting nutrient in eutrophic Lake

Mendota, Wisconsin (Gerloff & Skoog, 1954, 1957) and in Lake Kinneret, Israel (Serruya & Berman, 1969). Recently Lueschow *et al.* (1970) found that the mean monthly N (total organic plus inorganic) but not P levels of southern Wisconsin lakes provided a reliable index of their trophic status. Evidence is also accumulating to show that N rather than P is the critical factor governing algal growth and eutrophication in coastal waters (Shapiro, 1970; Ryther & Dunstan, 1971).

Recently the possibility that carbon dioxide (CO_2) can limit algal growth has been given renewed attention (Kuenztel, 1969, 1970; King, 1970). Whilst P and N levels in eutrophic waters are in general adequate to support an algal bloom, CO_2 may be limiting unless additional quantities can be supplied (Kuenztel, 1970). In his hypothesis Kuenztel suggests that heterotrophic bacteria supply the extra CO_2 required for massive algal blooms.

Shapiro (1970) doubts the importance of CO_2 limitation. He points out that (a) the recycling of P can only extend the duration of an algal bloom, not the standing crop; (b) a constant input of P is needed to maintain algal growth due to the P-absorbing properties of lake sediments; (c) low water concentration of P during algal blooms only means that most of the available P is immobilized by the algae, and (d) assuming an atomic P/C ratio of 100 in algae, large amounts of P as well as CO_2 are required to produce massive algal blooms. Unless evidence to the contrary is presented, the arguments presented by Shapiro (1970) leave little doubt, except perhaps in acid environments high in N and P, that CO_2 is not the limiting nutrient for biomass production.

(b) *Bacteria*

The limited data available suggest that *Pseudomonas, Achromobacter, Alcaligenes* and *Flavobacterium* spp., predominate (Collins, 1963; McCoy & Sarles, 1969). Collins (1963), working on Lake Windermere, found that >96% of the organisms isolated were Gram negative, asporogenous rods and that <1% were cocci. In the absence of a single suitable medium it is extremely difficult to determine the total numbers of viable bacteria occurring in freshwaters. Data from Lake Windermere (Collins, 1963) suggest that the maximum bacterial count (47,000 organisms/ml) occurs at the base of the thermocline. Henrici (1940) compared the bacterial populations of oligotrophic and eutrophic lakes and showed (Table 1) that the counts obtained in eutrophic waters and sediments were several-fold greater than those obtained from corresponding oligotrophic lakes. Maximum bacterial counts occur in lake waters during the winter when precipitation and stream flow are at a maximum (Collins, 1963). The significance of the increased bacterial populations observed in eutrophic lakes is still largely unknown in terms of the environmental conditions promoting the increase.

Table 1

*Comparison of bacterial populations in oligotrophic
and eutrophic lakes (after Henrici, 1940)*

Lake	Plate count/ml of*	
	Water	Mud
Eutrophic		
Boulder	–	47,000
Alexander	675	144,240
Little John	505	39,050
Mendota	–	609,000
Muskellunge	133	10,930
Mean	438	170,100
Oligotrophic		
Weber	132	2350
Trout	66	29,790
Crystal	80	2160
Mean	93	11,400

*Henrici & McCoy (1938).

3 Sources of Nitrogen and Phosphorus Compounds

When considering the roles played by N and P compounds controlling biomass production it is essential to determine the N and P budget of the lake system under investigation. This involves estimates of the total quantities of N and P entering the system from all sources and the outflow of nutrients from the lake system. In addition to the major point sources, e.g. sewage and industrial effluents, a variety of diffuse sources from which nutrients may be derived make accurate budgets difficult and expensive to obtain. A considerable effort has been made to determine the N budget for Lake Mendota, Wisconsin (Lee *et al.*, 1966; Brezonik & Lee, 1968; Keeney, 1971) (Table 2). Lee *et al.* (1966) were of the opinion that the principal N source was derived from ground water seepage, whilst *c.*67% of the N was considered to be immobilized in the sediment. More recently Keeney (1971) revised the budget for Lake Mendota and concluded that stream drainage and precipitation are the major sources of N. These data no doubt will be revised further as more information becomes available, but they serve to indicate the sources and probable contributions of N to a particular lake system.

Owens (1970) attempted to determine the proportion of N and P entering freshwaters in the United Kingdom from a variety of sources including agricultural and urban runoff, rainfall, sewage effluent and industrial waste waters. He concluded that although sewage effluents contribute the majority of the P input they account for only a small proportion of the total N entering the systems. The main source of N, with the exception of densely populated urban areas, appears to be from land drainage. In another study in Loch Leven, Scotland (Holden, 1969) N primarily in the form of NO_3-N constitutes 96% of the N input and is derived from streams draining the immediate land area, whilst almost 90% of the P input was supplied as industrial sewage and effluents. In certain of the rivers governed by the Lancashire River Authority, NO_3-N levels as high as 2 p/m have been observed (A. D. Buckley, pers. comm.), which in the absence of industrial and sewage effluents, are due to surface runoff from agricultural land and seepage of NO_3-N containing ground water. Although N and P fertilizer usage in the U.K. has increased several-fold during the last decade, there does not appear to have been a concomitant increase in NO_3-N levels in the rivers so far investigated (Tomlinson, 1970).

Table 2

Nitrogen budgets for Lake Mendota

| Nitrogen source | Annual contribution | | | |
| | Previous estimate[a] | | Revised estimate[b] | |
	kg	%	kg	%
Municpal & industrial wastes	21,200	8	21,200	8
Urban runoff	13,700	5	13,700	7
Rural runoff	23,500	9	23,500	12
Precipitation on lake surface	43,900	17	43,900	21
Nitrogen fixation	36,100	14*	36,100	18
Groundwater				
(a) streams	35,900	14	35,900	18
(b) seepage	77,000	31	28,500	14
Marsh drainage	+	+	+	+
Approximate total	251,000	100	203,000	100

[a]from Lee *et al.* (1966) and Brezonik & Lee (1968); [b]from Keeney (1971).
*Now considered to be 2% (Stewart *et al.,* 1968).
+ Considered important, but no data available.

4. Forms and Quantities of Nitrogen and Phosphorus in Lakewaters

The principal forms of N occurring in lakewaters, excluding dissolved N_2 gas, are ammonia (NH_4-N), nitrate (NO_3-N), nitrite (NO_2-N) and organic N. Nitrate levels can range 0–5 p/m in surface waters and to several p/m in anoxic waters (Hutchinson, 1957). Nitrate levels are in general low ($<$0·1 p/m), but up to 1 p/m has been recorded in sediment interstitial waters (Konrad et al., 1970).

The concentrations of the various forms of N in water depend not only on the rates of input, but also on biological processes such as denitrification, mineralization, nitrification and N immobilization. For example in Loch Leven, Scotland NO_3-N is the principal N form found in the water, reaching a maximum concentration of 1·8 p/m in winter and declining to $<$ 0·2 p/m in the summer, due to immobilization and denitrification. Many deep lakes exhibit a dichotomic NO_3-N distribution pattern, with maximum concentration at intermediate depths and low NO_3-N in the bottom and surface waters due to denitrification and immobilization, respectively (Hutchinson, 1957). In surface waters NH_4-N is generally at a maximum following the autumn overturn, whilst in deep waters highest NH_4-N levels occur during summer stratification (Hutchinson, 1957).

As with soils, the N in lake sediments is primarily in an organic form. This matter may arise either from material washed into the lake or can be formed within the lake system. The surface sediments of lakes generally contain 0·1–4% of N and have C:N ratios of c. 11:1 (Keeney et al., 1970). The same authors surveyed the forms and quantities of N in sediments from 13 Wisconsin lakes. They determined the distribution of hydrolysable (hot 6N HCl) N as well as total and inorganic forms of N and also that the forms of hydrolysable N did not differ significantly between sediment samples from lakes differing widely in physical and chemical characteristics. However, considerable differences in inorganic N distribution values in sediments were noted. For example, acid sediments contained greater amounts of exchangeable NH_4-N but lower levels of soluble NH_4-N than did calcareous sediments.

The amount of P in the lake water varies greatly, being influenced by a range of geochemical, climatic, morphological and cultural factors. Within a given lake, P concentrations vary with depth, time of day and season. Total P concentrations in surface waters range from 0·01–0·05 p/m in relatively unproductive lakes (Vollenwieder, 1968) to 0·5–1·0 p/m in highly productive hypolimnic waters (Lueschow et al., 1970). Higher P levels are found in sediments, particularly during summer stratification, presumably due to release of P from anoxic sediments. The proportions of different forms of P varies considerably: inorganic P generally dominates in deeper water and in surface waters during the winter (Armstrong et al., 1971).

5. Influence of Oxidation-Reduction Potential

In stratified waters or in sediments where the oxygen (O_2) diffusion rate is lower than the O_2 uptake by biological and chemical systems, the E_h declines, rapidly. Since many biological and chemical transformations of N, C, S, P, Fe and Mn are markedly influenced by the redox potential, it is of considerable importance in aquatic and terrestrial ecosystems.

From the viewpoint of pollution, the E_h plays an important role in controlling the levels of available nutrients within a given system. For example, at low E_h levels ferric oxide adsorbed phosphate may be released from sediments, or NO_3-N may be reduced to gaseous nitrogen (denitrification), causing a net loss of N from the ecosystem. The production of methane (CH_4) under highly reduced conditions is perhaps one of the major energy losses from the system.

The subject has been considered by Hutchinson (1957) and fully reviewed recently by Stumm & Morgan (1970) and Ponnamperuma (1971). The biological interpretation of E_h levels recorded in aquatic systems is complex due to naturally occurring interfering substances, the presence of numerous O/R systems and to problems of measurement. Thus, the values recorded must be treated with caution. However, the general agreement amongst different workers is that whilst the observed E_h values may not be theoretically correct, they can be used with reasonable confidence when interpreting data.

(a) E_h values in lake systems

Typical recorded E_h values for aerated surface water range from 420–520 mV, although values as low as 80 mV have been recorded in certain hypolimnic waters (Hutchinson, 1957). The E_h of aquatic systems is relatively insensitive to O_2 concentration provided O_2 is present. Decreasing the O_2 tension from 100–0.1% saturation reduces the E_h by only 30 mV. Thus the change from aerobic oxidizing to anaerobic reducing conditions does not occur until the O_2 concentration is depleted (Greenwood, 1962). Relatively few studies have been made of the E_h of sediments, but figures as low as -300 mV have been reported (Weijden et al., 1970). A figure of -250 mV has been recorded in calcareous lake sediments (Graetz & Keeney, unpublished). In oligotrophic water, E_h seems to be controlled by the O_2 concentration, but during summer stratification in hypolimnic waters of eutrophic lakes the reduction processes proceed to Fe, S and finally the CH_4 stage (Hutchinson, 1957).

(b) E_h values in flooded soils

The considerable literature on O/R processes occurring in soils, which in many respects are similar to those occurring in lake sediments, has shown qualitatively, and in some instances quantitatively, the sequence of events which take place (see Table 3). It seems probable that a similar sequence also occurs in freshwater lake sediments.

Initially, depending on the availability of energy sources, O_2 is removed from the system by aerobic respiratory processes. The net decrease of the O_2 content also depends on surface aeration effects, currents and algal and plant O_2 production. In addition to O_2 uptake by algae in the dark, the principal micro-organisms involved in oxygen-demanding respiratory processes are assumed to be the predominant aerobic bacteria found in lake systems (see section 2(b)). Following the depletion of the available O_2, a sequence of microbial processes appears to take place whereby various alternative inorganic compounds are reduced. It is universally accepted that NO_3-N is the first alternative electron acceptor utilized by natural microbial populations. This process of dissimilatory NO_3-reduction in which nitrogen oxides and finally N_2 gas is produced, is referred to as denitrification. The process occurs under controlled experimental conditions at E_h values $< 300–350$ mV (Turner & Patrick, 1968; Meek et al., 1969). It is important to note that NO_3-N additions to flooded soils stabilize and maintain an E_h value of $100–200$ mV until the NO_3-N has been utilized (Redman & Patrick, 1965; Bell, 1969).

Table 3

Suggested effect of E_h on microbial metabolism in lake sediments (after Parr, 1969)

Sequence of events	Process	E_h (mV)	Bacterial metabolism
1	normal oxidations	+600	Aerobic
2	NO_3-reduction	+500	Aerobic*
	Mn^{4+} reduction	+400	Aerobic
	Fe^{3+} reduction	+300	Aerobic
3	SO^{2-}_4 reduction	0	
4	H_2 production	-150†	Obligately
5	CH_4 production	-220	anaerobic

* Using anaerobic respiratory processes in presence of NO_3; † not demonstrated in lake sediments.

The O/R systems are complicated by the reduction of the Fe^{3+} and Mn^{4+} ions to soluble forms at c. 200 mV (Table 3). However, whilst there is considerable evidence that these reductions are biologically mediated (Broadbent & Clark, 1965; Parr, 1969; Chen et al., 1971b) no evidence is available to indicate that Fe^{3+} or Mn^{4+} can be used as alternative electron acceptors in the absence of molecular O_2. It is assumed that these metallic ions are reduced by the end products of the metabolism of a wide range of aerobic and facultatively anaerobic micro-organisms. According to Takai & Kamura (1966) these processes are followed by the growth of obligatory anaerobic bacteria which can reduce $SO_4{}^{2-}$ to S^{2-}. These processes commence at an E_h value of -150 mV but the release of H_2S may be prevented by the formation of ferrous sulphide (FeS) in high iron soils (Connell & Patrick, 1968). These workers also noted that no sulphide appeared until all the NO_3-N had been reduced. Similar observations have also been reported in anoxic basins (Gupta, 1969). Although comprehensive studies of the characteristics of sulphate-reducing bacteria occurring in flooded soils and fresh waters have not been reported it is usually assumed that they are similar to species of *Desulfovibrio* and closely related species (see Campbell & Postgate, 1966). The metabolism of S compounds by bacteria has been reviewed recently by Trudinger (1969).

The anaerobic fermentation of a range of simple and complex organic substrates leads to the formation of organic acids which accumulate within the E_h -100 to -200 mV. Although it seems likely that in freshwater sediments, as in many soils, *Clostridium* spp. play an important role in these processes, confirmatory studies are required. For example, in acidic anaerobic peat (pH 3·9) *Actinomyces*-like organisms are the predominant anaerobes (Siwasin, 1971).

These fermentations are followed by H_2 and CH_4 production which occurs in the E_h range -250 to -300 mV (Ponnamperuma *et al.*, 1966). Methane is produced by a highly specialized small group of bacteria which have only rarely been isolated in pure culture. The subject has been reviewed by Stadtman (1964) and more recently by Wolfe (1971), who concluded that H_2 and CO_2 or formate are the only substrates utilized by most of the strains available. A small amount of CH_4 has also been reported as a product of the metabolism of pyruvate in sulphate-reducing bacteria (Postgate, 1969). No data are available to indicate whether the free H_2 reported to be produced in soil (Parr, 1969) is also evolved from lake sediments.

(c) *Effects of E_h on eutrophication*

The E_h of sediments and probably to a less extent the overlying water can play a significant role in controlling the availability of particular microbial and plant nutrients. The occurrence of high concentrations of PO_4^{3+} in deoxygenated hypolimnic waters is well established. This $PO_4{}^{3+}$ is released from a previously immobilized pool of PO_4^{3+} in the sediments when reducing conditions develop (Mortimer, 1941, 1942). This release may exceed the total input from other sources and hence may be of particular importance. Patrick & Mahapatra (1968) have reviewed the influence of flooding soils in relation to PO_4^{3+} release. It is clear that in certain soils a lower E_h enables the reduction of ferric phosphate to the more soluble ferrous form. This increase in available PO_4 occurs when the E_h is $<+200$ mV. Between $+200$ mV and -200 mV the concentration of extractable PO_4 increased from 10 to 35 p/m.

The relationship between the redox potential of a sediment-water system and N transformations has been recently investigated in a laboratory model system (Graetz, Aspiras & Keeney, unpublished). A large plastic cylinder (600 x 25 mm) containing a sediment-water sample was flushed with helium and sealed. After incubation for 42 days the water fraction was aerated for 20 days before reflushing with helium and resealing. Data on the E_h of water and sediment and of the NH_4-N and the NO_3-N concentrations in the water are presented in Fig. 1.

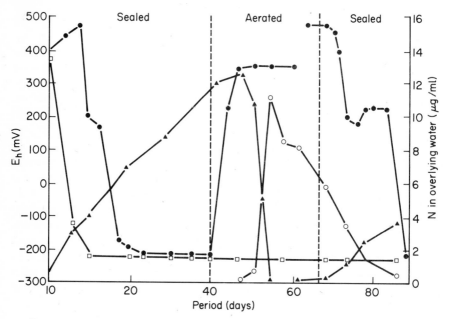

Fig. 1. The relationship between E_h and microbial N transformations in a lake sediment [illegible] ● E_h of water; □ E_h of sediment; ▲ NH_4-N; ○ NO_3-N.

As expected, the E_h of the sediment fell more rapidly than did that of the overlying water, which then increased during the aeration phase. In the water 90% of the NH_4-N produced during the anaerobic incubation period was oxidized to NO_3-N following aeration of the system, and the E_h rose to + 300 mV. It should be noted that the NO_3-N concentration declined in the aerobic system after all the NH_4-N had been nutrified. This was probably due to both assimilation and denitrification of the NO_3-N. In the second anaerobic phase considerable denitrification occurred but the E_h remained at + 200 mV until all the NO_3-N had been utilized, and then fell sharply.

6. Assimilation and Release of Nitrogen Compounds

Microbial processes leading to either an increase or decrease in the availability of N compounds to algae and other plant life are important in controlling the eutrophication process. In many systems micro-organisms are the primary agents responsible for N mobilization (organic–N to NH_4-N) thus producing available N. These organic N compounds are primarily proteinaceous and of course include those synthesized as a result of N fixation. Conversely, N can be at least temporarily immobilized. by the microbial populations in a carbon-rich system or permanently lost from the system as a result of denitrification. The relative rates of these processes determine whether or not there is a net increase of available N to the ecosystem. For heterotrophic organisms the availability of energy sources relative to N sources is important because N will only be assimilated into the microbial protoplasm until the available energy sources have been fully utilized. Excess N not required is then made available to the system, hence the importance of the available C:N ratio.

The relative roles played by the various biological components, microflora, fauna and flora, in the different N transformations clearly vary according to the morphological, physical and chemical characteristics of a particular environment (Johannes, 1968). In his review of mineral cycling in aquatic and terrestrial systems Pomeroy (1970) emphasizes that complex interactions probably more often limit productivity than does the availability of a single nutrient.

Differing views have been put forward concerning the relative role of bacteria and plankton in nutrient regeneration. One school of thought considers that bacteria and fungi are the primary agents for nutrient recycling (Golterman, 1960, 1964; Kuznetzov, 1968). Kuznetzov (1968) cited work suggesting that the bacterial biomass in certain Russian reservoirs was equal to or greater than that of the phytoplankton. However, Johannes (1968) whilst acknowledging that in terrestrial systems micro-organisms are dominant, considers that in freshwaters herbivorous zooplankton predominate, bacteria only becoming important for the decomposition of material not consumed by the zooplankton or their remains. Nitrogen is excreted as NH_4-N free amino acids and other órganic compounds which are then available for utilization by phytoplankton and bacteria (Dugdale & Goering, 1967). The 4 microbial processes which are likely to be most important in determining the availability of N are ammonification, nitrification, denitrification and nitrogen fixation.

(a) *Ammonification*

Comparatively little information is available on either the micro-organisms or the environmental characteristics controlling the formation of NH_4-N from organic matter in lake systems. Kuznetzov (1968) reported that a broad range of micro-organisms are involved, including *Bacillus, Pseudomonas* and *Micrococcus* spp. It can be assumed that the parameters which influence ammonification rates in soil, e.g. pH, temperature, inorganic ion concentrations and O_2 availability (Power, 1968) also influence the process in aquatic environments. Early studies in Lake Mendota, Wisconsin, by Domogalla *et al.* (1926) showed that NH_4-N production increases during stratification (both summer and winter) in hypolimnic waters due to decomposition of sinking detritus as well as release from the sediment. Yoshima (1932), however, found that whilst NH_4-N increased in deep waters during early summer the concentration fell in late summer, perhaps due to immobilization, even though the lake remained stratified. Recent work (Foree *et al.*, 1971) has shown that in fresh waters the net mineralization of N (and P) from decomposing algae is greater under anaerobic conditions than aerobic conditions, presumably due, in part at least, to the lower level of energy produced during fermentation.

(b) *Nitrification*

This process, resulting in the conversion of NH_4-N to NO_3-N by obligately aerobic autotrophic bacteria of the genera *Nitrosomonas* (NH_4-N $\rightarrow NO_2$-N) and *Nitrobacter* (NO_2-N $\rightarrow NO_3$-N) has been extensively studied in terrestrial systems (Alexander, 1965) and sewage purification processes (Downing, Patrick & Knowles, 1964). The large volume of data available on the biochemical and physiological characteristics of autotrophic nitrifying bacteria has been reviewed by Nicholas (1963). In addition to the autotrophic organisms, numerous heterotrophic micro-organisms including certain bacteria, fungi and algae have the ability to oxidize organic N at least as far as NO_2-N (Alexander, 1965). It is generally accepted that these organisms play a negligible role in soil nitrification but their occurrence and importance in aquatic systems remain to be assessed. In aquatic systems, the trophic state may determine whether the process is functional or not. This process can lead to N losses due to diffusion of the NO_3-N to anoxic zones (and subsequent denitrification) whilst the utilization of molecular O_2 results in an overall reduction in O_2 in waste laden waters (Courchaine, 1968; Butts *et al.*, 1970). The review by Vollenweider (1968) suggest that aquatic plants utilize either NH_4-N or NO_3-N, although in general algae are believed to prefer NH_4-N.

Measurements of nitrification rates *in situ* are difficult to determine due to the competing reactions of immobilization and denitrification, so that few data are available. Hutchinson (1957) considers that nitrification takes place whenever O_2 is present, and such increases in NO_3-N attributable to nitrification have been measured (Domogalla & Fred, 1926; Mortimer, 1941, 1942; Brezonik, 1968). Recent laboratory investigations show that this occurs readily in lake sediments which have been vigorously stirred to maintain aerobic conditions. Data from experiments using a well buffered calcareous sediment show high nitrification rates (up to 25 μg of NO_3-N of sediment/day). Nitrification rates were much lower in noncalcareous sediments. Addition of N-Serve (2-chloro-6 (trichloro-methyl) pyridine), a specific inhibitor of *Nitrosomonas* (Goring, 1962), stopped nitrification entirely for 7 days in the Lake Mendota sediment sample. An interesting observation was that under static conditions little NO_3-N accumulated but the NH_4-N levels declined, suggesting that nitrification and denitrification were occurring simultaneously. The data of Chen *et al.* (1971a) suggest that in well oxidized, stirred sediments, such as those found in shallow waters, nitrification may contribute appreciable quantities of NO_3-N to the overlying water. A similar observation has been made in soil (Greenwood, 1962) and in the surface layers of flooded paddy soils (Patnaik, 1965).

(c) *Denitrification*

The process whereby NO_3-N and NO_2-N are biologically reduced to gaseous N oxides (N_2O and NO) and molecular N_2 results in the release of N from aquatic systems and hence is of considerable importance in controlling N levels. Reactions leading to the nonenzymic losses of N in acid media (chemodenitrification) (Broadbent & Clark, 1965) and to assimilatory NO_3-N reduction are thus excluded.

The physiology of denitrification in pure cultures of soil micro-organisms has been studied in detail (Nicholas, 1963; Broadbent & Clark, 1965). All the organisms examined so far possess the normal respiratory chain enzymes and are unable to grow anaerobically except in the presence of NO_3- (or in certain circumstances NO_2-), in which case the inorganic N source functions as an alternative terminal electron acceptor. Bandurski (1965) has shown that except for the terminal respiratory enzymes, the electron transport systems are identical under both aerobic and anaerobic conditions. In terrestrial systems the commonly encountered genera are *Pseudomonas, Bacillus, Alcaligenes* and *Cytophaga* (Cook, 1961). These organisms, which utilize a wide range of simple organic compounds as energy sources, frequently are involved in other N transformations such as ammonification and proteolysis. Except for their denitrifying ability they do not appear to differ significantly from other representatives of the genera. Almost no information is available regarding population sizes or types of denitrifying bacteria found in aquatic systems. Preliminary data (Holding & Keeney, unpublished) on Wisconsin lake sediments show the dominant bacteria are Gram negative rods or coccoid rods of the genera *Pseudomonas* and *Alcaligenes*.

Several investigators have confirmed that considerable denitrification can occur in lake and river systems. These include studies in Switzerland (Vollenweider, 1968), the U.S.A. (Goering & Dugdale, 1966; Brezonik & Lee, 1968; Chen *et al.*, 1971b, c), the U.K. (Owens, 1970) and U.S.S.R. (Kuznetzov,

1968). Brezonik & Lee (1968), working on Lake Mendota, showed that $8-26\mu g$ of N/l/day were lost as a result of denitrification. Similar denitrification rates $(7-13\mu g$ of 1/day) have been observed in samples obtained from Lake Windermere and Loch Leven (Herbert & Holding, unpublished). Calculated on an annual basis, denitrification can account for significant losses of N. For example, Vollenweider (1968) considers that in certain Swiss lakes denitrification can account for up to 80% of the N losses. Similarly in Loch Leven, Scotland, up to 42% of the N budget cannot be accounted for, and denitrification is a possible mechanism whereby this N is lost from the system (Holden, 1969).

It is generally accepted that the reactions operating in true biochemical systems are: $NO_3 \rightarrow NO_2 \rightarrow NO \rightarrow NO_2 \rightarrow N_2$ (Cooper & Smith, 1963; Cady & Bartholomew, 1960). Nitric oxide (NO) is reduced rapidly to N_2 gas in closed systems, and the N_2 is the only gaseous product commonly observed in sewage sludges and wastes (Brezonik & Lee, 1966, 1968; Kuznetzov, 1968; Chen *et al.*, 1971*b*). Recently Chen *et al.* (1971*b, c*) confirmed that the sequence $NO_3 \rightarrow NO_2 \rightarrow [NO] \rightarrow N_2O \rightarrow N_2$ occurred during denitrification in lake sediments.

(d) *Nitrogen fixation*

A considerable volume of data is available to show that N fixation in lakewater is usually correlated with the occurrence of heterocystous blue-green algae. The process appears to be light dependent although certain blue-green algae can fix N in the dark (Fay, 1965). Hutchinson (1957) was probably one of the first workers to realize the possible significance of algal N fixation in lakewaters, yet it is only recently using the heavy isotope ^{15}N, that more definitive data have become available. Perhaps the most authoritative assessment of N fixation, based on ^{15}N studies, has been that of Horne & Fogg (1970) working on Lake Windermere and Esthwaite Water. Their data showed a positive correlation between N fixation and the concentration of organic N in the water. Nitrogen fixation generally occurred when the NO_3-N levels were low (0-0.1 mg/l). High NH_4-N levels had no direct effect on N fixation in the 2 lakes when stratified. However, no fixation could be demonstrated during the short period of high NH_4-N concentration occurring just after the autumnal overturn. In Windermere and Esthwaite Water it is estimated that N fixation probably constitutes < 1% of the total N input (Horne & Fogg, 1970). This is a considerably lower than that obtained by other workers. Brezonik & Lee (1968) estimated that in Lake Mendota N fixation may contribute up to 14% of the N input; however, recent data by Stewart *et al.* (1968) suggest that 2% would be a more accurate figure. The significance of N fixation in relation to the N budget of the lake systems is thus probably less than previously believed. Nevertheless it may be a major source of bound N at particular times of the year. Kuznetzov (1968) is of the opinion that N fixation contributes the bulk of the N input in certain Russian reservoirs.

Blue-green algae, identified species include *Anabaena, Nostoc, Calothrix, Cylindrospermum* and *Mastigocladus,* appear to be the dominant N fixers in surface waters (Lund, 1965; Stewart, 1968, 1970). Although photosynthetic bacteria are known to fix N in pure culture (Lindström *et al.,* 1950; Schick, 1971) little if anything is known regarding rates of N fixation in aquatic environments. Stewart (1968) observed that a green sulphur photosynthetic bacterium *Pelodictyon* was the dominant organism at the thermocline of a Norwegian Fjord where active N fixation, as measured by the acetylene reduction technique, was taking place. Organisms responsible for N fixation in anoxic waters and sediments, below the photolithotrophic zone appear to predominantly *Clostridium* spp. (Brezonik, 1971). Similar *Clostridium* spp. have been implicated as N fixers in flooded paddy soils (Rice *et al.,* 1967) and they are widely distributed (Brouzes *et al.,* 1969).

7. General Conclusions

Due to the many complex competing biological reactions occurring within a given ecosystem it is extremely difficult to determine the relative importance of each individual microbial process. The key N transformations occurring in lakewaters and sediments are shown in Fig. 2. The NO_3-N balance at any given time is governed by the relative rate of NO_3-N loss (as a result of denitrification and immobilization) to the rate of NO_3-N regeneration (from groundwater seepage, stream drainage and ammonification/nitrification). Nutritionally it is

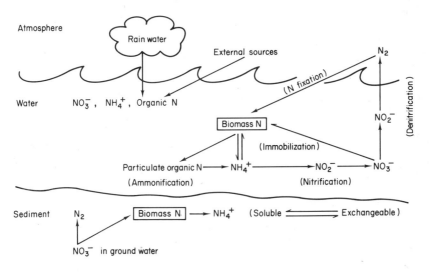

Fig. 2. Nitrogen transformations occurring in lakewaters and sediments.

now fairly well established that algal growth is largely governed by N and P availability. Nitrogen fixation by heterocystous blue-green algae probably constitutes only 1—2% of the input into the freshwaters so far examined, but it may well be important to biological productivity when inorganic N levels become depleted during the summer months. The small bacterial populations found in oligotrophic waters suggest their role in nutrient cycling is perhaps minimal. In eutrophic waters, however, O_2 depletion by micro-organisms can lead to a whole series of events, and these are to a large extent governed by the redox potential. For example, PO_4 release from lake sediments and losses of N from the ecosystem due to denitrification are of considerable importance. Micro-organisms are probably also involved in many reactions not considered here. Fogg (1968) has fully discussed the role of various organic compounds present in fresh waters on algal growth. Many of these compounds may be of bacterial origin and emphasize the lack of knowledge on the epiphytic algal microflora or its effect on algal growth. In addition, the possible importance of CO_2 in limiting algal growth requires careful investigation since there is very little information on the source of the carbon required by algae.

Although the importance of field studies should not be underestimated, the development of valid model systems, whereby individual parameters can be closely controlled will be invaluable in differentiating specific microbial processes. It may then be feasible to predict the future behaviour of fresh waters in response to increased or decreased nutrient loads. This review has also demonstrated the paucity of data available on the indigenous microbial population of fresh waters and lake sediments.

8. Acknowledgments

The Wisconsin contribution was supported by a grant from the College of Agriculture and Life Sciences, University of Wisconsin, and by research grants from the Federal Water Quality Administration No. 16010 EHR. For the Edinburgh contribution to the paper we wish to acknowledge the financial support of the Natural Environmental Research Council, London

9. References

Alexander, M. (1965. Nitrification. *Agronomy, U.S. Dept. Agric.* **10**, 307.
Armstrong, D. E. Spyridakis, D. E. & Lee, G. F. (1971). Cycling of nitrogen and phosphorus in the Great Lakes. *Wat. Res.*
Bandurski, R. S. (1965). Biological reduction of sulfate and nitrate. In *Plant Biochemistry.* (Eds J. Bonner & J. E. Varner). New York: Academic Press.
Bell, R. G. (1969). Studies on the decomposition of organic matter in flooded soil. *Soil Biol. Biochem.* **1**, 105.
Brezonik, P. L. (1968). The dynamics of the nitrogen cycle in natural waters. Ph.D. thesis. University of Wisconsin, Madison, Wisconsin.
Brezonik, P. L. (1970). Cultural eutrophication: the nature of the problem and some recommendations for its study and control. Presented at Workshop on Problems in Water Pollution, University of Colorado, Boulder, August 1970.

Brezonik, P. L. (1971). Nitrogen: sources and transformations in natural waters. Presented at the American Chemical Society Meetings, Los Angeles, April 1971.

Brezonik, P. L. & Lee, G. F. (1966). Sources of elemental nitrogen in fermentation gases. *Int. J. Air & Wat. Pollut.* **10**, 145.

Brezonik, P. L. & Lee, G. F. (1968). Denitrification as a nitrogen sink in Lake Mendota, Wisconsin. *Envir. Sci. Tech.* **2**, 120.

Broadbent, F. E. & Clark, F. (1965). Denitrification. *Agronomy, U.S. Dept. Agric.* **10**, 344.

Brouzes, R., Lasik, J. & Knowles, R. (1969). Effect of organic amendments, water content and oxygen on the incorporation of $^{15}N_2$ by some uncultivated forest soils. *Can. J. Microbiol.* **15**, 899.

Butts, T. A., Schnepper, D. H. & Evans, R. L. (1970). Dissolved oxygen resources and waste assimilation capacity of the LaGrange Pool, Illinois River. Report of Investigation 64, Illinois State Water Survey.

Cady, F. B. & Bartholomew, W. V. (1960). Sequential products of anaerobic denitrification in Norfolk soil material. *Proc. Soil Sci. Soc. Am.* **24**, 477.

Campbell, L. L. & Postgate, J. R. (1966). Classification of Desulfovibrio spp., the sulphate-reducing bacteria. *Bact. Rev.* **30**, 732.

Chen, R. L., Keeney, D. R. & Konrad, J. G. (1971a). Nitrification in lake sediments. *J. envir. Qual.* In press.

Chen, R. L., Keeney, D. R., Konrad, J. G., Holding, A. J. & Graetz, D. A. (1971b). Gas production in sediments of Lake Mendota, Wisconsin. *J. envir. Qual.* In press.

Chen, R. L., Keeney, D. R., Graetz, D. A. & Holding, A. J. (1971c). Denitrification and nitrate reduction in lake sediments. *J. envir. Qual.* In press.

Collins, V. G. (1963). Distribution and ecology of bacteria in freshwater. *Proc. Soc. Wat. Treat. Exam.* **12**, 40.

Connell, W. S. & Patrick, W. H., Jr. (1968). Sulfate reduction in soil: effects on redox potential and pH. *Science, N.Y.* **159**, 86.

Cook, F. D. (1961). The denitrifying bacteria of soil. Ph.D. Thesis, University of Edinburgh.

Cooper, G. S. & Smith, R. L. (1963). Sequence of products formed during denitrification in some diverse Western soils. *Proc. Soil Sci. Soc. Am.* **27**, 659.

Courchaine, R. J. (1968). Significance of nitrification in stream analysis—effects on the oxygen balance. *J. Wat. Pollut. Control Fed.* **40**, 835.

Domogalla, D. P. & Fred, E. B. (1926). Ammonia and nitrate studies of lakes near Madison, Wisconsin. *J. Am. Soc. Agron.* **18** (10), 897.

Domogalla, B. P., Fred, E. B. & Peterson, W. A. (1926). Seasonal variation in the ammonia and nitrate content of lake waters. *J. Am. Wat. Wks Ass.* **15**, 369.

Downing, A. L., Painter, H. A., & Knowles, G. (1964). Nitrification in the activated sludge process. *J. Proc. Inst. Sew. Purif.* **1**, 445.

Dugdale, R. C. & Goering, J. J. (1967). Uptake of new and regenerated nitrogen in primary productivity. *Limnol. Oceanogr.* **12**, 196.

Fay, P. (1965). Heterotrophy and nitrogen fixation in *Chlorogloea fritschii. J. gen. Microbiol.* **39**, 11.

Fogg, G. E. (1968). The physiology of an algal nuisance. *Proc. R. Soc. B.* **173**, 175.

Foree, E. G., Jewell, W. J. & McCarthy, P. L. (1971). The extent of nitrogen and phosphorus regeneration from decomposing algae. *Wat. Res.* In press.

Fruh, E. G. (1967). The overall picture of eutrophication. *J. Wat. Pollut. Control Fed.* **39**, 1149.

Gerloff, G. C. & Skoog, F. (1954). Cell contents of nitrogen and phosphorous as a measure of their availability for growth *Mocrocystic aeruginosa. J. Ecol.* **35**, 348.

Gerloff, G. C. & Skoog, F. (1957). Nitrogen as a limiting factor for productivity for the growth of *Macrocystis aeruginosa* in southern Wisconsin lakes. *J. Ecol.* **38**, 556.

Goering, J. J. & Dugdale, V. A. (1966). Estimate of rates of denitrification in subarctic lake. *Limnol. Ocenogr.* **9**, 448.

Goering, J. H. & Neess, J. C. (1964). Nitrogen fixation in two Wisconsin lakes. *Limnol. Oceanogr.* **9**, 530.

Golterman, H. L(1960). Studies on the cycle of elements in fresh water. *Acta bot. neerl.* **9**, 1 (cited in Johannes, 1968).

Golterman, H. L. (1964). Mineralization of algae under sterile conditions or by bacterial breakdown. *Verh. int. Verein theor. Agnew. Limnol.* **15**, 544 (cited in Johannes, 1968).

Goring, C. I. A. (1962) Control of nitrification by 2-chloro-6-(trichloro-methyl) pyridine. *Soil Sci.* **93**, 211.

Greenwood, D. J. (1962). Measurement of microbial metabolism in soil. In *The Ecology of Soil Bacteria.* (Eds T. R. G. Gray & D. Parkinson). Toronto, Canada: University of Toronto Press.

Gupta, R. S. (1969). Biochemical relationships and inorganic nitrogen equilibrium in semi-enclosed basins. *Tellus* **21**, 170.

Henrici, A. T. (1940). The distribution of bacteria in lakes. *Am. Ass. Adv. Sci. Pub.* **10**, 39.

Henrici, A. T. & McCoy, E. (1938). The distribution of heterotrophic bacteria in the bottom deposits of some lakes. *Wis. Acad. Sci. Let.* **31**, 323.

Holden, A. V. (1969). Chemical Investigations in Loch Leven, Scotland. Report to the Supervisory Committee for Loch Leven Research.

Horne, A. J. & Fogg, G. E. (1970). Nitrogen fixation in some English Lakes. *Proc. R. Soc.* B. **175**, 351.

Johannes, R. E. (1968). Nutrient regeneration in lakes and oceans. *Adv. Microbiol. Sea* **1**, 203.

Keeney, D. R. (1971). The nitrogen cycle in sediment-water systems. Paper presented before Div. S-2, Soil Science Society of America, New York, August 1971.

Keeney, D. R., Konrad, J. G. & Chesters, G. (1970). Nitrogen distribution in some Wisconsin lake sediments. *J. Wat. Pollut. Control Fed.* **42**, 411.

King, D. L. (1970). The role of carbon in eutrophication. *J. Wat. Pollut. Control Fed.* **42**, 2035.

Konrad, J. G., Keeney, D. R., Chesters, G. & Chen, K. (1970). Nitrogen and carbon distribution in sediment cores of selected Wisconsin lakes. *J. Wat. Pollut. Control Fed.* **42**, 2094.

Kuentzel, L. E. (1969). Bacteria, carbon dioxide, and algal blooms. *J. Wat. Pollut. Control Fed.* **41**, 1737.

Kuentzel, L. E. (1970). Phosphorous vs. carbon factor in algal blooms and deterioration of water quality. Presented at Division of Environmental Sciences, New York Academy, February, 1971.

Kuznetsov, S. I. (1968). Recent studies on the role of micro-organisms in the cycling of substances in lakes. *Limnol. Oceanogr.* **13**, 211.

Lee, G. F. (1966). Report on the nutrient sources of Lake Mendota. Madison, Wisconsin: Lake Mendota Problems Committee (revised, 1969).

Lindstrom, E. S., Trove, S. R., & Wilson, P. W. (1950). Nitrogen fixation by the green and purple sulphur bacteria. *Science, N.Y.* **112**, 197.

Lund, J. W. G. (1965). The ecology of fresh water phytoplankton. *Biol. Rev,* **40**, 231.

McCoy, E., & Sarles, W. B. (1969). Bacteria in lakes: populations and functional relations. *In Eutrophication: Causes, Consequences, Correctives.* Washington, D.C.: National Academy of Science.

Mackenthun, K. M. (1969). The practice of water pollution biology. Washington, D.C.: U.S. Dept. Interior, FWPCA.

Meek, B. D., Grass, L. B. & Mackenzie, A. J. (1969). Applied nitrogen losses in relation to oxygen status of soils. *Proc. Soil Sci. Soc. Am.* **33**, 575.

Mortimer, C. H. (1941). The exchange of dissolved substances between mud and water in lakes. *J. Ecol.* **29**, 280.

Mortimer, C. H. (1942). The exchange of dissolved substances between mud and water in lakes. *J. Ecol.* **30**, 147.

Nicholas, D. J. D. (1963). The metabolism of inorganic nitrogen and its compounds in micro-organisms. *Biol. Rev.* **38**, 350.

Nursall, J. R. (1969). The general analysis of an eutrophic system. *Verh. int. Verein. theor. angew. Limnol.* **17**, 109.

Owens, M. (1970). Nutrient balances in rivers. Paper presented as Society for Water Treatment and Examination Symposium on Eutrophication, March 1970.

Parr, J. F. (1969). Nature and significance of inorganic transformations in tile-drained soils. *Soils Fertil.* **32**, 411.

Patnaik, S. (1965). Nitrogen-15 tracer studies on the transformations of applied nitrogen in submerged rice soils. *Proc. Indian Acad. Sci.* **61B**, 25.

Patrick, W. H. & Mahapatra, I. C. (1968). Transformation and availability to rice of nitrogen and phosphorus in water-logged soils. *Adv. Agron.* **20**, 323.

Pomeroy, L. E. (1970). The strategy of mineral cycling. *A. Rev. Ecol. Systematics* **1**, 171.

Ponnamperuma, F. N. (1971). Oxidation-reduction reactions. In *Soil Chemistry*. (Eds G. Chesters & J. M. Bremmer). New York: Marcel Dekker.

Ponnamperuma, F. N., Martinez, E. & Loy, T. (1966). Influence of redox potential and partial pressure of carbon dioxide on pH values and the suspension effect of flooded soils. *Soil Sci.* **101**, 421.

Postgate, J. R. (1969). Methane as a minor product of pyruvate metabolism by sulphate-reducing and other bacteria. *J. gen. Microbiol.* **57**, 293.

Power, J. F. (1968). Mineralisation of nitrogen in grass roots. *Proc. Soil Sci. Soc. Am.* **32**, 673.

Redman, F. H. & Patrick, W. H. (1965). Effect of submergence on several biological and chemical soil properties. *Bull. La. agric. Exp. Stn* No. 592.

Rice, W. A., Paul, E. A. & Wettler, L. R. (1967). The role of anaerobiosis in asymbiotic nitrogen fixation. *Can. J. Microbiol.* **13**, 829.

Ryther, J. H. & Dunstan, W. M. (1971). Nitrogen, phosphorus, and eutrophication in the coastal marine environment. *Science, N.Y.* **1711**, 1008.

Schick, H. J. (1971). Substrate and light dependent fixation of molecular nitrogen in *Rhodospirillum rubrum*. *Arch. Mikrobiol.* **75**, 89.

Serruya, C. & Berman, T. (1969). The evolution of nitrogen compounds in Lake Kinneret. *Dev. Wat. Qual. Res.* (June) p. 73.

Shapiro, J. (1970). A statement on phosphorus. *J. Wat. Pollut. Control Fed.* **42**, 772.

Siwasin, J. (1971). Anaerobic micro-organisms in soil. M.Sc. thesis, University of Edinburgh.

Stadtman, T. (1967). Methane fermentation. *A. Rev. Microbiol.* **21**, 121.

Stewart, W. D. P. (1968). Nitrogen input into aquatic ecosystems. In *Algae, Man and the Environment*. (Ed. D. F. Jackson). New York: Syracuse University Press.

Stewart, W. D. P. (1970). Algal fixation of atmospheric nitrogen. *Pl. Soil* **32**, 555.

Stewart, W. D. P., Fitzgerald, G. P. & Burris, R. H. (1968). Acetylene reduction by nitrogen-fixing blue-green algae. *Arch. Mikrobiol.* **62**, 336.

Stumm, W. & Morgan, J. J. (1970). *Aquatic Chemistry*. New York: Wiley-Interscience.

Takai, Y. & Kamura, T. (1966). The mechanism of reduction in water-logged paddy soil. *Folia microbiol.* **11**, 304.

Thomas, E. A. (1953). Empirische und experimentelle Untersuchungen zur Kentnis der Minimumstoffe in 46 Seen der Schweiz und angronozonder Gobiete. *Schweiz. Ver Gas Wasserfachm* **2**, 1.

Thomas, E. A. (1963). Die Veralgung von Seen und Flüsen, deren ursache und abwehr. *Schweiz. Ver. Gas Wasserfachm* **6/7**, 3.

Tomlinson, T. E. (1970). Trends in nitrate concentrations in English rivers, and fertilizer use. Paper presented at the Society for Water Treatment and Examination Symposium on Eutrophication, March 1970.

Trudinger, P. A. (1969). Assimilatory and dissimilatory metabolism of inorganic sulphur compounds by micro-organisms. In *Advances in Microbial Physiology,* vol. 3. (Eds A. H. Rose & J. F. Wilkinson). London: Academic Press.

Turner, F. T. & Patrick, W. H., Jr. (1968). Chemical changes in water-logged soils as a result of oxygen depletion. *Trans. 9th Int. Congr. Soil Sci.* **4**, 53.

Van Der Weijden, C. H., Schulling, R. D. & Das, H. A. (1970). Some geochemical characteristics of sediments from the North Atlantic Ocean. *Mar. Geol.* **9**, 81.

Vollenweider, R. A. (1968). Scientific fundamentals of the eutrophication of lakes and flowing waters with particular reference to nitrogen and phosphorus as factors in eutrophication. DAS/CSI/68.27. Paris: Organisation Economic Cooperation and Development.

Wolfe, R. S. (1971). Microbial formation of methane. In *Advances in Microbial Physiology*, vol. 6. (Eds A. H. Rose & J. F. Wilkinson). London: Academic Press.

Yoshima, S. (1932). Seasonal variations of nitrogenous compounds and phosphate in the water of Takasuka pond, Saitama, Japan. *Arch. Hydrogeol.* 24, 155.

Factors Affecting Algal Blooms

J. A. STEEL

*Metropolitan Water Board, Water Examination Department,
Roseberry Avenue, London E.C.1, England*

CONTENTS

1. Introduction

IN ALMOST ANY body of water at least one of the lower forms of plant or animal life is found; in most a great many such representatives will be algae. Whether or not such algae form a major part of the natural biocoenosis will depend to some extent on man's intervention in it, and whether or not their presence is desirable will depend on man's requirement of the particular ecosystem. These algae may be undesirable in that, even when relatively sparse, they so change the character of the medium that it becomes unacceptable. For instance, one may cite the colonial chrysophyte *Synura uvella* which is associated in water treatment systems with a very noticeable and persistent taste of cucumber. Some members of the Cyanophyta have been implicated in causing death or distress to cattle allowed to drink waters in which such algae were present. *Microcystis aeruginosa*, for example, has yielded extremely rapid and efficient toxins in laboratory extractions (Bishop *et al.*, 1959). The well publicized "Red Tides" of a marine dinoflagellate are further instances of a primary producer creating considerable difficulties for other members of the ecosystem.

Mostly, of course, the difficulties occasioned to man are far less dramatic but nonetheless can be disrupting. The presence of large blooms of algae is clearly only a disturbance to those who want the water in which the algae are thriving.

Fish farmers may positively encourage algal blooms by deliberate fertilization in an attempt to increase the overall productivity of their lagoons and would presumably regard blooms of nontoxic algae as desirable. For those engaged in water supply, however, blooms are to be avoided whenever possible. The difficulty these blooms occasion water supply is primarily due to interference in the water treatment units attempting to separate the water and the particles suspended therein.

It is difficult to generalize about the desirability or otherwise of any particular alga. However, based on experience of sand filtering systems, some examples of these algae which may be deemed a nuisance can be attempted. The ubiquitous *Asterionella formosa,* a diatom which forms stellate colonies, can cause serious difficulty either by forming a surface mat on the sand or, if the colonies disrupt, by individual cells penetrating the filter. *Nitzschia acicularis,* a pennate diatom, is a difficult alga to filter because of its pore blocking and filter passing capabilities, due to its needle-like form. *Stephanodiscus astraea,* a centric diatom, is normally of such large dimensions, at least in the very eutrophic reservoirs of the lower Thames Valley, that considerable biomasses of this alga may be present in the water mass before serious filtration difficulties occur. *Tribonema bombycinum,* a filamentous xanthophyte, and various species of the genus *Melosira* (a filamentous diatom) also cause filtration difficulty by pore blockage and/or surface matting. The *Melosira* spp. may continue to cause such difficulty even after death due to the persistence of the silica shell, a characteristic common to all the diatoms. The members of the Cyanophyta most commonly experienced, *Aphanozomenon flos aquae, Anabaena circinalis* and *Microcystis aeruginosa,* are capable of producing very large standing crops during the late summer with obvious effects on the filtration process. *Microcystis aeruginosa* also produces an extracellular mucopolysaccharide, particularly when moribund, which causes consumer complaint due to later flocculation following soft drink manufacture or tea making (Windle Taylor, 1966).

The nuisance algae cause difficulties to water treatment processes because of their direct interference in any of the treatment units, either physical or chemical, or their later manifestation at the consumers' premises. The result has been that in the multiple stage treatment systems used to purify very eutrophic waters, some effort has been made to understand the various factors which seem to be of importance in producing the blooms of algae which cause the more serious difficulties.

2. Requirements for Growth

The algae at any given moment find themselves in an environment which may be more, or less, favourable to their growth. In any attempt to assess the relative merits of a situation we must be careful not to judge by what we as humans perceive as favourable. Hence we are bound to base judgements, in so far as is possible, on any known response of an alga to the factor in question. Although it is desirable that this response be obtained within the particular environment it is clearly not always possible, and resort must be made to laboratory investigations under controlled conditions. In the following the responses to the various factors to be detailed have been obtained from both field and laboratory studies.

(a) *Incident radiation*

Incident radiation is of importance to the alga in that it represents the energy which the plant requires for its photosynthetic processes. Both the right quality and quantity of light are required for any growth. The photosynthetic pigments of algae, in common with most plants, are such that they absorb radiation within the wave band *c*. 400–700 nm. This generally represents 46–48% of the total incoming radiation (Talling, 1957a). This fraction is becoming known as photosynthetically active radiation (PhAR). In so far as the absorption pigments have some wave length specificity, it is possible that the spectral quality of the light is also important. As radiations of differing wavelengths are not equally absorbed in the downward passage through water, the spectral quality of the light in the depths will be different from that nearer the surface. It is not known how important this fact is; probably only minor in terms of absolute production, but possibly of some consequence in determining the type of alga which will grow.

With respect to production the absolute quantity of the incident radiation rate seems to be of less importance than its relative magnitide to an intensity labelled by Talling (1957a) as I_k. This is the intensity within the light intensity/photosynthetic rate relationship of the alga, at which the photosynthetic rate is 0·70–0·75 of the maximal photosynthetic rate. Figure 1(*a*) illustrates the photosynthetic rate relative to the maximum rate as determined by a $^{14}0$ fixation method for *A. formosa* and a *Melosira* sp. during late winter. Figure 1(*b*) illustrates the relative photosynthetic rate plotted against the relative light intensity, from which the similarity of the response emerges. The I_k for *A. formosa*, is ~13 Kergs/cm²/sec, whereas the I_k for the *Melosira* is only ~4 Kergs cm⁻² s⁻¹. Thus the *Melosira* would seem much better equipped to deal with lower light intensities than *A. formosa*. Latterly the infrequent appearance of significant quantities of *Melosira* in the reservoirs of the lower Thames Valley has prevented any further investigation of this behaviour, but it is very similar to an observation by Talling (1957a) on *A. formosa* and *Melosira italica* subsp. *subarctica*.

As shown by Vollenweider (1965) this type of light response is very general amongst planktonic algae, and relatively simple mathematical formulations can

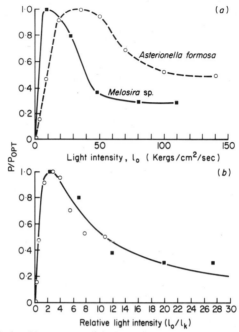

Fig. 1. The relationships between light intensity and photosynthetic rate in algae. (*a*)
Direct relationships in terms of the photosynthetic rate relative to the maximum
observed for 2 types of algae; (*b*) relationships in terms of relative values. In (*a*) the
curves were fitted by eye; in (*b*) the curve was calculated.

be used to describe it. Such expressions can be further expanded to give some
means of estimating the total daily photosynthesis per unit area (Steel, 1970).
This estimate is important in assessing the productive capability of any algae in
given circumstances.

The absolute light level is important in that inhibition of photosynthesis
occurs at the higher light levels, this inhibition being more or less reversible
(Goldman, Mason & Wood, 1963), and at high light levels photodestruction of
the pigmentation may take place.

Some algae are capable of growing heterotrophically, that is in the absence of
light. The process is, however, relatively so energetically inefficient that
extremely large biomasses are unlikely to be produced in any reasonably short
period, and one can state basically that the exclusion of light prevents algal
blooms.

(b) *Temperature*

Thermal effects on the growth processes of algae can be very misleading if
confined to their anabolic physiology. The photosynthetic rate maximum has a

Q_{10} of $c.$ 2·0—2·3 and so the algae in a warm environment has a photo-synthetic rate maximum far in excess of their midwinter maximum rate. Frequently, however, the algae are not able to manifest themselves in any great biomass even during summer. This may be due in part to sedimentation effects, but in a natural environment it is possible that at higher temperatures their catabolic processes are relatively greater. Some measure of this relationship is afforded by the relative respiration. This is the ratio between the respiration rate of an alga and its photosynthetic rate maximum, and normally has a value of 0·5—0:10. At first replication of overwintering diatoms, the relative respiration may be as low as 0·02—0:03. Although the inefficiency of experimental techniques makes it difficult to describe this figure as absolute, it is probably a reasonable indication that this early season relative respiration is lower than those more generally found later in the year. The result of this favourable situation is that the algae are able to grow to profusion notwithstanding the fact that the temperature may remain at or near the winter level of 2—3° throughout the period of growth. It appears that for many vernal diatoms such a temperature regime is a positive advantage.

It may be that the very low relative respiration of overwintering diatoms results from their having been through the low incident radiation levels of winter. In the lower Thames Valley autumnal diatom crops usually have a much less favourable relative respiration than have the vernal blooms. This would clearly result if the respiration rate had a positive temperature coefficient greater than that of the photosynthetic rate maximum, for then the relative respiration would also possess a positive temperature coefficient. This implies that the warmer the water, the less efficient would be the carbon retention of the algae, at least if those algae are diatoms. It seems that other factors also affect the relative respiration and, as suggested above, the previous light history may be important. Beyers (1965), for instance, studying photosynthesis and respiration of laboratory microcosms, consistently found significant depression of respiration during dark periods.

This behaviour would indicate how closely interwoven are the effects of light and temperature to an alga, and how the effects of one may confound those of the other. It is possible to distinguish groups of planktonic algae broadly by their apparent ability to exploit various combinations of light and temperature. Thus $M.$ $italica$ has the ability to reproduce efficiently in the low light and low temperature of winter, whereas at the other extreme $A.$ $circinalis$ has the same ability in the high light and high temperatures of summer.

The higher temperatures of summer allow sudden outbursts of growth should the situation become favourable and, as the absolute rates are much greater, any beneficial change in the relative respiration, for instance, implies an absolutely greater net gain; hence possibly greater growth rates.

(c) *Nutrients*

With the present day awareness of the potential problems associated with eutrophication, defined by Lund (1967) as "the process of becoming rich in dissolved nutrients", much work and discussion have been devoted to the role of nutrients in supporting large blooms of algae. In such considerations emphasis, in many instances perhaps over emphasis, has been accorded to dissolved nitrogen and phosphorous. Lund (1970) puts the matter in perspective with his apposite observations of the effects of nutrient concentrations on primary production.

From Lund's definition of eutrophication the inevitable question arises "how rich is rich?", and the answer must begin, equally inevitably with "it all depends . . .". In many areas perhaps undue importance has been placed on nutrient removal in eutrophic situations because of an insufficient consideration of this question and its answer. Notwithstanding these remarks nutrient concentrations are, of course, a fundamental factor in allowing any algal growth, and their total exclusion, as with light, precludes any algal bloom. In lowland surface waters this desirable state of affairs at present cannot be achieved economically, hence the need to find out how closely it must be approximated. In such investigation it is important to appreciate how minute are the nutrient requirements of the algae even when forming appreciable blooms.

(i) *Essential elements*

Nicholas (1963) has listed 16 elements apart from C, H and O_2, essential for growth: N, P, Ca, Mg, Na, K, S, Fe, Mn, Cu, Zn, Mo, B, Cl, Co and V. Although other elements, such as Au and Pb, are found in ash residues, an absolute requirement has not yet been demonstrated. These elements are subdivided into macronutrients, e.g. N and P, and micronutrients, e.g. Mo and Mn. Many algae also seem to require trace organic materials such as vitamin B_{12}. Occasionally there may be a specific nutrient requirement, the most noticeable in freshwater being the dissolved silica that the diatoms require for their shells. Lund (1950) has shown that if this dissolved chemical be decreased to a concentration <0 5 mg SiO_2/l further diatom growth is prevented.

(ii) *Nitrogen and phosphorus requirements*

Some consideration of N and P may serve to indicate the difficulties in assessing the role of nutrients in production. Figure 2 shows an observed relationship between winter maximum NO_3-N and PO_4-P concentrations and maximum summer standing crops expressed as chlorophyll-a concentration. From a waterworks point of view, taking into account chlorophyll-a concentrations which cause treatment difficulty, this figure suggests that the undesirable chlorophyll-a concentrations begin to occur with N concentration of c. 525 μg of NO_3-N/l and P concentrations of c. 9 μg of PO_4-P/l. Normally the N content of algae is some 10–15 times greater than the P content. An apparent uptake ratio of this order is only approached in the large biomasses (Fig. 2). Thus, it may be suggested that the smaller crops are being P limited, the N concentration always being >10 times the P concentration. One has to associate

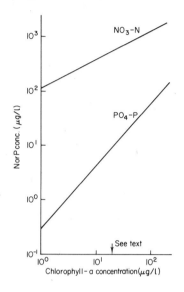

Fig. 2. The relationships between winter NO_3-N and PO_4-P concentrations and summer standing crops expressed as chlorophyll-a concentration (after Lund, 1970).

with this concentration, however, the findings of Mackereth (1953) on the absolute P requirements and uptake behaviour of *A. formosa* in the English Lake District. He found that this alga used P more or less abundantly, depending upon the ambient levels. Thus at low levels of dissolved PO_4-P, during the relatively nonproductive periods, PO_4-P is stored in the cell so that during the productive periods that reserve is available for a bloom. This results in higher chlorophyll-a/PO_4-P ratios at low PO_4-P concentrations. In high ambient PO_4-P concentrations the cells tend to conserve their internal P and so immediately begin to deplete the environment when they actively grow and reproduce. These cells should then be able to utilize their stored P for further growth. However, the great efficiency with which P is used is such that in many environments other factors limit the growth before this occurs. Mackereth found that 1 μg of P was sufficient to support 16×10^6 *A. formosa* cells. Thus a concentration of 1 μg of PO_4-P/l, i.e. 1/700 of the M.W.B. (Metropolitan Water Board) norm, allows an *A. formosa* cell concentration only slightly less than the greatest observed in the M.W.B. reservoirs during the past few years. PO_4-P is also extremely mobile (Golterman, 1960), a large percentage of the P content being released shortly after death; thus there would seem relatively little locking up of

P, in contrast to the relatively more stable SiO_2 which is taken from the system when the diatoms sediment to the bottom.

Consideration of N also has inherent difficulty in that many members of the blue-green algae can fix dissolved elemental N (Dugdale & Dugdale, 1962). This means that they are relatively independent of the dissolved combined N supply, although they still require a supply of other essential nutrients. Zooplankton N excretion may also provide N for utilization by algae. Rainwater, particularly where the prevailing winds blow over large urban or industrial areas, may also be a relatively unconsidered source of nutrients. A regular analysis of rainwater collected in London shows quite considerable concentrations of some of the nutrients required by algae (Windle Taylor, 1966). Taking these figures as being an indication of the order of concentrations to be expected, it is possible that 1 week of normal rainfall early in the year could supply $2 \cdot 5$ mg of PO_4-P/m^2 and $21 \cdot 8$ mg of NH_3- and NO_3- N/m^2. From Mackereth's data, this would support $c.\ 40 \times 10^9$ *A. formosa* m^2. In a reservoir such as the Queen Mary (volume 30×10^6 m^3, 12 m deep) this represents a concentration of $3 \cdot 33 \times 10^6$ cells/l, a sufficient concentration to cause some concern to waterworks. As the relatively nonproductive period of winter is longer than 1 week, a considerable N and P build up could take place and support quite reasonable vernal blooms.

Studies using mathematical models of the primary production systems in the lower Thames Valley reservoirs allow some estimation of the nutrient requirements for the potential maximum cell concentrations (Steel, unpublished). These suggest that even if the N and P concentration in those reservoirs could be decreased to slightly $<10\%$ of the present day levels, no great, if any, reduction in the maximum algal crops would result. This may be viewed in conjunction with observations of Ridley (1970) who illustrated an increase in N and P concentrations in the River Thames at Walton from 1905 to 1968. Since 1928 the Queen Mary reservoir has been in use and the indications are that the algal problems in the 1930's were much as have been experienced some 30 years later, despite a 72% increase in NO_3-N and a 728% increase in PO_4-P in the water during this period.

3. Algae in Natural Environments

The environment within which the algae exist affects the capacity of those algae to exploit their productive capabilities. Within that environment the algae are affected by "biological neighbours" and the manner in which the environment affects parameters of importance to algal production.

(a) Biotic interference

The presence of the algae themselves affects their growth capability in that they cause a decrease in the depth of the euphotic zone by their light absorption. This selfshading effect can be critical in determining the size of the blooms which may be formed in given water masses and will be more fully discussed elsewhere (Steel, in preparation). It is possible that the dense populations which form the blooms also have their growth processes reduced by "neighbour nearness". This may be due to substances released by the algae during the formation of the bloom. For instance, the percentage saturation of dissolved O_2 can affect algal physiology (Gessner & Pannier, 1958) and large blooms are normally associated with water supersaturated with respect to dissolved O_2 produced as a byproduct of photosynthesis. In culture, some algae can produce extracellular substances which prevent other species of algae growing in that particular medium. The most noted of these products is probably "Chlorellin" produced by *Chlorella* spp. Whilst it is possible that such products have some influence in near unispecific dominance, other reasons are more probably the cause (Talling, 1957b).

Most algae seem to leak some of their fixed carbon products, the leakage being particularly as glycolates (Fogg *et al.*, 1965). Such leakage may represent a fairly considerable fraction of the total net production to have taken place. Interpreting the difference between O_2 produced and ^{14}C fixed in *in situ* production profiles as some indication of the release of fixed carbon, it could be suggested that during a spring bloom of diatoms, some 4–5 g of C/m^2 (as organic carbon) may be released. It is possible that this C source is utilized by algae and/or bacteria (Watt, 1965).

Many blooms seem to have a relatively higher respiratory rate than less concentrated algae, but it cannot as yet be said that the measured O_2 utilization is exclusively algal respiration. Figure 3 illustrates the chlorophyll-a concentration in the Queen Elizabeth II reservoir and the coliform bacteria concentration in the reservoir and the Thames at Walton during July–August 1970. The alga was predominantly *A. circinalis*. During the final stages of this bloom there was an enormous increase in the coliform content of the reservoir. Presumably this increase must be to some extent a reflection of the previous river peak, but from physical considerations it is unlikely that the observed reservoir concentrations would have resulted from this cause alone. It is probable therefore that blooms of blue-green algae at least, and particularly if moribund have a much larger associated bacterial flora. The response of coliform bacteria to such blooms may occasionally be so great as to cause considerable concern in water supply (MacKenzie, 1952). The indigenous bacterial flora also shows a similar response (Price, pers. comm.). These bacteria may have some part to play in nutrient recycling and availability. *M. aeruginosa* for instance, frequently contains large numbers of bacteria in its mucilage (Gerloff & Skoog, 1957).

Zooplankton exert an influence on the algae, predominantly by grazing, but also possibly by the release of their secretions (Nauwerk, 1963; Steel *et al.*, 1970). Some algae, particularly the smaller forms, are actively grazed and can be removed by reasonable standing crops of zooplankton. Algae such as *Chlorella*

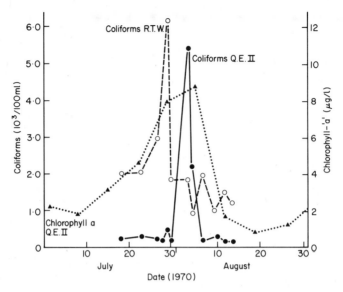

Fig. 3. Coliform bacteria and chlorophyll-a concentrations in River Thames water at Walton during July–August 1970.

sp. and *Oocystis* spp. are within this category. Many of the larger algae, for instance *S. aestraea*, do not seem to be willingly ingested for similar algae have been observed being apparently actively rejected. The rejection procedure seems to be incapable of dealing with large concentrations of such algae and they may then be ingested. When this happens the algae are not always digested and pass through the gut apparently undamaged. As the algae experience large pH changes in their passage through the gut it is possible that the effect of this on their photosynthetic pigments is such that the cells are no longer productive once ~~ingested. If this is so, the population loss is all the potential desirintants of any~~ ingested algae.

Canter & Lund (1951) have shown that parasitism on phytoplankton is a widespread phenomenon, and can markedly influence the course of a bloom. This influence is usually exerted once the incidence of infection is *c.* 25%; at <10% no apparent effects are observable.

(b) *Physical influences*

Physical interference in the growth capability of algae may come about by high concentrations of silt particles suspended in amongst the algal population. This silt, by its light absorption, competes with the algae for the radiation of the underwater light field. The degree of this competition clearly depends on the quantity and quality of that silt. The turbulence of the medium has a great influence in determining both the amount and the nature of the silt and so exerts a considerable affect on the light conditions experienced by the algae and, therefore, their production rate.

The water itself, and any dissolved colouring material, also absorbs the subsurface light. This absorbtion usually exhibits general Beer-Lambert characteristics in that the light absorbed within a small depth increment is proportional to the intensity of the radiation at the depth of the increment. This results in an exponentially decreasing radiation intensity with increasing depth. If the turbulence is capable of circulating the algae through this underwater light field, the algae experience great variation in both light intensity and gradient, particularly in optically dense water. If the depth of water through which the algae are being circulated is greater than that of the euphotic column, then even during the hours of daylight the algae may spend some time in darkness, thereby losing carbon by respiration. It is possible to construct models, at least for diatoms, which seem to represent reasonably such factors, and from which can be calculated the required depth of circulation to offset the growth capability of a given diatom within the circumstances (Steel, 1970).

Much effort has been, and is being, expended to investigate the role of turbulence in producing patterns of algal distribution which impose so great an energetic cost on them that they can no longer produce bloom. In such efforts, however, care must be taken to ensure that the turbulence does not maintain in suspension a population which would otherwise settle out of the water column. *Melosira* spp. seem to benefit greatly from periods of higher turbulence (Lund, 1966). Figure 4 illustrates the movement of a *M. granulata* population in the

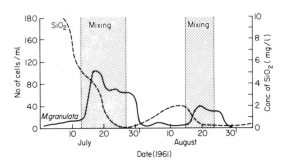

Fig. 4 The mixing of *M. granulata* in the Queen Mary reservoir due to water movement (after Ridley, 1962).

Queen Mary reservoir during a period in which the water column was fully circulating. The first mixing coincided with high dissolved SiO_2 concentration and a larger crop resulted than in the second mixed period which coincided with much lower SiO_2 concentrations.

MAP—8

The temperature of the water must also be considered in relation to sedimentation and flotation, for at lower temperatures the viscosity of the water materially assists in supporting sinking algae. The buoyancy of blue-green algae seems usually to be a favourable characteristic in allowing them to survive in the lower viscosity and turbulence of summer.

Turbulence also affects the phytoplankton/zooplankton relationship by possible differential distribution. The zooplankton may also be influenced in their distribution by changes in the light caused by the turbulence which prevents algal aggregation.

In reservoirs such as those of the Metropolitan Water Board with high relative throughputs of river-derived water the algal growth which may take place is affected by that throughput. In reservoirs in constant circulation the outflow imposes a continuous cropping of the algal standing crop. In rivers where considerable flow fluctuation may take place, a high flow rate flushes out any crop of algae established during periods of lower flow.

4. Conclusions

It will be appreciated that there are many factors which can influence the ability of the algae to form blooms. These factors may also interact so as to give an impressive array of environmental/physiological situations. It is of interest to those to whom algae are of concern to be able to discriminate amongst these possibilities so as to decide the most important factors allowing, or preventing, an algal bloom at any given moment. Such information is clearly important for prediction both for initial design and for management of existing resources. It is also important for objective evaluation of the results of systems which allow massive artificial influence within the ecosystem (e.g. Ridley, Cooley & Steel, 1966). Where such systems are available it is necessary that they produce their desired effects without unforeseen present and/or future deterioration in the desirable aspects of the ecosystem. Such insights are also necessary to maintain system optimization even if great changes in environment occur, perhaps taking the system outside the original design concepts. It is also clearly necessary to understand the relative importance of factors important in primary production, many of which may not always be apparent, so that energy is not expended on investigating factors which may not be of great consequence in a particular situation.

5. Acknowledgments

This contribution is published by permission of Dr E. Windle Taylor, C.B.E., Director of Water Examination, Metropolitan Water Board. The views expressed are not necessarily those of the Metropolitan Water Board. My thanks are due to Dr J. E. Ridley and Dr N. P. Burman for much helpful comment.

6. References

Beyers, R. J. (1965). The pattern of photosynthesis and respiration in laboratory microecosystems. *Memorie Ist. Ital. Idrobiol.* **18** (suppl.), 61.

Bishop, C. T., Anet, E. F. L. T. & Gorham, P. R. (1959). Isolation and identification of the fast-death factor in *Microcystis aeruginosa NRC-1*. *Can. J. Biochem. Physiol.* **37**, 453.

Canter, H. M. & Lund, J. W. G. (1951). Studies on plankton parasites. III. Examples of the interaction between parasitism and other factors determining the growth of diatoms. *Ann. Bot., N.Z.* **15** 359.

Dugdale, V. A. & Dugdale, R. C. (1962). Nitrogen metabolism in lakes. II. Role of nitrogen fixation in Sanctuary Lake, Pennsylvania. *Limnol. Oceanogr.* **7**, 170.

Fogg, E. G., Nalewajko, C. & Watt, W. D. (1965). Extracellular products of phytoplankton photosynthesis. *Proc. R. Soc.* B. **162**, 517.

Gerloff, G. C. & Skoog, F., (1957). Nitrogen as a limiting factor for the growth of *Microcystis aeruginosa* in southern Wisconsin lakes. *Ecology* **38**, 556.

Gessner, F. & Pannier, F., (1958). Influence of oxygen tension on respiration of phytoplankton. *Limnol. Oceanogr.* **3**, 478.

Goldman, C. R., Mason D. T. & Wood, B. J. B. (1963). Light injury and inhibition in Antarctic freshwater phytoplankton. *Limnol. Oceanogr.* **8**, 313.

Golterman, H. L. (1960). Studies on the cycle of elements in fresh water. *Acta bot. neerl.* **9**, 1.

Lund, J. W. G. (1950). Studies on *Asterionella formosa* Hass. II. Nutrient depletion and Spring maximum *J. Ecol.* **38**, 1.

Lund, J. W. G. (1966). The importance of turbulence in the periodicity of certain freshwater species of the genus *Melosira*. (Translation) *Zh. bio-bot. Tsyklu, Kyev*, **51**, 176.

Lund, J. W. G. (1967). Eutrophication. *Nature, Lond.* **214**, 557.

Lund, J. W. G. (1970). Primary production. *Proc. Soc. Wat. Treat. Exam.* **19**, 332.

MacKenzie, E. F. W. (1952). *Rep. Results chem. bact. Exam. Lond. Wat.* Vol. 43.

Mackereth, F. J. (1953). Phosphorous utilization by *Asterionella formosa* Hass. *J. exp. Bot.* **4**, 296.

Nauwerk, A., (1963). Die Beziehungen.zwischen Zooplankton und Phytoplankton in See Erken. *Sym. bot. upsal.* **17**(5), 1.

Nicholas, D. J. D. (1963). Inorganic nutrient nutrition of micro-organisms. In *Plant Physiology* III. (Ed. F. C. Stewart). London: Academic Press.

Ridley, J. E. A. (1962). Thermal Stratification in relation to phytoplankton production in Thames Valley storage reservoirs. Ph.D. Thesis University of London.

Ridley, J. E. A., Cooley, P. & Steel, J. A. (1966). Control of thermal stratification in Thames Valley reservoirs. *Proc. Soc. Wat. Treat. Exam.* **15**, 225.

Ridley, J. E. A. (1970). The biology and management of eutrophic reservoirs. *Proc. Soc. Wat. Treat. Exam.* **19**, 374.

Steel, J. A. (1972). The application of fundamental limnological research in water supply system design and management. *Zoo. Soc. Lond. Symposium on Conversion and Productivity of Natural Waters.* In press.

Steel, J. A. Duncan, A. & Andrew, T. E. (1970). The daily carbon gains and losses in the seston of Queen Mary Reservoir, England, during 1968. UNESCO—*IBP Symposium on Productivity Problems of Freshwaters, Kazimierz Dolny, Poland.* In press.

Talling, J. F. (1957a). Photosynthetic characteristics of some freshwater plankton diatoms in relation to underwater radiation. *New Phytol.* **56**, 29.

Talling, J. F. (1957b). The growth of two plankton diatoms in mixed cultures. *Physiologia Pl.* **10**, 215.

Vollenweider, R. A. (1965). Calculation models of photosynthesis-depth curves and some implications regarding day rate estimates in primary production measurements. *Memorie Ist. Ital. Idrobiol.* **18**, (suppl), 425.

Watt, W. D. (1965). Release of dissolved organic material from the cells of phytoplankton populations. *Proc. R. Soc.* B. **164**, 521.

Windle Taylor, E. (1966). *Rep. Results chem. bact. Exam. Lond. Wat.* Vol. 43.

Nitrogen Removal from Wastewaters
by Biological Nitrification and Denitrification

P. L. McCarty

Department of Environmental Engineering, Stanford University, Stanford, California, U.S.A.

AND

R. T. Haug

Department of Environmental Engineering, Loyola University, Los Angeles, California, U.S.A.

CONTENTS

1. Introduction

COMPOUNDS OF nitrogen contained in industrial, agricultural and municipal wastewaters can be detrimental to the quality of receiving waters. In particular, reduced compounds of N can have an adverse effect on the dissolved O_2 resources of receiving waters through bacterial nitrification. Nitrification of secondary effluents has been traditionally practised in England to serve as an indicator of a well oxidized effluent. By contrast, the practice in the United States until the late 1960s was to minimize nitrification of wastewaters during treatment in order to save on capital and operating costs. However, evidence of adverse effects on the dissolved O_2 balance of receiving waters due to nitrification has recently led several States to consider standards to limit the discharge of reduced N compounds.

Compounds of N may also serve as fertilizers which stimulate the growth of algae and other aquatic plants, and in turn can lead to the acceleration of eutrophication. This is a very real problem in many of the inland waters of the U.S.A. Complete removal of nitrogen from wastewaters is now being proposed in several cases to help limit the rate of eutrophication.

Figure 1 illustrates a simplified view of the N cycle. Organic compounds of N, for the most part, are in particulate form and can be removed from water by plain sedimentation or chemical precipitation. Inorganic N compounds such as ammonia (NH_3), nitrite (NO_2), and nitrate (NO_3) can also be removed after transformation into the organic form by bacteria or algae. The N cycle suggests an additional removal method for inorganic N, i.e. nitrification and denitrification. Ammonia can first be nitrified to either the NO_2^- or NO_3^- form by aerobic biological processes, and then denitrified in the absence of dissolved O_2 to molecular N, a normal and harmless constituent of all natural waters.

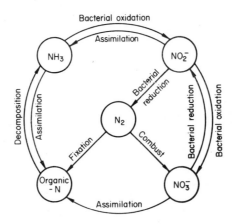

Fig. 1. Transformation of nitrogen compounds.

A summary of the normal range of N compounds contained in domestic wastewaters in the United States and the usual removals obtained during both primary and secondary treatment is given in Table 1. The nitrogen removals are ⁝⁝⁝ transformation from the NH_4^+ to NO_2 or NO_3 form takes place. Because of a smaller water usage, domestic wastewaters in England have total N concentrations which are considerably higher, 50–60 mg/l (Downing, Painter & Knowles, 1964; Downing & Hopwood, 1964; Downing, Tomlinson & Truesdale, 1964). During primary and secondary treatment, some biological conversions of N from one form to another occur, as already discussed, and the removals observed result largely from sedimentation of the particulate forms which may have been in the original wastewater or else formed during treatment. Many particulate N compounds are converted to soluble forms during anaerobic sludge digestion and lower overall removals in a treatment plant are obtained if the resulting supernatant liquors are returned to the primary or secondary portions of the treatment plant.

Bacterial nitrification–denitrification is perhaps the most generally promising biological method for N removal because of its moderate cost, reasonable reliability, high potential removal efficiency and economical area requirements. With certain waters, such as groundwaters and agricultural drainage waters, the major portion of the N is in the NO_3^- form so that the nitrification stage is not required. The requirements for the nitrification and denitrification stages are quite different and are discussed separately below.

2. Biological Nitrification

Nitrification is achieved in 2 steps by different species of autotrophic nitrifying bacteria.

Step 1:

$$NH_4^+ + \tfrac{3}{2}\, O_2 \xrightarrow{\text{\textit{Nitrosomonas} spp.}} NO_2^- + 2H^+ + H_2O \qquad (1)$$

Table 1

Nitrogen concentration and removals during treatment of municiple wastewaters in U.S.A.

		Content in			
		Effluent primary Settling natural		Effluent from secondary biological treatment†	
Nitrogen type	Raw waste (mg/l)	mg/l	% removal	mg/l	% removal
Organic total	10–25	7–20	10–40	3–6	50–80
dissolved	4–15	4–15	0	1–3	50–80
suspended	4–15	2–9	40–70	1–5	50–80
Ammonia	10–30	10–30	0	10–30	0
Nitrite	0–0·1	0–0·1	0	0–0·1	0
Nitrate	0–0·5	0–0·5	0	0–0·5	0
Total:	20–50	20–40	5–25	15–40	25–55

* Data from Johnson, 1958; Heukelekian & Balmat, 1959; Wuhrmann, 1964; Barth, Mulbarger, Salotto & Ettinger, 1966; Barth, Brenner & Lewis, 1968.
 † No nitrification.

Step 2:

$$NO_2^- + \tfrac{1}{2} O_2 \xrightarrow{\text{\textit{Nitrobacter} spp.}} NO_3^- \qquad (2)$$

Overall

$$NH_4^+ + 2O_2 \rightarrow NO_3^- + 2H^+ + H_2O. \tag{3}$$

Both *Nitrosomonas and Nitrobacter* are obligate autotrophs and cannot oxidize substrates other than NH_4^+ or NO_3^- (Lees, 1952; Lozinov & Ermachenko, 1957; Delwiche & Finstein, 1965).

Numerous experiments have been made to determine whether other organisms might also possess the ability to nitrify (Quastel & Scholefield, 1949; Schmidt, 1954; Fisher, Fisher & Appleman, 1956; Eylar & Schmidt, 1959 Doxtoder & Roviera, 1968). A few species of fungi produced NO_2 and NO_3 in limited amounts. However, in comparison with the nitrifying ability of the autotrophic bacteria, nitrification was not an important energy yielding mechanism for these heterotrophic organisms.

The above reactions furnish energy for the growth of the nitrifying bacteria, during which some of the N is assimilated into bacterial protoplasm, CO_2 being used as a source of cell carbon. With a cell formulation empirically of $C_5H_7O_2N$ the assimilation reaction can be written as

$$5CO_2 + NH_4^+ + 2H_2O \rightarrow C_5H_7O_2N + 5O_2 + H^+ \tag{4}$$

Equations describing the overall reactions of nitrification and assimilation (*Notes on Water Pollution,* 1971) can be approximated as:

by *Nitrosomonas* spp.

$$55NH_4^+ + 5CO_2 + 76O_2 \rightarrow C_5H_7O_2N + 54NO_2^- + 52H_2O + 109H^+ \tag{5}$$

by *Nitrobacter* spp.

$$400NO_2^- + 5CO_2 + NH_4^+ + 195O_2 + 2H_2O \rightarrow C_5H_7O_2N + 400NO_3^- + H^+. \tag{6}$$

On this basis, the nitrification of 20 mg/l of ammonium nitrogen (NH_3-N) would involve only c.2 mg/l of Nitrosomonas and c.1/1 mg/l of Nitrobacter, while consuming 83 mg/l of dissolved O_2 and producing 2.2 moles of H for each mole of NH_3-N oxidized. The cells would contain c.2% of the original NH_3-N.

The H^+ produced during nitrification is neutralized by the HCO_3^- in the water (assuming the pH to be <8.5) according to the equation:

$$H^+ + HCO_3^- \rightarrow CO_2 + H_2O. \tag{7}$$

This neutralization results in a decrease in the HCO_3 alkalinity and an increase in the CO_2 concentration, both of which tend to lower the pH value. Approximately 7·13 mg of HCO_3^- alkalinity as $CaCO_3$ are required to neutralize the H^+ produced during oxidation of 1 mg of NH_3-N. Calculations involving carbonic acid equilibria show that if all CO_2 produced during neutralization remains in solution, the amount of NH_3-N which can be oxidized, while maintaining the final $pH > 6\cdot 0$, $\simeq 1/10$ the alkalinity expressed as $CaCO_3$. For a water with an alkalinity of 200 mg/l as $CaCO_3$, $c.$ 20 mg of NH_3-N could be oxidized before the pH fell to $<6\cdot 0$ if all produced CO_2 stayed in solution. Fortunately, in most nitrifying reactors the CO_2 is stripped from solution as it is produced which tends to maintain a more neutral pH level.

The nitrifying bacteria they can maintain active metabolism over a wide range of pH value. Maximum rates of reaction for both *Nitrosomonas* and *Nitrobacter* are reported to occur over the pH range 7–9 (Winogradsky & Winogradsky, 1933; Hofman & Lees, 1953; Boon & Laudelout, 1962; Engel & Alexander, 1958). While reduced rates are generally observed at lower pH levels, several investigators (Frederick, 1956; McCarty & Broderson, 1962; Morrill & Dawson, 1967; Brar & Giddens, 1968) have observed slow nitrification in soils and in extended aeration activated sludge plants at pH values as low as 4·5–5·0. At the Stanford University Water Quality Laboratory an acclimatized nitrifying culture was able to maintain a maximum rate of nitrification at a pH value as low as 6, with greatly reduced activity occurring at a pH of 5·5.

(a) *Nitrification in the activated sludge process*

Nitrification may take place during the normal aerobic treatment of organic wastes in the activated sludge process (Fig. 2). To assure stable nitrification a significant population of nitrifying bacteria must be maintained in the activated sludge.

Downing and his coworkers (Downing, Painter & Knowles, 1964; Downing & Hopwood, 1964) and Wuhrmann (1964, 1968) pioneered efforts to determine the factors affecting nitrification in the activated sludge process. Downing, Painter & Knowles (1964) showed that there is a minimum period of aeration, t_m, which must be exceeded to maintain the nitrifying culture. On any single passage through the activated sludge unit the proportionate increase of the nitrifying population must at least equal that of the heterotrophic population, otherwise the nitrifying organisms will eventually be lost from the system. A sufficiently long period of aeration assures that the organic oxidizers do not outproduce on a percentage basis the slower growing nitrifying bacteria. These investigators have shown that t_m increases with decreasing temperature and concentration of activated sludge and with increasing concentration of organic matter in the waste. A required t_m for nitrification of $c.9$ h was calculated for a sewage with a 5 days BOD of 250 mg/l, an activated sludge concentration of 6000–7000 mg/l and a temperature of 7°. The value of t_m increased to $c.20$ h if the activated sludge suspended solids concentration was only 3000 mg/l.

Another parameter which can be used successfully to control nitrification is the solids retention time, θ_c (frequently called the sludge age), which is defined (Lawrence & McCarty, 1970) as:

$$\theta_c = \frac{\text{wt of suspended solids in treatment system}}{\text{wt of suspended solids removed/day}}.$$

The solids retention time is a measure of the average retention time of the cells in the system. If the solids retention time is maintained below the maximum generation time of the organisms, the bacteria are washed from the system and the process fails. The minimum solids retention time required to maintain the nitrifying population is closely related to t_m (Downing *et al.*, 1964). Either parameter can be successfully used to control nitrification.

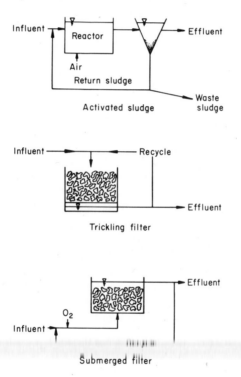

Fig. 2. Possible processes for bacterial nitrification.

The relationship between θ_c and NH_3-N oxidation by nitrification is illustrated in Fig. 3 from data by Wuhrmann (1968). Values of $\theta_c > 2-4$ days were required at temperatures of $c.20°$ for efficient nitrification to occur. Wuhrmann reported that a θ_c value of > 4 days was necessary to consistently attain a high degree of nitrification. This corresponds with a value of 4–5 days reported by Johnson & Schroepfer (1964). These values are just greater than the minimum generation times for the nitrifying bacteria. At reduced temperatures, the generation time is increased and a longer θ_c is required.

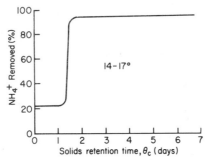

Fig. 3. Effect of solids retention time on removal of ammonia nitrogen by nitrification (after Wuhrmann, 1964).

In the design of a biological system it is advisable to apply a safety factor to ensure maximum efficiency and reliability. A design θ_c of 10 days for nitrification was recommended by Jenkins & Garrison (1968) and should give sufficient safeguard for most purposes: this should allow nitrification at 15°. It may not be sufficient however, at temperatures of 10° or lower, which perhaps occurs in wastewaters only during the coldest winter months.

To affect greater process stability, Barth, Brenner & Lewis (1968) recommended that nitrification be conducted in a separate aeration tank following that used for organic removal. This ensures greater reliability as each unit can be operated under more optimal conditions. Also, several potential inhibitors to the nitrification reaction are removed from the waste during the organic removal stage. In their pilot plant studies over a temperature range 12–22°, an average of 87% conversion of reduced N compounds to NO_3^- was obtained with a detention time of 3 h, a mixed liquor suspended solid concentration of 1195 mg/l and a θ_c value of 22 days. Nitrification was carried out efficiently and reliably during a study period of 6 months.

In addition to an adequate solids retention time and minimum period of aeration the dissolved O_2 concentration in the aeration tank must be kept $> c$. 2 mg/l (Garrett, 1961; Downing & Hopwood, 1964; Water Pollution Research Laboratory, 1964) and biological inhibitors must be absent (Downing, Tomlinson & Truesdale, 1964).

(b) *Nitrification with the trickling filter*

Some nitrification may take place during the treatment of organic wastes by the trickling filter process (Fig. 2). Organisms attach themselves to the surfaces of the stones and utilize the substrates in the waste flowing over the biological film. If organic oxidation and nitrification are conducted in the same filter, organic oxidation is usually observed in the top of the filter and nitrification near the bottom (Heukelekian, 1942; Wuhrmann, 1964). Other investigators (Sorrels & Zeller, 1956, 1963) have observed that in dual filter arrangements most of the carbonaceous O_2 demand is exerted in the first filter while nitrification becomes significant in the second. Process variables for the trickling filter include the depth and size of filter, type of media, hydraulic loading, liquid temperature, strength of organic material and the presence of inhibitors.

Because of the attached growth, the trickling filter is efficient in retaining biological solids and can maintain a long θ_c. However, nitrification is usually only observed at warm temperatures and at low hydraulic loading rates of < 5 m³ of waste/m³ of filter volume/day (Heukelekian, 1945; Grantham et al., 1950; Sorrels & Zeller, 1956, 1963; Wuhrmann, 1964). Heukelekian (1945) concluded that the lack of nitrification in a trickling filter was usually caused by the limited contact time between the waste and the biological film, nitrifying bacteria being usually firmly established in the film. Because increased hydraulic loading leads to increased sloughing of solids it is also possible that θ_c is less than that necessary to establish nitrification at high organic loadings. The biological solids production from organic oxidation is considerably greater than the solids production from nitrification. Therefore, there should be a greater rate of solids sloughing during organic oxidation than during NH_4^+ oxidation in the absence of organic material. The increased sloughing may mean that the solids retention time is not sufficient to maintain an adequate nitrifying population in those sections of a filter where organic oxidation is occurring. Nitrification would then be observed in the bottom of a trickling filter or in the second filter of a dual filter system where organic oxidation is less pronounced.

The nitrifying efficiency of a trickling filter can be increased by increasing the depth of the filter, by decreasing the hydraulic loading and by prior removal of the carbonaceous oxygen demand. For example, Balakrishnan & Eckenfelder (1969) indicated that for a trickling filter using a synthetic medium with a specific surface area of 70 ft²/ft³ and a loading of 20×10^6 gal/acre/day, 80% nitrification of a secondary effluent could be obtained at 30° in a filter 10 ft deep, at 20° in a filter 25 ft deep and at 15° in one 40 ft deep.

In comparison with the activated sludge system, the trickling filter is not as nitrification is usually only observed at high concentration of nitrification. By comparison the activated sludge system is the more desirable for nitrification . (Barth, Brenner & Lewis, 1968).

(c) Nitrification with the submerged filter

The submerged filter (Fig. 2) is a new reactor designed for nitrifying secondary effluents and is currently under investigation at the Stanford University Water Quality Laboratory. It is patterned after the anaerobic filter developed by Young & McCarty (1969) and provides an upward flow of fluid through a fixed medium bed to which the nitrifying bacteria are attached. The fixed film growth provides a long solids retention time of hundreds of days, significantly longer than that obtained with the activated sludge process. The submerged flow operation allows direct control of the hydraulic detention time which is not possible with the trickling filter. These advantages enable the achievement of efficient, stable nitrification at temperatures as low as $1°$.

Table 2

Summary of the submerged filter performance

Temperature ($°C$)	Detention time (min)*	Influent NH_3-N (mg/l)	Recycle flow Waste flow	Amount % oxidized
Single passage operation				
25	7·5	8		70
	30	8		95
15	15	10		87
	30	10		91
10	15	10		71
	30	10		88
5	60	12		75
	120	12		92
1	120	14		86
Recycle operation				
25	30	20	1·7	90
	60	20	1·6	95
	60	20	3.0	95
16	60	30	1·1	91

*Based on raw waste flow.

From equation (3) $c.4·5$ mg of O_2 are required to oxidize 1 mg of NH_3-N. Several methods are available to supply this large O_2 demand. For a single passage through the submerged filter, oxygenation to saturation at a partial pressure of 1 atm can provide sufficient dissolved O_2 to oxidize $8-12$ mg/l of NH_3-N, depending on the temperature. A recycle system can then be used to oxidize higher NH_3 concentrations. Oxygenation at partial pressures > 1 atm can supply all the dissolved O_2 required for nitrification with a single passage through the system. This method, however, requires a pressurized reactor. Air or O_2 can also be directly bubbled through the filter. The rising bubbles create some turbulence and tend to decrease the solids capture efficiency of the filter.

The retention time, based on the filter void volume, and the waste temperature are the main parameters affecting filter performance. A summary of submerged filter performance with oxygenation at 1 atm partial pressure using single passage and recycle systems is given in Table 2. Efficiency of nitrification of $c.90\%$ can be obtained with a detention time of $c.15$ min at $25°$ and can be maintained down to a temperature just above freezing by increasing the detention time to $c.2$ h.

The H^+ produced during the oxidation of NH_4^+ to NO_2^- causes the pH value to fall as the waste passes through the submerged filter. The CO_2 produced during the neutralization of the H^+ is not stripped from solution so that the maximum pH fall, as previously discussed, is realized. However, studies have shown that biological adaptation was possible to the extent that pH values as low as 6 were not inhibitory: below pH 5·5 adaptation was not so complete.

A recycle system was successfully used for oxidation of higher NH_3-N concentrations in the typical range 20–30 mg/l of NH_3-N. Ninety % oxidation of 20 mg/l of NH_3-N was obtained with a detention time based on raw waste flow of 30 min at $25°$ and 60 min at $16°$ (Table 2). In general, over the temperature range $5–25°$, 90% nitrification was readily obtained in the submerged filter with detention times of 30–120 min. This same general performance was observed when O_2 was supplied by direct bubbling through the filter.

Efficiencies $> 90–95\%$ require greatly extended detention times and are probably not practical to achieve with the filter. This suggests that for required efficiencies $> 90\%$, a more economical system may be nitrification with the submerged filter, followed by break point chlorination to remove the remaining NH_3-N residuals.

Nitrification in the submerged filter has proven to be a stable process. The upflow system selects organisms which either attach to a surface or form a flocculant and well settling mass. Under these conditions suspended material was almost never observed in the effluent. The system responded well to changing conditions. The detention time and the temperature could be rapidly changed without causing interruption of service.

3. Biological Denitrification

In the U.S.A. an increasing degree of emphasis is being placed on complete removal of N forms from wastewater effluents to protect lakes, rivers and reservoirs from excessive growths of aquatic plants. The most promising general method of accomplishing this to date is through biological denitrification of previously nitrified effluents.

Denitrification is a microbial process by which NO_3^- and NO_2^- are reduced to molecular N (Alexander, 1961). Nitrous oxide is sometimes formed during denitrification, but this does not appear as a significant intermediate during denitrification in aqueous solutions (Johnson & Schoepfer, 1964). The molecular N endproduct is an inert gas which, unlike other compounds of N, is relatively unavailable for biological growth. Thus, denitrification converts N from objectionable forms to a nonobjectionable one. Denitrification is a respiratory mechanism in which NO_3^- replaces molecular O_2. This is in contrast to assimulation in which the NO_3' is reduced to NH_4-N and is used for the synthesis of cell protein.

The ability to bring about denitrification is a characteristic of a wide variety of common facultative bacteria including the genera *Pseudomonas, Achromobacter* and *Bacillus.* In the absence of molecular O_2, these organisms use nitrates or nitrites as terminal electron acceptors while oxidizing organic matter for energy. There are also certain autotrophic bacteria that can bring about denitrification while oxidizing an inorganic energy source.

The denitrifying abilities of micro-organisms differ (Alexander, 1961). Some can reduce NO_3^- to NO_2^- only, some can reduce NO_2^- to molecular N only, and some can bring about the reduction of both NO_3^- and NO_2^- to molecular N. With a naturally occurring heterogeneous population, species capable of carrying out denitrification to different degrees will undoubtedly be present and nitrites may or may not appear as an intermediate. However, it is convenient to consider denitrification as a 2 step process, the first representing reduction of nitrate to nitrite and the second a reduction of NO_2^- to N gas.

The organic matter in municipal wastewaters or the microbial cells formed during waste treatment have been used as electron donors in many studies on denitrification (Johnson & Schroepfer, 1964; Wuhrmann, 1968; Johnson & Vania, 1971). However, the amount and nature of this organic matter is not readily controllable and this has led to some degree of unreliability and inefficiency in the process (Barth, Brenner & Lewis, 1968; Johnson & Vania, 1971). In addition, many industrial and agricultural wastewaters do not have suitable electron donors present. The addition of organic material under carefully controlled conditions to serve as an electron donor is one way to overcome these limitations.

Various organic materials, such as sugar, acetic acid, ethanol, acetone and methanol, have been found satisfactory for use as electron donors, but methanol was found to be the least expensive (McCarty, Beck & St. Amant, 1969). Considering denitrification as 2 step process the following reactions can be written using methanol as an electron donor:

Step 1:
$$NO_3^- + 1/3 \ CH_3OH \rightarrow NO_2^- + 1/3 \ CO_2 + 2/3 \ H_2O \qquad (8)$$
Step 2:
$$NO_2^- + 1/2 \ CH_3OH \rightarrow 1/2 \ N_2 + 1/2 \ CO_2 + 1/2 \ H_2O + OH^- \qquad (9)$$

Overall:

$$NO_3^- + 5/6\ CH_3OH \rightarrow 1/2\ N_2 + 5/6\ CO_2 + 7/6\ H_2O + OH^- \qquad (10)$$

Thus, 5/6 mole of methanol are required for the denitrification reaction alone to reduce one mole of NO_3^- completely to molecular N. If only 1/3 moles of methanol were added, it is possible that the NO_3^- would only be reduced to NO_2^- so that no effective N removal would result. This consideration indicates that a treatment process should be designed for nearly complete denitrification of the portion to be treated, rather than for partial treatment as the latter would be unpredictable and probably wasteful of chemicals.

Electron donor must be added not only to satisfy the denitrification reactions of equations (9)–(11), but also an additional amount must be added for bacterial growth. In order to evaluate this additional need, the consumptive ratio has been determined, and is defined as the ratio of the total quantity of an organic chemical consumed during denitrification to the stoichiometric requirement for denitrification and deoxygenation alone.

When methanol was used as the electron donor, a consumptive ratio of 1:3 was found (McCarty, Beck & St. Amant, 1969). Methanol can also be added to biologically lower the dissolved oxygen level as is necessary for denitrification. From these considerations, the following formulae were developed to estimate methanol requirements for denitrification and to estimate the resulting biomass production:

methanol requirement:

$$C_m = 2{\cdot}47\,N_o + 1{\cdot}53\,N_1 + 0{\cdot}87\,D_0 \qquad (11)$$

biomass production:

$$C_b = 0.53\,N_o + 0{\cdot}32\,N_1 + 0{\cdot}19\,D_o \qquad (12)$$

where C_m = required methanol concentration (mg/l), C_b = biomass production (mg/l), N_o = initial NO_3-N concentration (mg/l), N_1 = initial NO_2-N concentration (mg/l); D_o = initial dissolved O_2 concentration (mg/l).

If O_2 enters the denitrifying unit during treatment, the value for D_o should be increased to account for this effect. This is likely to be of some significance in an uncovered pond or in an activated sludge-type process.

The biomass formed during denitrification has an empirical formula similar to that for other bacteria as given in equation (4). From this consideration and from equations (11) and (12) it can be calculated that the reduction of 30 mg/l of NO_2-N (assuming D_o and N_1 are zero) requires 74 mg/l of methanol and results in the production of 16 mg/l of bacteria containing 2 mg/l of assimilated N. Thus, a little over 6% of the inorganic N would be assimilated into cell protein during denitrification.

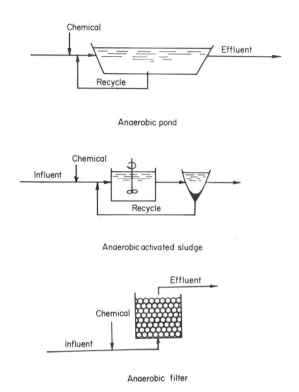

Fig. 4. Possible processes for bacterial denitrification.

Figure 4 illustrates 3 processes which may be used to carry out denitrification: anaerobic ponds, anaerobic activated sludge and an anaerobic submerged filter. Both the anaerobic pond and the anaerobic filter have been used for denitrification of agricultural wastewaters during a 3-years field scale study at Firebaugh, Calif. (Tamblyn & Sword, 1969). These studies were conducted to determine the feasibility of various processes for removal of an estimated 20 mg/l of NO_3-N from 500 mg of future agricultural wastewaters to be collected in subsurface drainage fields under $c.\,1.7 \times 10^6$ acres of irrigated land in the San Joaquin Valley. The removal of the drainage water is necessary to carry away excess salts from the soil. The salinity of the water will vary from 6800 mg/l with present flows to 2500 mg/l for flows in the year 2020. Nitrogen removal is considered necessary since the nutrient-rich drainage water will be collected and transported along the proposed 250 miles San Luis Drain and discharged into San Francisco Bay. These investigations have been conducted by an interagency group composed of the California Department of Water Resources, the U.S. Environmental Protection Agency and the U.S. Bureau of Reclamation.

(a) *Denitrification in the anaerobic pond*

In the Firebaugh studies an uncovered pond, 60 × 60 × 15ft, and a covered pond, 60 × 200 × 15 ft, were evaluated (St. Amant, Beck & Tamblyn, unpublished). Recycle from the pond bottom at a flow rate of 25–50% of the influent flow rate was used to provide a bacterial seed to the incoming water. The uncovered pond proved unsatisfactory for high efficiencies of denitrification as resulting algal growth produced unwanted O_2 which increased methanol requirements and hindered the denitrification reaction. A pond covered with large styrofoam blocks, however, performed efficiently, removing up to 90% of the 20 mg/l of NO_3-N at a detention time of 10 days and a temperature of 20°. As temperatures approached 10°, the need for a detention time of up to 20 days was indicated. In the pond system, no provision was made for removal of biomass synthesized during denitrification. One–2 mg/l of organic plus NH_3-N resulted from this synthesis and was normally presented in the effluent along with 10–20 mg/l of microbial solids.

(b) *Denitrification in the anaerobic filter*

Several pilot plant denitrification filters similar in construction and operation to the submerged filter for nitrification were operated at the Firebaugh site for over a year (Tamblyn & Sword, 1969). A 10 × 10 × 6 ft filter and several 18 in and 36 in diam filters were evaluated. Media investigated included sand, activated carbon, volcanic cinders, gravel, coal and a commercially produced plastic trickling filter media. Various media sizes from coarse sand to 3 in diam gravels were used. The detention times studied ranged from 0·5 to 4 h, representing flow rates from c.0·6–0·15 g/ft^2.

Of the media evaluated, the most satisfactory was the 1 in diam gravel. A filter with this medium was in operation for >1 year at a detention time of 1 h and gave an average removal of 90% of the 20 mg/l of influent NO_3-N. This efficiency was maintained even though the influent water reached a low temperature of 10°. The head loss through the filter gradually increased with use and after 2·1 year of operation reached values as high as 8 lb/in². Investigations for methods of correcting this gradual filter plugging were conducted. Periodic flushing to remove excess accumulations of microbial solids seemed to be indicated.

Parkhurst, Dryden, McDermott & English (1967) have reported on the use of an anaerobic filter containing activated carbon as the filter medium and methanol as an electron donor. Their filter was operated with downflow as with a normal high-rate filter for water treatment. They backflushed the filter daily to prevent clogging.

(c) *Denitrification in the anaerobic activated sludge system*

Anaerobic activated sludge systems have been used both in the laboratory and in the field for denitrification of municipal waste treatment plant effluents. Wuhrmann (1964), Ludzack & Ettinger (1962), and Johnson & Schroepfer (1964) used the normal organic materials present in municipal wastewaters as electron donors for denitrification and suggested various schemes and flowsheets for this process. In all cases, high efficiencies of nitrogen removal were prevented as NH_3-N was liberated during the oxidation of the electron donors.

Christianson, Rex, Webster & Virgil (1956) added both sugar and methanol in an activated sludge-type system and obtained efficient removal of NO_3 at concentrations of 250–1100 mg/l. Barth, Brenner & Lewis (1968) used methanol for the denitrification of municipal wastewaters in pilot plant activated sludge systems. The methanol requirements were somewhat higher than that given by equation (11), perhaps due to the introduction of oxygen during mixing of the open tanks. They used a detention time of 3 h, a mixed liquor suspended solids concentration of 2010 mg/l and a solids retention time of 38 days. In a 1 gal/min pilot plant, Johnson & Vania (1971) obtained an effluent with < 1 mg/l of NO_3-N using a mixed liquor suspended solids concentration of 1715 mg/l and a detention time of 1·3 h at $20°$ or 2·6 h at $10°$.

The solids retention times in activated sludge systems or in ponds should probably be about the same and at least 10 days. This gives a factor of safety over the minimum retention time for a high efficiency of denitrification of $c.$ 4 days as found by Moore & Schroeder (unpublished).

4. Nitrification – Denitrification Processes

Suggested flowsheets for the combined nitrification-denitrification system to be used in conjunction with municipal wastewater treatment for nitrogen removal are illustrated in Fig. 5. In the first system (Ludzack & Ettinger, 1962;

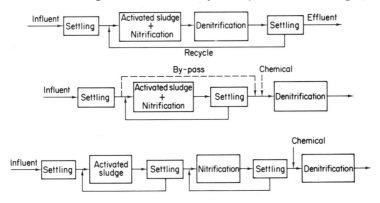

Fig. 5. Flow sheet for various nitrification–denitrification biological systems.

Wuhrmann, 1964; Johnson & Schroepfer, 1964) the nitrification occurs in the same tank as organic removal and the mixed liquor then flows into a separate non-aerated tank where denitrification takes place. The mixed liquor volatile suspended solids are oxidized while serving as the electron donor for the denitrification reactions. While this system has been operated successfully, it requires close control and knowledgeable operation. Some ammonia is released during oxidation of the suspended solids in the denitrification tank so that removal efficiencies no higher than 50–70% are obtained.

In the second system, organic oxidation and nitrification are separated from the denitrification step (Johnson & Vania, 1971). Either raw wastewater or an organic such as methanol is used as an electron donor in the denitrification tank. When wastewater is used for this purpose, efficiency of nitrogen removal is usually no more than 40–60% because of the high concentration of NH_3 and organic N carried over in the wastewater. Addition of methanol, however, is much more easily controlled and does not add N to the denitrification tank so that N removal efficiencies as high as 80–90% can be obtained.

In the last system, the 3 processes of organic removal, nitrification and denitrification are separated to give better control. While the tank volumes here may be somewhat greater than for either of the first 2 systems, the increased dependability and high removal efficiencies obtained are distinct advantages (Barth, Brenner & Lewis, 1968).

For these systems there is a choice as to the type of nitrification and denitrification process employed. For a particular situation the decision usually depends on the required efficiency of N removal, the cost of each process and the reliability desired in the system. Each of the processes has its advantages and disadvantages. Experience with full scale operation in some cases is limited and accurate cost data are not yet available. A reliable determination of the optimum system for N oxidation or removal will require more time and experience with full scale operation.

Acknowledgement

This study was supported in part by Research Fellowship No. 1-F1-WP-26, 451-01 and Research Grant No. WP-17010 from the United States Environmental Protection Agency.

6. References

Alexander, M. (1961). *Soil Microbiology*. New York: Wiley.

Balakrishnan, S. & Eckenfelder, W. W. (1969). Nitrogen relationships in biological treatment processes – II. Nitrification in trickling filters. *Wat. Res.* **3**, 167.

Barth, E. F., Brenner, R. C. & Lewis, R. F. (1968). Chemical-biological control of nitrogen and phosphorus in wastewater effluents. *J. Wat. Pollut. Control Fed.* **40**, 2040.

Barth, E. F., Mulbarger, M., Salotto, B. V. & Ettinger, M. B. (1966). Removal of nitrogen by municipal wastewater treatment plants. *J. Wat. Pollut. Control Fed.* **38**, 1208.

Boon, B. & Laudelout, H. (1962). Kinetics of nitrite oxidation by *Nitrobacter winogradskyki. Biochem. J.* **85**, 440.

Brar, S.S. & Giddens, J. (1968). Inhibition of nitrification in Bladen grassland soil. *Soil Sci. Soc. Am.* **32**, 821.

Christianson, C. W., Rex, E. H., Webster, W. M. & Virgil, F. A. (1956). Reduction of nitrate nitrogen by modified activated sludge. U.S. Atomic Energy Commission, TID-7517 (part ia), p.264.

Delwiche, C. C. & Finstein, M. S. (1965). Carbon and energy sources for the nitrifying autotroph *Nitrobacter. J. Bact.* **90**, 102.

Downing, A. L. & Hopwood, A. P. (1964). Some observations on the kinetics of nitrifying activated-sludge plants. *Schweiz. Z. Hydrol.* **26**, 271.

Downing, A. L., Painter, H. A. & Knowles, G. (1964). Nitrification in the activated-sludge process. *J. Proc. Inst. Sewage Purif.* **3**, 130.

Downing, A. L., Tomlinson, T. G. & Truesdale, G. A. (1964). Effect of inhibitors on nitrification in the activated-sludge process. *J. Proc. Inst. Sewage Purif.* **3**, 537.

Doxtoder, K. G., Rovira, A. D. (1968). Nitrification by *Aspergillus flavus* in sterilized soil. *Aust. J. Soil Res.* **6**, 141.

Engel, M. S. & Alexander, M. (1958). Growth and autotrophic metabolism of *Nitrosomonas europaea. J. Bact.* **76**, 217.

Eylar, O. R. & Schmidt, E. L. (1959). A survey of heterotrophic micro-organisms from soil for ability to form nitrite and nitrate. *J. gen. Microbiol.* **20**, 473.

Fisher, T., Fisher, E. & Appleman, (1956). Nitrite production by heterotrophic bacteria. *J. gen. Microbiol.* **14**, 238.

Frederick, L. R. (1956). The formation of nitrate from ammonium nitrogen in soils. I. Effects of temperature. *Soil Sci. Soc. Am.* **20**, 496.

Garrett, M. T. (1961). Significance of growth rate in the control and operation of bio-oxidation treatment plants. *Ind. Wat. & Wastes Conf., Rice University, Houston, Texas.*

Grantham, G. R., Phelps, E. B., Calaway, W. T. & Emerson, D. L. (1950). Progress report on trickling filter studies. *Sewage Wks J.* **22** (7), 867.

Heukelekian, J. (1942). The influence of nitrifying flora, oxygen and ammonia supply on the nitrification of sewage. *Sewage Wks J.* **14**, 964.

Heukelekian, H. (1945). The relationship between accumulation biochemical and biological characteristics of film and purification capacity of a biofilter and a standard filter. III. Nitrification and nitrifying capacity of the film. *Sewage Wks J.* **17**, 516.

Heukelekian, H. & Balmat, J. L. (1959). Chemical composition of the particulate fractions of domestic sewage. *Sewage ind. Wastes* **31**, 413.

Hofman, T. & Lees, H. (1953). The biochemistry of nitrifying organisms. 4. The respiration and intermediary metabolism of *Nitrosomonas. Biochem. J.* **54**, 579.

Jenkins, D. & Garrison, W. E. (1968). Control of activated sludge by mean cell residence time. *J. Wat. Pollut. Control Fed.* **40**, 1905.

Johnson, W. K. (1958). Nutrient removals by conventional treatment processes. *Proc. 13th Indust. Waste Conf.* Purdue Univ. Ext. ser. 96, p. 151.

Johnson, W. K. & Schroepfer, G. J. (1964). Nitrogen removal by nitrification and denitrification. *J. Wat. Pollut. Control Fed.* **36**, 1015.

Johnson, W. K. & Vania, G. B. (1971). Nitrification and denitrification of wastewater. *Sanit. Engng Rep. No. 175 S.* University of Minnesota.

Lawrence, A. W. & McCarthy, P. L. (1970). Unified basis for biological treatment design and operation. *J. Am. Soc. civil Engrs, San. engng Div.* SA3, 757.

Lees, H. (1952). The biochemistry of the nitrifying organisms. 1. The ammonia-oxidizing system of *Nitrosomonas. Biochem. J.* **52**, 134.

Lozinov, A. B. & Ermachenco, V. A. (1957). Accumulation of organic matter by *Nitrosomonas europoea* cultures grown on Winogradsky medium. *Microbiologiya (Transl.)* **26**(2), 166.

Ludzack, F. J. & Ettinger, M. B. (1962). Controlling operation to minimize activated sludge effluent nitrogen. *J. Wat. Pollut. Control Fed.* **24**, 920.
McCarty, P. L., Beck, L. & Amant, P. St. (1969). Biological denitrification of wastewaters by addition of organic materials. *Proc. 24th Purdue Ind. Waste Conf. Lafayette, Ind.* p. 1261.
McCarty, P. L. & Broderson, C. F. (1962). Theory of extended aeration activated sludge. *J. Wat. Pollut. Control Fed.* **34**, 1095.
Morrill, L. & Dawson, J. E. (1967). Patterns observed for oxidation of ammonium to nitrate by soil organisms. *Soil Sci. Soc. Am.* **31**, 757.
Notes on Water Pollution. (1971). Nitrification in the BOD test. Water Pollution Research Laboratory, Stevenage, England.
Parkshurst, J. D., Dryden, F. D., McDermott, G. N. & English, J. (1967). Pomona activated carbon pilot plant. *J. Wat. Pollut. Control Fed.* **39**, R70.
Quastel, J. & Scholefield, P. (1949). Influence of organic nitrogen compounds on nitrification in soil. *Nature, Lond.* **164**, 1068.
Schmidt, E. L. (1954). Nitrate formation by a soil fungus. *Science, N.Y.* **119**, 187.
Sorrels, J. H. & Zeller, P. J. A. (1956). Two-stage trickling filter performance. *Sewage Wks J.* **28** (8), 943.
Sorrels, J. H. & Zeller, P. J. A. (1963). Supernatant on trickling filters. *J. Wat. Pollut. Control Fed.* **35**, 11, 1419.
Tamblyn, T. A. & Sword, B. R. (1969). The anaerobic filter for the denitrification of agricultural subsurface drainage. *Proc. 24th Purdue Ind. Waste Conf. Lafayette, Ind.* p. 1135.
Water Pollution Research Laboratory (1964). Effect of dissolved oxygen on nitrification. London: H.M.S.O.
Winogradsky, S. & Winogradsky, H. (1933). Etudes sur la microbiologie du sol. Nouvelles recherches sur les organisms de la nitrification. *Ann. inst. Pasteur, Lille* **50**, 350.
Wuhrmann, K. (1964). Nitrogen removal in sewage treatment processes. *Verh. Internat. Verein. Limnol.* **XV**, 580.
Wuhrmann, K. (1968). Objectives, technology and results of nitrogen and phosphorus removal processes. In *Advances in Water Quality Improvement.* (Eds E. F. Gloyana, W. W. Eckenfelder). Austin: University of Texas Press.
Young, J. C. & McCarty, P. L. (1969). The anaerobic filter for waste treatment. *J. Wat. Pollut. Control Fed.* **41**, R160.

Degradation of Herbicides by Soil Micro-organisms

S. J. L. WRIGHT

*School of Biological Sciences,
University of Bath,
Bath, Somerset, England*

CONTENTS

1. Introduction

MORE THAN 600 weed species are known to compete with crops (Shaw, 1965); weed control therefore becomes an important factor in efficient and intensive agriculture.

Chemical weed killers, herbicides, may be applied to the soil or directly to the foliage according to type. Either way the soil and soil drainage water become the ultimate recipients of at least part of all herbicide applications and so are polluted to some extent. The potential scale of such pollution can be gauged by considering the vast quantities of these chemicals which are applied globally each year. It is estimated that the foliage and soils of the U.S.A. alone are doused with 1×10^9 lb of synthetic organic pesticides yearly, there being $> 60,000$ registered formulations containing > 800 active substances (Bartha & Pramer, 1970): many of them are herbicides.

In Britain the number of Ministry-approved herbicide products increased from 241 in 1964 to 386 in 1969, including 24 new active chemicals. In the same period sales of herbicides at home and abroad by British manufacturers rose from £7·8 to £22m (Third Report of the Research Committee on Toxic Chemicals, 1970).

Chemical weed control has progressed a long way since the adoption in 1898 of copper sulphate as a means of controlling broadleafed weeds in cereals. A spur was provided by the discovery in 1942 that the salts of the chlorinated phenoxyacetic acids were selectively herbicidal allowing, for the first time, efficiency and safety in use. Many widely used herbicides were developed during the 1940s and these became the forerunners of the present array of selective and nonselective compounds, a chemical arsenal that continually expands as manufacturers battle with each other as well as the weeds. At the same time there has been developmental work on application methods and formulations so that we find the active ingredients may be applied as dusts, powders, pellets or liquids, often with adjuvants and dyes.

It is important that a herbicide remains in the active state long enough to achieve the desired effect. However, where herbicides are persistent or where application rates exceed dissipation, injury is likely to sensitive plants grown in rotation with treated crops and accumulated residues may directly enter food chains. With increasing quantities of herbicides entering the soil each year concern is aroused as to possible effects on soil fertility through deleterious effects on the activities of certain microbial groups. Fletcher (1960) and Bollen (1961) reviewed data on the effects of herbicides on soil microbes.

It is plainly important that the fate of herbicides in soil should be extensively examined so that aspects of soil science, agronomy, chemistry, toxicology, plant physiology, public health and microbiology are covered. The microbiologist can contribute in essentially 2 ways. First, by investigating the effects of herbicides on the activities and interactions in the microbial community; secondly, by examining the effects of soil microbes on herbicides to determine how and to what extent the compounds may be transformed, detoxified or degraded.

This contribution attempts to show the essential role of microbes in the detoxication and degradation of some herbicides and to indicate general principles and methods which have hitherto been adopted in these studies.

2. The Fate and Persistence of Herbicides in the Soil

Once in the soil a herbicide is exposed to a variety of physical, chemical and biological influences which individually or collectively may affect the structural integrity and persistence of the compound. These factors, which have been considered in more detail by Aldrich (1953) and Sheets (1962), are leaching, adsorption by soil particles, decomposition and volatilization (Table 1). It is sometimes difficult to distinguish between chemical and microbial decompo-

Table 1

Factors influencing herbicide persistence in soil

Factor	Determined by notes
Leaching	(i) Water solubility
	(ii) Soil permeability and texture
	(iii) Amount and intensity of rainfall
Adsorption	(i) Physicochemical properties of the herbicide
	(ii) Number and type of adsorbing sites in soil
Decomposition	
(a) Microbial	Temperature, pH, moisture, aeration, nutrient status of soil
(b) Chemical	Oxidation, hydrolysis, complexing
(c) Photochemical	Possibly important in arid sunny areas
Volatilization	Important cause of loss of some surface-applied herbicides, especially if soil is moist at time of application

sition of a herbicide in soil because it is not always possible to claim that soil sterilization merely inactivates microbes without causing any other changes. Even so, studies with sterile soils and model systems have shown that in some cases strictly chemical systems participate in herbicide degradation. Reactions have been elucidated which have underlying similarities from chemical and biochemical standpoints (Kearney & Helling, 1969) and in some cases an understanding of such reactions has allowed prediction of the degradability of compounds and of the likely degradation products. Considerations of the relative importance of chemical and biological degradation have seldom been reported and are outside the scope of the present paper.

Herbicides vary considerably in persistence. Variations can be attributed to application rate and the inherent susceptibility or resistance to microbial attack which is conferred by certain molecular structures and substitutions (Alexander & Aleem, 1961; Alexander, 1965a). The simpler phenoxyalkanoates and phenylcarbamates at field application rates persist for only a few weeks while residues of some complex triazines, ureas and benzoates remain for more than a year after application. Some authors have reported on the persistence of several different herbicides (Hocombe, Holly & Parker, 1966) but comparison of data obtained in several laboratories is generally invalid because of variations in experimental conditions and assay sensitivity.

3. The Experimental Approach

In examining reports on the microbial degradation of herbicides it is often possible to see a fairly well defined sequence of events in the experimentation, ranging from studies on persistence in soil to investigation of degradative routes at the enzymic level. The first step involves comparison of persistence and interpretation of dissipation patterns in natural soil and in soil which has been treated to inactivate microbes, usually by autoclaving, Tyndallization or addition of an inhibitor such as azide. Residual herbicide is assayed frequently to obtain dissipation curves (Fig. 1), the interpretation of which is considered in a later section. Biological assays, using appropriate sensitive plant species, are preferred to chemical assays. By biological assay the important residual component, namely toxicity, is determined. Chemical assays on the other hand do not necessarily indicate changes in phytotoxicity.

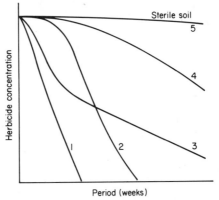

Fig. 1. Generalized herbicide persistence patterns in soil. For interpretation, see text [section 4 (a)].

The main step, the isolation of "active" organisms, is usually attempted after a period of prior enrichment by herbicide addition. Soil percolation techniques (Wright & Clark, 1969) have been used widely for this purpose and some authors favour the use of sludge or faecal material as a source of organisms. An elective process is used, designed to favour organisms which can utilize the herbicide as a source of carbon, nitrogen or energy for growth. Ideally this process permits easy recognition of the active organisms, for example by a specific reaction with the medium (Kaufman & Kearney, 1965; Clark & Wright, 1970a). There is no guarantee that the accepted isolation procedures will be successful. Failure may be attributed to inadequate simulation of the organism's natural environment in the isolation medium, possibly by omission of growth factors, or the degraded compound may provide insufficient energy to sustain growth of the organism. It seems probable that the exacting nature of the final isolation procedure will preclude the demonstration of the possible mixed microbial culture or successional degradation of a herbicide.

The microbiologist usually aims to study herbicide degradation under defined conditions, using pure microbial cultures and pure chemicals rather than commercial formulations. Such conditions are often so far removed from those in the soil environment as to render laboratory studies quite artificial. This is an unfortunate admission for physiological-ecological investigations and indicates that information obtained from *in vitro* experiments cannot necessarily be projected with certainty to the field where the microbial population and environment are infinitely more complex. It is desirable that investigations of herbicide degradation should be extended to include formulated compounds. Our own investigations (Clark & Wright, 1970b) on the herbicide IPC indicated that while petrochemical solvents used in the liquid formulation where somewhat inhibitory to the IPC-degrading bacteria in pure culture, degradation was unaffected to soil where other organisms removed the petrochemicals.

4. Microbial Metabolism of Herbicides

(a) *General considerations*

The catabolic and biochemical versatility of the bacteria, fungi and actino-mycetes in soil serve notice that these are the organisms which are likely to be involved in the alteration or degradation of an organic herbicide. Represen-tatives of the major microbial groups have been found capable of effecting at least certain degradative steps in the decomposition of many herbicides (Alexander, 1965b). Protozoa have not been directly connected with herbicide degradation although they may contribute to the dispersion of such compounds in the soil. However, recent work suggests that microscopic algae are able to detoxify certain herbicides. Kruglov & Paromenskaya (1970) observed consider-able detoxication of the triazine herbicide Simazine in soil inoculated with *Chlorosarcina*, the compound being partially metabolized by the algae. Total detoxication of the benzenesulphonylcarbamate Asulam at 50 p/m was obtained when incubated with a *Chlorella* sp. isolated from Asulam-treated soil (Wright, unpublished).

Alexander (1965b) considered that 5 conditions should be met for a pesticide to be subject to biodegradation: (i) organisms effective in metabolizing the compound must exist in the soil or be capable of developing therein; (ii) the compound must be in a degradable form; (iii) the compound must reach the organisms; (iv) the compound must induce the formation of enzymes concerned in the degradation, few of which are likely to be constitutive; (v) environmental conditions must favour the proliferation of the organisms and operation of the enzymes.

Table 2

Some examples of herbicides and organisms which degrade them

Type of compound	Chemical name (common name)	Structural formula	Active organisms*
Phenoxyacetate	2, 4-dichlorophenoxyacetic acid (2, 4-D)	(2,4-dichlorophenyl)–O–CH$_2$–COOH	Flavobacterium, Corynebacterium, Achromobacter, Pseudomonas, Arthrobacter, Aspergillus
	2-methyl-4-chlorophenoxyacetic acid (MCPA)	(4-chloro-2-methylphenyl)–O–CH$_2$–COOH	Corynebacterium, Achromobacter, Aspergillus.
Phenylcarbamate	isopropyl N-phenylcarbamate (IPC)	$\text{phenyl–NH–C(=O)–O–CH(CH}_3\text{)}_2$	Arthrobacter, Achromobacter, Pseudomonas, Flavobacterium, Agrobacterium.
	isopropyl N-3-chlorophenylcarbamate (CIPC)	$\text{(3-chlorophenyl)–NH–C(=O)–O–CH(CH}_3\text{)}_2$	As above
	4-chloro-2-butynyl-N-3-chlorophenyl carbamate (Barban)	$\text{(3-chlorophenyl)–NH–C(=O)–O–CH}_2\text{–C≡C–CH}_2\text{Cl}$	Penicillium
Phenylurea	N (4-chlorophenyl)-1-1dimethyl urea (Monuron)	$\text{(4-chlorophenyl)–NH–C(=O)–N(CH}_3\text{)}_2$	Pseudomonas, Xanthomonas, Penicillium, Aspergillus.
	N (4-chlorophenyl)-1-methoxy-1-methylurea (Monolinuron)	$\text{(4-chlorophenyl)–NH–C(=O)–N(OCH}_3\text{)(CH}_3\text{)}$	Bacillus
Acylanilide	3, 4-dichloropropionanilide (Propanil)	$\text{(3,4-dichlorophenyl)–NH–C(=O)–CH}_2\text{–CH}_3$	Fusarium
	N-3,4-dichlorophenyl-2-methylpentanamide (Karsil)	$\text{(3,4-dichlorophenyl)–NH–C(=O)–CH(CH}_3\text{)–CH}_2\text{–CH}_2\text{–CH}_3$	Penicillium, Pullularia

Class	Compound	Structure	Organisms
Chlorinated aliphatic acid	2,2-dichloropropionic acid (Dalapon)	CH_3-CCl_2-COOH	Arthrobacter, Bacillus, Agrobacterium, Nocardia, Alkaligenes, Pseudomonas
	trichloroacetic acid (TCA)	CCl_3-COOH	Pseudomonas, Arthrobacter, Trichoderma
Substituted benzoic acid	3-amino-2,5-dichlorobenzoic acid (Amiben)	(dichloro-amino-benzoic acid: ring with Cl, $COOH$, Cl, H_2N)	Biologically degraded but no organisms identified
	2,6-dichlorobenzonitrile (Dichlobenil)	(benzonitrile ring with Cl, CN, Cl)	Fusarium, Penicillium, other fungi
Triazine	2-chloro-4,6-bis(ethylamino)-s-triazine (Simazine)	$H_5C_2-HN-C=N-C(Cl)=N-C(N=)-NH-C_2H_5$ (s-triazine ring)	Aspergillus
	2-chloro-4-ethylamino-6-isopropylamino-s-triazine (Atrazine)	$H_5C_2-HN-C=N-C(Cl)=N-C(N=)-NH-C_3H_7$ (s-triazine ring)	Aspergillus, Rhizopus, Fusarium, Penicillium
Dipyridyl	1,1'-dimenthyl-4,4'-bipyridylium cation (Paraquat)	$[H_3C-N\text{(pyridyl)}-\text{(pyridyl)}N-CH_3]^{++}$	Corynebacterium, Clostridium, Lipomyces

* From references cited in text. Genera only are indicated here.

How do some members of the soil microbial community acquire the ability to degrade a herbicide? This is not clearly understood generally, though there appear to be 3 possibilities. First, there are organisms which require little or no adaption before attacking a herbicide, having degradative enzymes of a constitutive type. Secondly, through random mutation occasional organisms acquire the ability to metabolize the compounds. Thirdly, organisms may have adaptive enzymes which are induced in the presence of the herbicide. The latter possibilities were considered by Audus (1960). Persistence patterns can give an indication of which of these schemes is involved, so that by considering the hypothetical results in Fig. 1 the following interpretations may be applied. Case (1)—there being no obvious time lag implies either that large numbers of organisms are present with constitutive enzymes to degrade the compound or organisms rapidly adapt. Case (2)—a distinct lag indicates either that time is required for adaption by many organisms or the proliferation of active organisms to a suitable population, and hence enzyme, level. Case (3)—herbicide is dissipated in 2 distinct phases, due either to rapid initial degradation leading to the accumulation of a more recalcitrant metabolite or to the complete decomposition of the compound involving different microbes acting in succession where degradation rates are not necessarily the same. Case (4)—a long delay before herbicide concentration begins to decline suggests that the compound is inherently resistant to attack and the eventual degradation, at a slow rate, is either due to a long adaption period or to the emergence of mutants. Case (5)—little or no degradation is seen, implying that degradation in the other systems is biological.

Undoubtedly physiochemical factors would affect the herbicide persistence but these have been excluded from the present summary in order to present interpretations on solely biological lines.

Whichever process operates, microbes with the ability to metabolize a herbicide and use it as carbon or energy source are at an ecological advantage in the absence of competition for the substrate. There are, however, cases where microbes may utilize a herbicide though not necessarily in preference to other energy sources (Sheets, 1964).

Examples of herbicides belonging to various groups, their structures and organisms which have been found to degrade them are given in Table 2. Clearly, common soil genera such as *Arthrobacter, Pseudomonas, Penicillium* and *Aspergillus* often feature as herbicide degraders and in many cases diverse organisms may be involved.

(b) *Phenoxyalkanoic acids*

The phenoxyalkanoic acids, or phenoxyalkyl carboxylic acids, comprise an aliphatic acid joined by ether linkage to a benzene ring. Chloro and methyl substitution of the aromatic ring and variation of the aliphatic chain length gives a number of herbicidal compounds which are used to control broadleafed weeds. Many phenoxyalkanoates are degraded in soil, 2,4-dichlorophenoxyacetic acid, 2,4-D, being particularly susceptible. Following the observations of Audus (1949) and Newman & Thomas (1949) that 2,4-D is detoxified by soil microbes there have been a number of isolations of phenoxyalkanoate-degrading organisms. The isolates included an organism designated *Bacterium globiforme* (Audus, 1950); *Flavobacterium* spp. (Jensen & Petersen, 1952; Burger, MacRae & Alexander, 1962); *Corynebacterium* spp. (Jensen & Petersen, 1952; Rogoff & Reid, 1956); *Achromobacter* sp. (Bell, 1957); *Nocardia* spp. (Webley, Duff & Farmer, 1957; Taylor & Wain, 1962); *Arthrobacter* sp. (Loos, Roberts & Alexander, 1967a).

Several pathways, reviewed by Loos (1969), have been suggested for the metabolism of phenoxyalkanoates. Mechanisms include β-oxidation of compounds with long fatty acid moieties, cleavage of the ether link between the side chain and the aromatic ring and ring hydroxylation prior to attack on the side chain.

(i) β-oxidation

Webley, Duff & Farmer (1957) showed that monochlorophenoxybutyric acids were β-oxidized to the corresponding chlorophenoxyacetic acids *via* the intermediate β-hydroxy acids. It has since been confirmed that the higher phenoxyalkanoates, with side chains having >2 carbon atoms, may be β-oxidized by soil microbes (Byrde & Woodcock, 1957; Taylor & Wain, 1962). The mixed microflora initiated degradation of these compounds in soil by β-oxidation (Gutenmann, Loos, Alexander & Lisk, 1964) and further degraded the β-oxidation products (Loos, 1969).

(ii) Ring hydroxylation

A Gram negative soil bacterium formed 2-hydroxy-4-chlorophenoxyacetate and 4-chlorocatechol when growing on 4-chlorophenoxyacetate, 4-CPA (Evans & Smith, 1954). Evans & Moss (1957) later showed that the 4-chlorocatechol was metabolized to β-chloromuconic acid. Evans & Smith (1954) and Bell (1960) proposed the 6-hydroxy derivative of 2,4-D as a metabolite. However, this route has not been conclusively demonstrated and Evans (1958) instead suggested that 2,4-dichlorophenol was formed. Either way, subsequent metabolities would be 3,5-dichlorocatechol and α-chloromuconic acid (Fernley & Evans, 1959). Proposed routes are summarized in Fig. 2.

Aspergillus niger both β-oxidized the side chain and hydroxylated the ring of some higher phenoxyalkanoates, an example being the conversion of phenoxybutyrate to 4-hydroxyphenoxyacetate (Byrde & Woodcock, 1957).

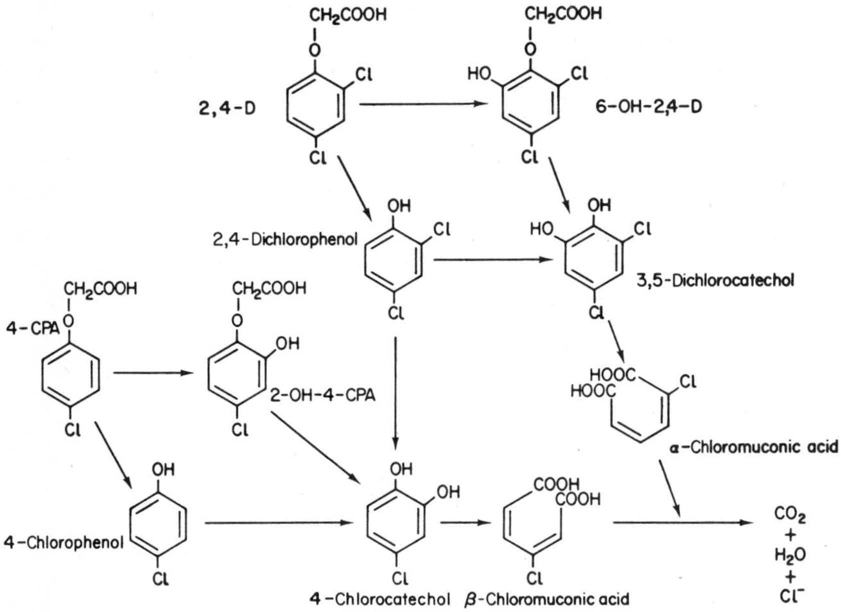

Fig. 2. Proposed pathways for microbial metabolism of 2,4-D and 4-CPA. Collected from data cited in the text.

(iii) *Ether link cleavage*

This process generates the corresponding phenolic compound. Steenson & Walker (1957) proposed that 2,4-dichlorophenol and thence 4-chlorocatechol were intermediates in 2,4-D metabolism and that 2-methyl-4-chlorophenol was formed from 2-methyl-4-chlorophenoxyacetate, MCPA. Several authors suggested that 2,4-D was converted to 2,4-dichlorophenol without giving direct evidence. However, Loos, Roberts and Alexander (1967a,b) and Loos, Bollag & Alexander (1967) using an *Arthrobacter* sp. confirmed that phenoxyacetates are metabolized via corresponding phenols. The phenolic compounds are in turn metabolized; Bollag, Helling & Alexander (1968) demonstrated the enzymic conversion of 2,4-dichlorophenol and 4-chlorophenol to 3,5-dichlorocatechol and 4-chlorocatechol respectively (Fig. 2). Degradation of 2,4-dichlorophenoxybutyric acid, 2,4-DB, proceeded via ether link cleavage giving 2,4-dichlorophenol and 4-chlorocatechol (MacRae, Alexander & Rovira, 1963), which in contrast to the β-oxidation side chain attack led to immediate detoxication of the herbicide.

(iv) *Ring cleavage*

The formation of chloromuconic acids on ring fission of the phenoxyacetate-derived chlorocatechols is mentioned above. Bollag, Briggs, Dawson & Alexander

(1968) observed that chlorocatechols are cleaved *via* chloromuconic acids to carboxymethylenebutenolides and thence to chlorosubstituted β-ketoadipate. Whilst many plant and microbial degration mechanisms are similar, aromatic ring cleavage is a microbial property (Loos, 1969).

Metabolism of sodium 2,4-dichlorophenoxyethyl sulphate (Sesone) provides an example of the microbial formation of a herbicide. The compound is not herbicidal but is rapidly coverted by microbes in soil to 2,4-D (Audus, 1952).

(c) *Phenylcarbamates*

Carbamic acid, H_2N-COOH, is unstable and spontaneously decomposes to CO_2 and NH_3. However, substitution of an amino proton with a phenyl group and esterification of the resulting N-phenylcarbamic acid, carbanilic acid, with different alcohols gives stable phytotoxic compounds. The phenylcarbamates have the basic structure C_6H_5-NH-CO-R which is shared by other types of herbicides. In the phenylcarbamates R is an alkoxy group; in acylanilides R is an alkyl group and in phenylureas it is an alkylamino group.

Isopropyl N-phenylcarbamate, IPC, was the first widely used phenylcarbamate herbicide and its success led to the development of others such as isopropyl N-3-chlorophenylcarbamate, CIPC, and 4-chloro-2-butyryl N-3-chlorophenylcarbamate, Barban. Microbial degradation of IPC and CIPC had been indicated in several soil persistence studies (see Clark & Wright, 1970*a*), and Kaufman & Kearney (1965) first showed phenylcarbamate degradation by pure bacterial isolates which included *Pseudomonas, Flavobacterium, Achromobacter* and *Arthrobacter* spp. Clark & Wright (1970*a*) isolated *Arthrobacter* and *Achromobacter* spp. which utilized IPC as the sole source of carbon for growth, even when supplied as particles: IPC was rapidly dissipated from soil when suspensions of these organisms were added. A *Penicillium* sp. which metabolizes Barban has recently been isolated (Wright & Forey, unpublished).

Kaufman & Kearney (1965) detected 3-chloroaniline during early stages of CIPC degradation in soil and a cell-free preparation of a soil pseudomonad which hydrolysed CIPC to 3-chloroaniline was described (Kearney & Kaufman, 1965). The purified enzyme (Kearney, 1965) converted several N-phenylcarbamates to the respective anilines. Similarly IPC and CIPC were metabolized by bacteria *via* aniline and 3-chloroaniline respectively (Clark & Wright, 1970*b*), and *Penicillium jenseni* coverted Barban to 3-chloroaniline (Wright & Forey, unpublished). All phenylcarbamate-degrading organisms isolated in our laboratory hydrolyse IPC, CIPC and Barban, and in each organism the enzyme was found to be closely associated with the microbial cell surface. Apparently the ability to initiate phenylcarbamate metabolism by liberating the aniline compound is a broad specificity process widely distributed among soil microbes. The anilines are not phytotoxic and their formation represents herbicide detoxication (Clark & Wright, 1970*b*).

MAP–9

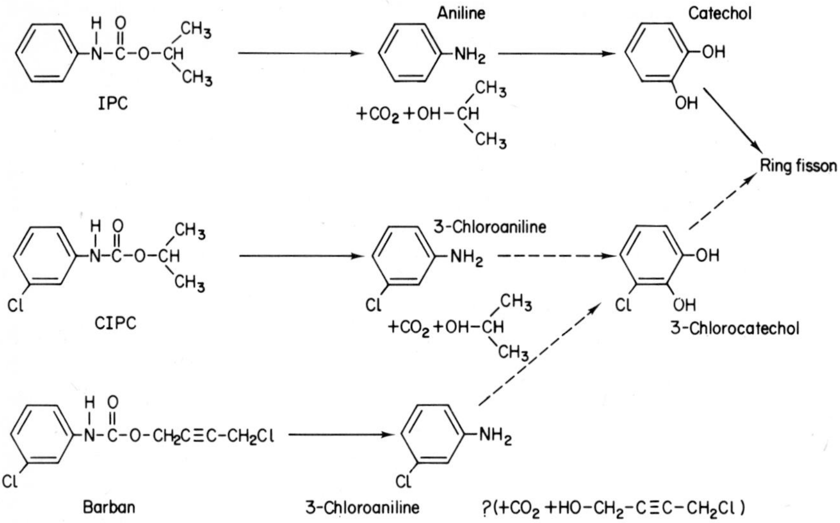

Fig. 3. Microbial metabolism of phenylcarbamates. ⟶ established; - -→ possible routes.

It is not clear whether the conversion of phenylcarbamates to anilines is due to esterase or amidase activity. Amidase was implied by the observation that acylanilides were also attacked (Kearney, 1965). Regardless of which bond is broken the products would be the same owing to instability of the initial metabolities (Kearney, 1965), so that in addition to the aniline compound, CO_2 and *iso*propanol would be formed from IPC and CIPC (Fig. 3). Recent studies (Wright & Spragg, unpublished) on an aniline metabolism by an IPC-degrading *Arthrobacier* indicate that catechol is formed and this undergoes fission, probably ortho fission. It seems likely that 3-chloroaniline would follow a similar degradative route.

Certain insecticidal methylcarbamates increase the herbicidal persistence of CIPC in soil by acting as competitive inhibitors of the phenylcarbamate hydrolysing enzyme (Kaufman & Kearney, 1966; Kaufman, Kearney, von Endt & Miller, 1970).

(d) *Phenylurea compounds*

The basic phenylurea herbicides are chloro-substituted dimethyl derivatives and compounds with more complex ring structures, substitutents and alkyl moieties add to the range (Geisbühler, 1969).

Inactivation of phenylurea herbicides occurs in soil conditions favouring microbial growth and though the rate of disappearance from soil is strongly influenced by adsorption-desorption equilibria microbes are at least partly responsible (Sheets, 1964). A *Pseudomonas* sp. utilized 4-chlorophenyl-1,

1-dimethylurea, Monuron, as a sole carbon source (Hill *et al.*, 1955) and Wallnöfer (1969) described a strain of *Bacillus sphaericus* which decomposed *N*-methoxyphenylurea compounds.

Degradation of the methyl- and methoxy-phenylurea compounds by soil microbes proceeds by stepwise demethylation and demethoxylation, respectively, to form the corresponding phenylurea which is then hydrolysed to the aniline compound with liberation of CO_2 and NH_3 (Geissbühler, Haselbach, Aebi & Ebner, 1963; Dalton, Evans & Rhodes, 1966; Tweedy, Loeppky & Ross, 1970). The stepwise reactions gradually decrease herbicidal activity and so represent detoxication processes (Geissbühler, 1969). A cell-free extract of *B. sphaericus* degraded *N'*-methoxyphenylureas to halogen-substituted anilines and CO_2, but demethylation and demethoxylation were apparently not involved (Wallnöfer & Bader, 1970).

(e) *Acylanilides*

Propanil, 3,4-dichloropropionanilide (DCPA), is an important herbicide widely used for weed control in rice (Bartha & Pramer, 1970).

Bartha & Pramer (1967) concluded that in soil a microbial acylamidase catalysed the cleavage of DCPA into 3,4-dichloroaniline (DCA) and propionic acid. The propionate was utilized by soil organisms while 2 molecules of DCA condensed to form 3,3',4,4'-tetrachloroazobenzene, TCAB (Fig. 4). Similar observations were recorded for the related herbicides Dicryl and Karsil (Bartha,

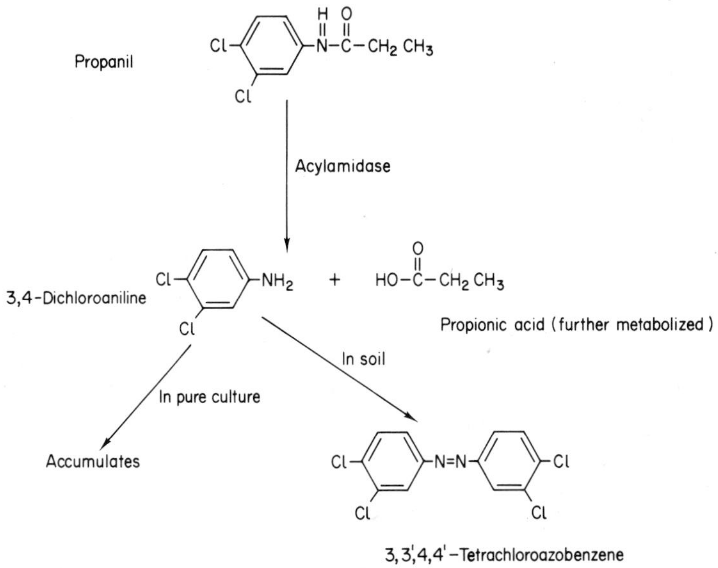

Fig. 4. Metabolism of 3,4-dichloropropionanilide, Propanil, in soil and in pure culture (after Bartha & Pramer, 1967, and Lanzilotta & Pramer, 1970*a,b*).

Linke & Pramer, 1968a). However, pure culture studies (Lanzilotta & Pramer, 1970a) with *Fusarium solani* showed that though the organism used Propanil as a sole carbon and energy source, DCA accumulated in the medium without undergoing condensation to form the azo compound. A specific acylamidase hydrolysing Propanil was extracted from the *Fusarium* (Lanzilotta & Pramer, 1970b). Sharabi & Bordeleau (1969) obtained fungi which hydrolysed Karsil forming DCA as a major product which accumulated without forming the azo compound.

Anilines occupy an important position in the metabolism of phenylcarbamates, phenylureas and acylanilides (Fig. 5). Bartha, Linke & Pramer (1968b) found that unsubstituted aniline was decomposed in soil but chloroanilines were transformed to the corresponding chloroazobenzenes, a peroxidatic mechanism being implicated. Bordeleau & Bartha (1970) proposed that phenylhydroxylamines formed from anilines condense with unchanged anilines to form the azobenzenes. Investigations of chloroaniline and azobenzene formation in soil relate to environmental pollution and public health since some azo-compounds are known to be carcinogenic (Bartha *et al.*, 1968b) and DCA and TCAB are mutagenic (Prasad, 1970). The results of Bartha *et al.* were obtained using aniline concentrations which could be considered abnormally high. Analysis of soils after exposure to urea herbicides under field conditions revealed no azo derivatives (Guth & Boyd, unpublished cited by Geissbühler, 1969). Further,

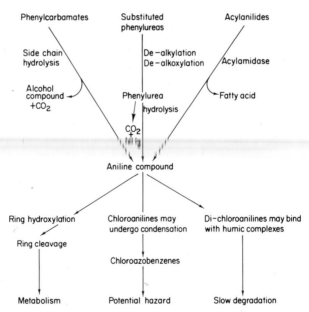

Fig. 5. Pathways of herbicide degradation via aniline compounds, indicating alternative subsequent routes.

Sprott & Corke (1971) showed that in soil only a small percentage of DCA was converted to TCAB. It is likely that at low concentrations chloroanilines may bind with humic materials in soil, so increasing their persistence owing to the slow turnover rate of humic complexes (Chisaka & Kearney, 1970; Bartha, 1971).

(f) Chlorinated aliphatic acids

Dalapon, 2,2-dichloropropionate, and TCA (trichloroacetate) have been used widely in weed control. At normal application rates Dalapon disappears from soil in 2–4 weeks and TCA may require up to 12 weeks (Kearney, Kaufman & Alexander, 1967). Both compounds are subject to microbial degradation (Foy, 1969) which is accompanied by release of chloride ions.

A variety of "active" organisms has been isolated (Jensen, 1957a,b; Magee & Colmer, 1959; Hirsch & Alexander, 1960; Kearney, Kaufman & Beall, 1964).

Jensen (1957a) suggested that the first step in metabolism of these compounds involves hydrolytic dechlorination, giving the corresponding hydroxy or keto acids which can then be utilized as an energy source. The results of Beall, Kearney & Kaufman (1964) who used [^{14}C] Dalapon and Kearney, Kaufman & Beall (1964) who used cell-free extracts to dehalogenate Dalapon, confirmed Jensen's proposed scheme. The likely precursor of pyruvate was 2-chloro-2-hydroxypropionate (Fig. 6). Hirsch & Alexander (1960) obtained partial degradation of Dalapon by a streptomycete without liberation of chloride.

The microbial degradation of Dalapon in soil was inhibited by the triazole herbicide Amitrole though Dalapon had no effect on Amitrole persistence (Kaufman & Sheets, 1965).

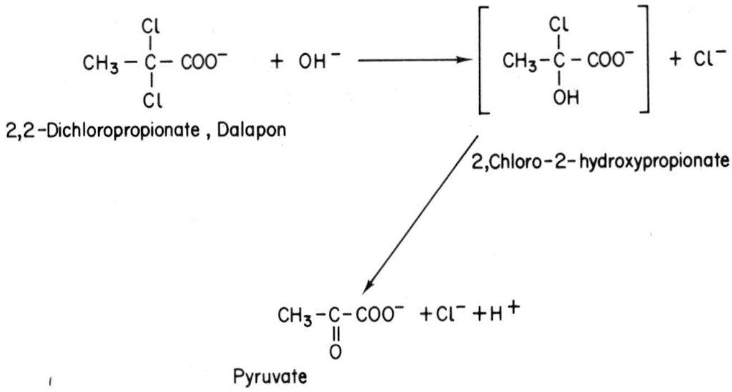

Fig. 6. Enzymic dehalogenation of 2,2-dichloropropionate, Dalapon, to pyruvate (after Kearney et al., 1964).

(g) Substituted benzoates

Commonly used herbicidal benzoates include Amiben, 3-amino-2,5-dichloro-benzoic acid; Dinoben, 3-nitro-2,5-dichlorobenzoic acid; TBA, 2,3,6-trichloro-benzoic acid; Diclobenil, 2,6-dichlorobenzonitrile and Dicamba; 2-methoxy-3,6-dichlorobenzoic acid. Swanson (1969) reviewed the fate and degradation of these compounds. Though some microbial degradation is implied, none of the compounds is readily attacked, a feature which can be related to the extent of chlorine substitution of the aromatic ring. Amiben is degraded more readily than TBA or Dicamba. Kaufman & Sheets (1965) observed that Dicamba decomposed more rapidly in soils treated simultaneously with 2,4-D.

(h) Triazine compounds

These herbicides, based on the triazine ring, usually have alkylamino substituents at positions 4 and 6 with chloro or alkoxy groups at position 2. Simazine, 2-chlor-4,6-bis(ethylamino)-s-triazine, and Atrazine, 2-chloro-4-ethylamino-6-isopropylamino-s-triazine, are particularly widely used.

The s-triazine ring remains intact for long periods in the soil and is quite resistant to microbial attack (Kearney et al., 1967). Ten months were required for dissipation of 90% of the toxicity of 4 p/m of Simazine in soil (Burnside, Schmidt & Behrens, 1961).

Soil fungi, notably Aspergillus fumigatus, metabolize the ethylamino side chains of Simazine leaving the less phytotoxic chlorine-substituted triazine ring intact (Kaufman, Kearney & Sheets, 1965; Kearney, Kaufman & Sheets, 1965). Hydroxysimazine, a metabolite in plants and soil, was not involved in this pathway.

Several soil fungi attacked Atrazine by N-dealkylation, a process not necessarily ensuring detoxication (Kaufman & Blake, 1970). Dehalogenation, with formation of nontoxic hydroxyatrazine was considered possible in some organisms and was probably the major degradative route in soil.

(i) Dipyridyl compounds

The comparatively recently discovered herbicides Diquat and Paraquat are extremely rapid general weed killers used as crop defoliants and desiccants and for aquatic weed control. The herbicides are very strongly adsorbed by certain soil particles rendering them relatively unavailable to biological systems, a factor which confers several agronomic advantages (Funderburk, 1969). Commonly formulated as the dichlorides, the cationic portion of the molecule is the active part in each case (Table 2). Paraquat is more persistent than Diquat (Funderburk & Bozarth, 1967).

Baldwin, Bray & Geoghegan (1966) obtained a strain of Corynebacterium fascians which gave 30–40% decomposition of Paraquat; unidentified anaerobes and a strain of Clostridium pasteurianum were similarly active. The same authors

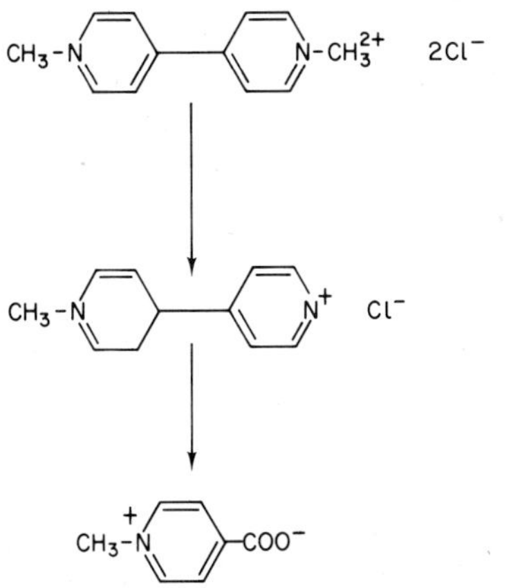

Fig. 7. Proposed pathway of Paraquat degradation by an unindentified soil bacterium (after Funderburk & Bozarth, 1967).
(1) − 1-methyl-4, 4′-dipyridinium ion
(2) − 1 methyl-4-carboxypyridinium ion

isolated *Lipomyces starkeyi* which utilized Paraquat as a preferred nitrogen source and was able to completely decompose the herbicide. *Lipomyces starkeyi* also degraded Diquat (Funderburk, 1969).

An unidentified bacterium degraded [^{14}C] Paraquat (Funderburk & Bozarth, 1967) forming a 1-methyl-4,4′-dipyridinium ion (1), and a 1-methyl-4-carboxypyridinium ion (2) (see Fig. 7). The proposed degradative pathway involved demethylation followed by cleavage of one of the heterocyclic rings.

5. Conclusions

The toxicity of some herbicides in soil declines in a manner suggesting the involvement of biological agencies and in some cases effective micro-organisms have been isolated and degradative pathways established. Information relating to the biodegradation of these compounds has been collected in this paper. It is pertinent, however, to draw attention to the need for more research in this field. We have little knowledge of the true situation in soil where several factors operate to affect the herbicide and its degradation products.

Investigation of interactions between herbicides and soil microbes offers a challenging area of research. The isolation of herbicide-degrading organisms is in itself rarely a simple operation. New approaches are required, in particular the experimentation should be extended so as to isolate and test algae, protozoa and auxotrophic organisms. Hitherto most isolates, by virtue of the techniques used have been prototrophic bacteria, fungi and actinomycetes. Algae especially should receive more attention, the soil surface horizon being the site of both the majority of soil algal activity and deposition of most herbicides.

Often technical problems arise due to the insolubility or volatility of the herbicides and added complications are likely when studying the fate of mixtures of herbicides. Some herbicides affect the persistence of others (Kaufman & Sheets, 1965). Organisms which degrade a herbicide in pure culture conditions are not necessarily those responsible for its degradation in soil. Organisms which can metabolize a herbicide to some extent are not necessarily able to use it as a carbon source for growth, accounting possibly for some failures to isolate organisms by elective culture techniques.

Will information on the microbial degradation of herbicides be of any practical use? In answer it can be said that there appear to be distinct possibilities for controlling herbicide persistence. Inoculation of herbicide formulations of the soil with capsules of active organisms or enzymes, and the addition of factors to promote microbial activity, are possible devices for reducing persistence or promoting rapid degradation after a desired time. Alternatively, extending the persistence of certain herbicides by adding competitive inhibitors of the herbicide-degrading enzymes holds considerable promise (Kaufman *et al.,* 1970). The foregoing observations are based on results of small scale experiments. Such processes, though feasible from biological standpoints, would however be difficult to implement for financial reasons. A promising development was the formulation of herbicides in water-degradable polymers which allowed control of the rate of release of herbicides and gave increased herbicidal efficiency (Beasley & Collins, 1970).

Even apparently "safe" compounds may also show marked nontarget inhibitory activity. The degradable herbicides IPC and CIPC are known to inhibit growth in algae, fungi and protozoa, and the observation (Timson, 1970) that these compounds reduce the rate of mitosis in human cells raises questions as to their future use. Whilst it is desirable that the soil should not contain persistent herbicide residues, lack of persistence in some instances would be an agronomic embarrassment. The task facing manufacturers is to design herbicides which exert the required effect after which they do not persist. Desirable features of such compounds include the fact that nontarget organisms are affected for the shortest possible time, target species are less likely to develop resistance and real control with economic advantages is obtained.

The need for efficient, intensive cropping of the land currently requires that vast quantities of herbicides and other pesticides are administered directly or indirectly to the soil. We are obliged to maintain the fertility and attendant natural balances in the soil, and fully co-ordinated, long term research on all aspects of the interactions between agricultural chemicals and the soil is already overdue.

Among aspects of herbicide evaluation where the research effort has been deficient and is now required to secure safe exploitation of their potentialities, the Third Report of the Research Committee on Toxic Chemicals (1970) included: (1) behaviour and persistence of herbicides in the soil; (2) influence of environmental factors on the activity of herbicides; (3) toxicity of new herbicides to soil micro-organisms and fauna; (4) long term consequences of repeated usage of herbicides on the same site. Similarly, the Royal Commission on Environmental Pollution (1971) emphasized the importance of long term research into the ecological effects of herbicides and suggested as a matter of urgency the phased replacement of persistent pesticides by those that are less persistent.

6. Acknowledgments

The support of the Agricultural Research Council and the assistance of Mrs A. Forey during the preparation of this paper are gratefully acknowledged.

7. References

Aldrich, R. J. (1953). Herbicides. Residues in soil. *J. agric. Fd Chem.* **1**, 257.

Alexander, M. (1965*a*). Persistence and biological reactions of pesticides in soils. *Proc. Soil Sci. Soc. Am.* **29**, 1.

Alexander, M. (1965*b*). Biodegradtion of pesticides. In *Pesticides and their Effects on Soil and Water.* Symp. Soil. Sci. Soc. Am. Columbus: Spec. Publ. A.S.A.

Alexander, M. & Aleem, M. I. H. (1961). Effect of chemical structure on microbial decomposition of aromatic herbicides. *J. agric. Fd Chem.* **9**, 44.

Audus, L. J. (1949). The biological detoxication of 2,4-dichlorophenoxyacetic acid in soil. *Pl. Soil* **2**, 31.

Audus, L. J. (1950). Biological detoxication. of 2,4-dichlorophenoxyacetic acid in soils: isolation of an effective organism. *Nature, Lond.* **166**, 356.

Audus, L. J. (1952). Fate of sodium 2,4-dichlorophenoxyethyl-sulphate in soil. *Nature, Lond.* **170**, 886.

Audus, L. J. (1960). Microbiological breakdown of herbicides in soils. In *Herbicides and the Soil.* (Eds E. K. Woodford & G. R. Sagar). Oxford: Blackwells.

Baldwin, B. C., Bray, M. F. & Geoghegan, M. J. (1966). The microbial decomposition of paraquat. *Biochem, J.* **101**, 15P.

Bartha, R. (1971). Fate of herbicide-derived chloroanilines in soil. *J. agric. Fd Chem.* **19**, 385.

Bartha, R., Linke, H. & Pramer, D. (1968*a*). Transformation of anilide herbicides and chloroanilines in soil. *Bact. Proc.* A**26**, 5.

Bartha, R., Linke, H. & Pramer, D. (1968*b*). Pesticide transformations: production of chloroazobenzenes from chloroanilines. *Science, N.Y.* **161**, 582.

Bartha, R. & Pramer, D. (1967). Pesticide transformation to aniline and azo compounds in soil. *Science, N.Y.* **156**, 1617.

Bartha, R. J. & Pramer, D. (1970). Metabolism of acylanilide herbicides. *Adv. appl. Microbiol.* **13**, 317.

Beall, M. L., Kearney, P. C. & Kaufman, D. D. (1964). Comparative metabolism of 1-^{14}C- and 2-^{14}C-labelled dalapon by soil micro-organisms. *Weed Soc. Am. Abstr.* p.11.

Beasley, M. L. & Collins, R. L. (1970). Water-degradable polymers for controlled release of herbicides and other agents. *Science, N.Y.* **169**, 769.

Bell, G. R. (1957). Some morphological and biochemical characteristics of a soil bacterium which decomposes 2,4-dicholorophenoxyacetic acid. *Can. J. Microbiol.* **3**, 821.

Bell, G. R. (1960). Studies on a soil *Achromobacter* which degrades 2,4-dichlorophenoxy-acetic acid. *Can. J. Microbiol.* **6**, 325.

Bollag, J. M., Briggs, G. G., Dawson, J. E. & Alexander, M. (1968). 2,4-D metabolism. Enzymatic degradation of chlorocatechols. *J. agric. Fd Chem.* **16**, 829.

Bollag, J. M., Helling, C. S. & Alexander, M. (1968). 2,4-D metabolism. Enzymatic hydroxylation of chlorinated phenols. *J. agric. Fd Chem.* **16**, 826.

Bollen, W. B. (1961). Interactions between pesticides and soil micro-organisms. *A. Rev. Microbiol.* **15**, 69.

Bordeleau, L. M. & Bartha, R. (1970). Azobenzene residues from aniline-based herbicides; evidence for labile intermediates. *Bull. Envir. Cont. Toxicol.* **5**, 34.

Burger, K., Macrae, I. C. & Alexander, M. (1962). Decomposition of phenoxyalkyl carboxylic acids. *Proc. Soil Sci. Soc. Am.* **26**, 243.

Burnside, O. C., Schmidt, E. L. & Behrens, R. (1961). Dissipation of simazine from the soil. *Weeds* **9**, 477.

Byrde, R. J. W. & Woodcock, D. (1957). The metabolism of some phenoxy-*n*-alkylcarboxylic acids by *Aspergillus niger*. *Biochem. J.* **65**, 682.

Chisaka, H. & Kearney, P. C. (1970). Metabolism of propanil in soils. *J. agric. Fd Chem.* **18**, 854.

Clark, C. G. & Wright, S. J. L. (1970*a*). Detoxication of isopropyl *N*-phenyl-carbamate (IPC) and isopropyl *N*-3-chlorophenylcarbamate (CIPC) in soil, and isolation of IPC-metabolizing bacteria. *Soil Biol. Biochem.* **2**, 19.

Clark, C. G. & Wright, S. J. L. (1970*b*). Degradation of the herbicide isopropyl *N*-phenylcarbamate by *Arthrobacter* and *Achromobacter* spp. from soil. *Soil Biol. Biochem.* **2**, 217.

Dalton, R. L., Evans, A. W. & Rhodes, R. C. (1966). Disappearance of Diuron from cotton field soils. *Weeds* **14**, 31.

Evans, W. C. (1958). The chlorophenoxyacetic acid herbicides. In *Encyclopedia of Plant Physiology,* Vol. 10. (Ed. W. Ruhland). Berlin: Springer.

Evans, W. C. & Moss, P. (1957). The metabolism of the herbicide *p*-chlorophenoxyacetic acid by a soil micro-organism-the formation of a β-chloromuconic acid on ring fission. *Biochem. J.* **65**, 8P.

Evans, W. C. & Smith, B. S. W. (1954). The photochemical inactivation and microbial metabolism of the chlorophenoxyacetic acid herbicides. *Biochem. J.* **57**.

Fernley, H. N. & Evans, W. C. (1959). Metabolism of 2,4-dichlorophenoxyacetic acid by soil *Pseudomonas* isolation of α-chloromuconic acid as an intermediate. *Biochem. J.* **73**, 22P.

Fletcher, W. W. (1960). The effect of herbicides on soil micro-organisms. In *Herbicides and the Soil.* (Eds E. K. Woodford & G. R. Sagar). Oxford: Blackwells.

Foy, C. L. (1969). The chlorinated aliphatic acids. In *Degradation of Herbicides.* (Eds P. C. Kearney & D. D. Kaufman). New York: Marcel Dekker.

Funderburk, H. H. (1969). Diquat and paraquat. In *Degradation of Herbicides.* (Eds P. C. Kearney & D. D. Kaufman). New York: Marcel Dekker.

of diquat and paraquat. *J. agric. Fd Chem.* **15**, 563.

Geissbühler, H. (1969). The substituted ureas. In *Degradation of Herbicides.* (Eds P. C. Kearney & D. D. Kaufman). New York: Marcel Dekker.

Geissbühler, H., Haselbach, C., Aebi, H., & Ebner, L. (1963). The fate of *N'*-(4-chlorophenoxy)-phenyl-*N N'*-dimethylurea (C-1983) in soils and plants. III Breakdown in soils and plants. *Weed Res.* **3**, 277.

Gutenmann, W. H., Loos, M. A., Alexander, M. & Lisk, D. J. (1964). Beta oxidation of phenoxyalkanoic acids in soil. *Proc. Soil Sci. Soc. Am.* **28**, 205.

Hill, G. D., McGahen, J. W., Baker, H. M., Finnerty, D. W. & Bingeman, C. W. (1955). The fate of substituted urea herbicides in agricultural soils. *Agron. J.* **47**, 93.

Hirsch, P. & Alexander, M. (1960). Microbial decomposition of halogenated propionic and acetic acids. *Can. J. Microbiol.* **6**, 241.

Hocombe, S. D., Holly, K. & Parker, C. (1966). The persistence of some new herbicides in the soil. *Proc. 8th Br. Weed Control Conf.* **2**, 605.

Jensen, H. L. (1957*a*). Decomposition of chloro-substituted aliphatic acids by soil bacteria. *Can. J. Microbiol.* **3**, 151.

Jensen, H. L. (1957*b*). Decomposition of chloro-organic acids by fungi. *Nature, Lond.* **180**, 1416.

Jensen, H. L. & Petersen, H. I. (1952). Detoxication of hormone herbicides by soil bacteria. *Nature, Lond.* **170**, 39.

Kaufman, D. D. & Blake, J. (1970). Degradation of Atrazine by soil fungi. *Soil Biol. Biochem.* **2**, 73.

Kaufman, D. D. & Kearney, P. C. (1965). Microbial degradation of isopropyl-*N*-3-chlorophenylcarbamate and 2-chloroethyl-*N*-3-chlorophenylcarbamate. *Appl. Microbiol.* **13**, 443.

Kaufman, D. D. & Kearney, P. C. (1966). Microbial degradation of carbamate pesticide combinations in soils. *Division of Agricultural and Food Chemistry, ACS Abstracts,* A45.

Kaufman, D. D., Kearney, P. C. & Sheets, T. J. (1965). Microbial degradation of Simazine. *J. agric. Fd Chem.* **13**, 238.

Kaufman, D. D., Kearney, P. C., Von Endt, D. W. & Miller, D. E. (1970). Methylcarbamate inhibition of phenylcarbamate metabolism in soil. *J. agric. Fd Chem.* **18**, 513.

Kaufman, D.D. & Sheets, T. J. (1965). Microbial decomposition of pesticide combinations. *Agron. Abstr.* **57**, 85.

Kearney, P. C. (1965). Purification and properties of an enzyme responsible for hydrolyzing phenylcarbamates. *J. agric. Fd. Chem.* **13**, 561.

Kearney, P. C. & Helling, C. S. (1969). Reactions of pesticides in soils. *Residue Rev.* **25**, 25.

Kearney, P. C. & Kaufman, D. D. (1965). Enzyme from soil bacterium hydrolyses phenylcarbamate herbicides. *Science, N.Y.* **147**, 740.

Kearney, P. C., Kaufman, D. D. & Alexander, M. (1967). Biochemistry of herbicide decomposition in soils. In *Soil Biochemistry.* (Eds A. D. McLaren & G. H. Peterson). New York: Marcel Dekker.

Kearney, P. C., Kaufman, D. D. & Beall, M. L. (1964). Enzymatic dehalogenation of 2,2-dichloropropionate. *Biochem. biophys. Res. Commun.* **14**, 29.

Kearney, P. C., Kaufman, D. D. & Sheets, T. J. (1965). Metabolites of Simazine by *Aspergillus fumigatus. J. agric. Fd Chem.* **13**, 369.

Kruglov, Yu. V. & Paromenskaya, L. N. (1970). Detoxication of Simazine by microscopic algae. *Microbiology* **39**(I), 139.

Lanzilotta, R. P. & Pramer, D. (1970*a*). Herbicide transformation. I. Studies with whole cells of *Fusarium solani. Appl. Microbiol.* **19**, 301.

Lanzilotta, R. P. & Pramer, D. (1970*b*). Herbicide transformation. II. Studies with an acylamidase of *Fusarium solani. Appl. Microbiol.* **19**, 307.

Loos, M. A. (1969). Phenoxyalkanoic acids. In *Degradation of Herbicides.* (Eds P. C. Kearney & D. D. Kaufman). New York: Marcel Dekker.

Loos, M. A., Bollag, J. M. & Alexander, M. (1967). Phenoxyacetate herbicide detoxication by bacterial enzymes. *J. agric. Fd Chem.* **15**, 858.

Loos, M. A., Roberts, R. N. & Alexander, M. (1967*a*). Phenols as intermediates in the decomposition of phenoxyacetates by an *Arthrobacter* species. *Can. J. Microbiol.* **13**, 679.

Loos, M. A., Roberts, R. N. & Alexander, M. (1967*b*). Formation of 2,4-dichlorophenol and 2,4-dichloroanisole from 2,4-dichlorophenoxyacetate by an *Arthrobacter* sp. *Can. J. Microbiol.* **13**, 691.

Macrae, I. C., Alexander, M. & Rovira, A. D. (1963). The decomposition of 4-(2,4-dichlorophenoxy) butyric acid by a *Flavobacterium* sp. *J. gen. Microbiol.* **32**, 69.

Magee, L. A. & Colmer, A. R. (1959). Decomposition of 2,2-dichloropropionic acid by soil bacteria. *Can. J. Microbiol.* **5**, 255.

Newman, A. S. & Thomas, J. R. (1949). Decomposition of 2,4-dichlorophenoxyacetic acid in soil and liquid media. *Proc. Soil Sci. Soc. Am.* **14**, 160.

Prasad, I. (1970). Mutagenic effects of the herbicide 3,4-dichloropropionanilide and its degradation products. *Can. J. Microbiol.* **16**, 369.

Rogoff, M. H. & Reid, J. J. (1956). Bacterial decomposition of 2,4-dichlorophenoxyacetic acid. *J. Bact.* **71**, 303.

Royal Commission on Environmental Pollution, First Report (1971). London: H.M.S.O.

Sharabi, N. E. & Bordeleau, L. M. (1969). Biochemical decomposition of the herbicide N-(3,4-dichlorophenyl)-2-methylpentanamide and related compounds. *Appl. Microbiol.* **18**, 369.

Shaw, W. C. (1965). Research and education needs in the use of pesticides. In *Pesticides and their Effects on Soil and Water.* Symp. Soil Sci. Soc. Am. Columbus: Spec. Publ. A.S.A.

Sheets, T. J. (1962). Persistence of herbicides in soils. *Proc. west. Can. Weed Control Conf.* **19**, 37.

Sheets, T. J. (1964). Review of disappearance of substituted urea herbicides from soil. *J. agric. Fd Chem.* **12**, 30.

Sprott, G. D. & Corke, C. T. (1971). Formation of 3,3',4,4'- tetrachloroazobenzene from 3,4-dichloroaniline in Ontario soils. *Can. J. Microbiol.* **17**, 235.

Steenson, T. I. & Walker, N. (1957). The pathway of breakdown of 2,4-dichloro-and 4-chloro-2-methyl-phenoxyacetic acid by bacteria. *J. gen. Microbiol.* **16**, 146.

Swanson, C. R. (1969). The benzoic acid herbicides. In *Degradation of Herbicides.* (Eds. P.C. Kearney & D. D. Kaufman). New York: Marcel Dekker.

Taylor, H. F. & Wain, R. L. (1962). Side-chain degradation of certain ω-phenoxyalkane carboxylic acids by *Nocardia coeliaca* and other micro-organisms isolated from soil. *Proc. R. Soc.* B **156**, 172.

Third Report of The Research Committee on Toxic Chemicals (1970). London: Agricultural Research Council.

Timson, J. (1970). Effect of the herbicides propham and chlorpropham on the rate of mitosis of human lymphocytes in culture. *Pest. Sci.* **1**, 191.

Tweedy, B. G., Loeppky, C. & Ross, J. A. (1970). Metabolism of 3-(p-bromophenyl) -l-methoxy-l-methylurea (metobromuron) by selected soil micro-organisms. *J. agric. Fd Chem.* **18**, 851.

Wallnöfer, P. (1969). The decomposition of urea herbicides by *Bacillus sphaericus,* isolated from soil. *Weed Res.* **9**, 333.

Wallnöfer, P. R. & Bader, J. (1970). Degradation of urea herbicides by cell-free extracts of *Bacillus sphaericus. Appl. Microbiol.* **19**, 714.

Webley, D. M., Duff, R. B. & Farmer, V. C. (1957). Formation of a β-hydroxy acid as an intermediate in the microbiological conversion of monochlorophenoxybutyric acids to the corresponding substituted acetic acids. *Nature, Lond.* **179**, 1130.

Wright, S. J. L. & Clark, C. G. (1969). Controlled soil perfusion with a multi-channel peristaltic pump. *Weed Res.* **9**, 65.

The Microbial Breakdown of Pesticides

R. E. Cripps

Shell Research Limited,
Borden Microbiological Laboratory,
Sittingbourne, Kent, England

CONTENTS

1. Introduction

THE PERSISTENCE in the environment for extended periods of many synthetic compounds has highlighted the fact that microbes, once considered to be capable of degrading almost every organic molecule (Gale, 1952), are not omnivorous and that some chemical structures are not palatable substrates for the numerous and diverse micro-organisms found in nature. This realization that microbes are fallible has led to investigations into the reasons why some molecules are nonbiodegradable, and attempts have been made to relate biodegradibility to chemical structure. It is the intention of this paper to summarize the available evidence and to relate this, as far as it is possible, to the observed persistence of some pesticidal chemicals.

To obtain comprehensive data on the subject, a vast number of different chemical species needs to be examined for susceptibility to degradation by microbial attack in order to find out which structures are broken down and which are not. The results which are presently available have arisen mainly from laboratory studies and it is apparent that a major difficulty facing all attempts to relate chemical structure to biodegradability is to decide how relevant are studies with pure cultures and isolated enzymes in the laboratory to actual experience with pesticidal molecules and their residues in the field.

An understanding of the factors involved in structure-biodegradability relationships should be of great assistance to the organic chemist because it could facilitate the design of those biologically active molecules in which the additional property of biodegradability is desirable. Such information would also be of great fundamental interest.

2. Problems in the Study of Biodegradation

Lack of biodegradation of a chemical in a particular environment can usually be attributed to one of two circumstances: (i) the environment itself is not conducive to microbial proliferation or even the performance of biochemical activities without proliferation, so that a potentially degradable compound remains unmetabolized, and (ii) the chemical possesses some property which prevents it being attacked by micro-organisms under any conditions. Such molecules have been termed recalcitrant (Alexander, 1965). The environment could be unsuitable for several reasons, such as extremes of temperature, the absence of water or of oxygen or the presence of materials toxic to the organism. Such situations are well known. In addition, the potential substrate may be protected from microbial attack by, for example, being bound to soil particles and thus preventing substrate-microbe contact. Intrinsic resistance to biodegradation can usually be attributed to the structure of the particular chemical.

There have been few systematic studies designed to obtain data relating the structure of a compound to its ease of biodegradation. The few results that are available are somewhat confusing because different workers have reported widely different susceptibilities to microbial attack on a single compound (see Heukelekian & Rand, 1955). These probably are due to different ways in which degradation was measured, e.g. biochemical oxygen demand or carbon dioxide disappearance, the absence or the inoculum of mixed organisms, e.g. soil or sewage, and the conditions of the experiments. Also, a compound may often be fairly readily degraded under laboratory conditions but may not be attacked in other, less suitable situations.

The paucity of systematic studies necessitates the examination of data from a range of work in order to arrive at some correlation between structure and degradability. Relevant data are found in studies in which microbes have been isolated by their ability to use organic molecules as sole sources of carbon and energy for growth. This approach yields valuable information but naturally only indicates those componds which are degradable, not those which are not. Workers in this area are, perhaps understandably, loath to publish negative results, that is, that a particular compound did not act as a suitable carbon and energy source in an enrichment culture experiment. The enrichment culture technique is limited, moreover, because, although in nature some organic compounds are undoubtedly completely transformed into CO_2, H_2O and cellular matter by a single organism, many compounds are mineralized by the co-operative action of more than one microbe. In this context, the phenomenon of co-metabolism, where a compound is transformed by an organism without the release of utilizable carbon to that organism, must be considered. That is to say, the organism cannot utilize a particular compound for growth and must obtain its biosynthetic precursors from some other carbon source. This phenomenon could well be very common in pesticide metabolism but it is often difficult to study in the laboratory because the selection of the microbes which can perform these transformations is beset with difficulties. The conventional and widely used enrichment culture technique is suitable only for the isolation of prototrophic organisms which can use a specific chemical as a source of carbon, nitrogen, phosphorus, sulphur or energy or as a combination of these. If the chemical under study is metabolized only by auxotrophic organisms, having a specific requirement, or by a process of co-metabolism, the technique is not suitable. Hence, many organisms which conceivably could be potent pesticide metabolizers cannot be isolated and studied in the laboratory.

3. Naturally Recalcitrant Molecules

It has been pointed out (Alexander, 1965) that besides synthetic organic chemicals many naturally occurring compounds are resistant to microbial attack and persist for extended periods in the environment. The lengths of time involved in these cases are not the few years usually quoted for some pesticides (although these compounds have often not been in existence for more than a few years) but centuries and, in some instances, thousands of years. Thus humus, fossilized materials, lignin, pollen exine and some proteinaceous substances can exist for very long periods, often due to protection conferred by the particular situation in which they occur as well as to the chemical properties of the compounds. An example of economic importance is the nonbiodegradability of the vast reserves of crude oil which occur in many areas of the world. This material, comprising many individual chemicals which are known to be degradable by the action of micro-organisms, e.g. hydrocarbons (van der Linden & Thijsse, 1965) is preserved because its decomposition is dependent on the presence of oxygen, an element absent from the anaerobic zones in which the crude oil occurs. The evidence suggests that oxygen, in this case, is required not only an an electron acceptor but also as a component of the biochemical reactions leading to the dissimilation of hydrocarbons so that other possible electron acceptors, such as sulphate and nitrate, are not effective alternatives.

4. Structural Factors Influencing Biodegradability

(a) *Effects of substitution in straight chain fatty acids and benzene derivatives*

A vast number of compounds has been shown to be degraded by pure cultures of micro-organisms and studies of the pathways of breakdown have revealed that similar metabolic routes are employed by different genera. In addition, it has been shown that different compounds do not usually have individual metabolic pathways for their dissimilation; in general only a few distinct pathways are employed and related compounds are often metabolized either by a series of reactions leading to the formation of a common intermediate or by analogous reaction sequences. A common observation is that a particular compound elicits the formation of a few specific enzymes which transform it to an intermediate whose further metabolism is catalysed by enzyme reactions common to the degradation of other compounds. The metabolism of straight-chain fatty acids by β-oxidation and the oxidation of n-alkanes by reactions leading to the formation of the homologous alcohol, aldehyde and acid (Fig. 1) are examples of analogous reaction sequences. No matter how many carbon atoms there are in the molecule, the same transformations are carried out by almost all microbes which have been studied. The metabolism of related compounds to a common

$$CH_3 (CH_2)_n COOH \longrightarrow \quad \longrightarrow \quad \longrightarrow CH_3 CH_2)_{n-2} COOH + \text{acetyl CoA}$$

$$CH_3 CH_2)_n CH_3 \longrightarrow \quad CH_3 (CH_2)_n CH_2 OH \longrightarrow CH_3 (CH_2)_n CHO \longrightarrow$$

$$CH_3 (CH_2)_n CHOH$$

Fig. 1. Analogous reaction sequences.

intermediate has been shown by studies on the breakdown of aromatic compounds. Figure 2 shows the structures of several benzene derivatives which are usually broken down by routes involving the formation of catechol. The metabolism of catechol is known to proceed by only two metabolic routes (Ribbons, 1965; Feist & Hegeman, 1969). Because so few metabolic pathways exist for the catabolism of related molecules, it is not surprising that even slight variations in the structure of chemicals which are easily metabolized can result in a loss of biodegradability by interfering with the normal metabolic sequences. This has been shown to be the case in studies with both pure and mixed cultures. Thus Heukelekian & Rand (1955), using sewage microflora, showed that the biochemical oxygen demand (BOD) of many pure organic compounds was affected by chemical structure. Some of their data (Table 1) make it apparent that the introduction of certain chemical modifications into a normally degradable

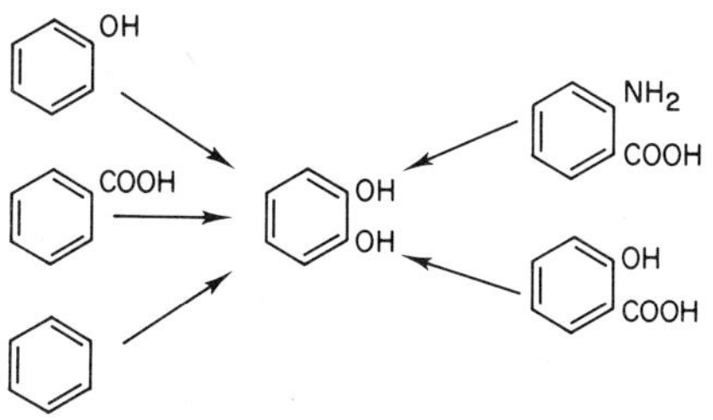

Fig. 2. Benzene derivatives which are degraded via catechol

Table 1

Biochemical oxygen demands of some degradable compounds and their nondegradable derivatives by sewage micro-organisms (from Heukelekian & Rand, 1955)

Degradable compound	BOD (g/g)	Nondegradable compound	BOD (g/g)
Acetic acid	0·34	Dichloracetic acid	0
Benzene	1·20	Chlorobenzene	0·03
		Nitrobenzene	0
Aniline	1·47	Ethylaniline	0·048
		Diethylanine	0

molecule results in a lowered BOD. In addition to these examples, this study revealed that ethers and chlorinated aliphatic compounds in general were resistant to degradation. In a similar investigation (Ludzack & Ettinger, 1960) in which sewage organisms were adapted to the compound before the biodegradability test, in an attempt to encourage the development of competent organisms, some compounds were still found to be resistant. The data (Table 2) reveal that ethers and highly branched structures are not degraded even by acclimatized organisms.

These two studies bring out some general rules in the relationship between chemical structure and biodegradability, namely, that the presence in a molecule of an ether linkage, chlorine atoms, branched carbon chains or substituted amino groups often confers increased resistance to microbial attack. These groupings are often found in pesticidal molecules and their residues.

Further evidence that branched-chain compounds present difficulties to microbial attack has been obtained from field experience with detergents (Swisher, 1963). The oft-quoted situation in which the early detergent compounds, which contained branched alkyl chains, produced undesirable effects due to the absence of biodegradation and the alleviation of these effects

MAP—10

Table 2

Degradation of some ethers and branched constructive compounds by an adapted sewage microflora (from Ludzack & Ettinger, 1960)

Compound	Theoretical oxidation (%)	% removed
Toluene	47	
n-Butyl benzene	14	
t-Butyl benzene	1	
n-Butyl benzene sulphonate	39	
sec-Butyl benzene sulphonate	2	
t-Butyl benzene sulphonate	0	
n-Butanol	44	
t-Butanol	2	
n-Amyl alcohol		84
t-Amyl alcohol		0
Diethyl ether		0
Dioxane		0
Ethanolamine		81
Triethanoloamine		0

by the substitution of linear alkyl chains provides an excellent example of the practical advantages of a knowledge of structure—degradability relationships.

Data on the degradation of substituted benzenes have been obtained by Alexander & Lustigman (1966). These workers followed the disappearance of several derivatives of benzene which were supplied as a sole source of carbon to a mixed culture of soil organisms. The results with compounds containing only one substituent in the benzene ring showed (Table 3) that the rate of degradation followed the order: benzoate = phenol > aniline > anisole > benzenesulphonate > nitrobenzene, indicating that the carboxy and hydroxy substituted

Table 3

Degradation of benzene derivatives by a mixed culture of soil organisms (from Alexander & Lustigman, 1966)

Compound	Decomposition period (days)
Benzoate	1
Phenol	1
Aniline	4
Anisole	8
Benzene sulphonate	16
Nitrobenzene	> 64

compounds were easily broken down but that amino, methoxy, sulphonic acid and nitro groups caused increasing resistance.

With benzene derivatives containing two substituents, it was shown that the presence of either a sulphonic acid or a chloro group resulted in greatly reduced rates of metabolism. Also, methoxy, nitro and amino groups were not conducive to biodegradability except in the case of some benzoic acid and phenol derivatives. Generally speaking, toluene derivatives were easily degradable unless a nitro or sulphonate grouping was present. From this it can be inferred that, as far as benzene compounds are concerned, the substituents most likely to reduce susceptibility to microbial attack are sulphonate, chloro and nitro, whether singly or with other substituents. On the other hand, phenols and benzoates are usually degradable. Some resistance can result from the presence of amino, methoxy and methyl groupings.

This study also revealed another important factor, namely that the position of the substituents has a marked effect on the rate of degradation of a molecule. Thus, *meta*-substituted benzene derivatives were almost invariably degraded more slowly than were the *ortho-* or *para*-substituted analogues.

The test conditions employed, however, suffer from certain limitations. The compounds were normally presented to the mixed cultures as sole sources of carbon and energy, but when glucose was included with one compound (*o*-anisidine) which was apparently not degradable, the concentration of the compound in the test system rapidly decreased. Thus the presence of a second carbon-energy source might well have revealed other degradable compounds, and any inferences derived from these results must be assessed with these limitations in mind. However, the results were in agreement with other studies since several workers have found that *meta*-substituted compounds are difficult to degrade. For example, Kameda, Toyoura ·& Kimina (1957) showed that of 32 pseudomonads isolated from soil, none could utilize *m*-methoxy, *m*-nitro- or *m*-amino-benzoates, although some of the strains grew with the *o*- and *p*-amino-benzoates. Also, Cartwright & Cain (1959) reported that organisms could easily be isolated for growth on *o*- and *p*-nitrobenzoic acids but with difficulty on the *meta*-substituted derivative, and other studies (Cripps, unpublished) showed that only 1 of 13 soil isolates could utilize *m*-hydroxybenzoate for growth whereas all grew on *p*-hydroxybenzoate. It would appear generally true, therefore, that with pure and mixed cultures *meta*-substituted isomers of synthetic organic materials are much more resistant to microbial attack. This work was carried out with compounds which were not necessarily pesticidal, but it has been shown (Alexander & Aleem, 1961) that the same generalizations can be applied to herbicides containing a substituted benzene nucleus.

The number of substituents on a benzene nucleus is also an important factor. MacRae & Alexander (1965) in a study of the breakdown by mixed populations of several halogenated benzoic acids, showed that the 3 isomeric mono-chlorobenzoates were all fairly readily degradable, the *meta* isomer being the least susceptible, but that 2,4-, 3,4- and 2,5-dichlorobenzoic acids and tri- and tetra-chlorobenzoic acids were not attacked. Hence, in this case, the number of substituents, rather than their position, determines the ease of microbial metabolism. Multiple substitutions of a benzene nucleus, of course, reduces the number of points at which microbial attack is possible.

Another group of compounds which has been studied is the *s*-triazines. These form the basis of a number of herbicides whose residues have been shown to be resistant to breakdown. The structures of the individual members of this class of pesticides are very similar, consisting of the triazine nucleus carrying usually either a chloro- or methoxy- group and two alkylamino substituents. In a study of the factors responsible for the resistance to degradation of the triazine nucleus, Hauck & Stephenson (1944) showed that the successive replacement of the hydroxyl groups of cyanuric acid (2,4,6-trihydroxy-1,3,5-triazine) by amino groups caused increasing resistance to the release of nitrogen during nutrification tests. Cyanuric acid itself was degradable but when all 3 hydroxyl groups were replaced by amino groups, no release of nitrogen was detected over a period of 4 weeks. These results suggest that the slow breakdown of triazine herbicides may be due to the presence of the substituted amino groupings and they reveal that the triazine ring itself, like the benzene nucleus, is biodegradable. This has been confirmed by the isolation of pure cultures which are able to use cyanuric acid as sole source of nitrogen (Jensen & Abdel-Ghaffar, 1969; Cripps, unpublished).

No clear reason can as yet be given for these effects. It may be that the addition of certain substituents into certain positions of the aromatic ring of a degradable compound could cause the molecule to be toxic to an organism or prevent it entering the cell. The answer might simply be that some molecular arrangements confer resistance on a compound due to steric effects preventing the necessary close association of the molecule with an enzyme.

(b) Pure vs mixed cultures

Although the work with mixed cultures revealed that certain molecular arrangements are likely to decrease susceptibility to a microbial attack, it has proved possible to isolate organisms in pure culture which are capable of utilizing some of the less easily degradable compounds after the induction of suitable enzymes. For example, the branched-chain compound *p-iso*propyl-toluene acts as a growth substrate for a pseudomonad (Leavitt, 1967), and organisms have been obtained which grow on arylsulphonates (Cain & Farr, 1968) and on 3,5-dinitro-*o*-cresol (Jensen & Lautrup-Larson, 1967; Hamdi & Tewfik, 1970). In addition, after growth on benzoic acid, organisms have been shown to be able to oxidize chlorine-substituted benzoic acids (Hughes, 1965;

Walker & Harris, 1970; Horvath & Alexander, 1970). These few examples show that the enzyme systems, which might be responsible for the breakdown of these compounds in biodegradability tests, can be studied *in vitro*, and these studies usually reveal that the transformations occur at easily measurable rates. This suggests that other factors besides an unfavourable chemical structure might be operating to decrease the rate of breakdown under other conditions, e.g. in mixed cultures or in the field. It may well be that the observed slow rates of microbial degradation of these compounds in studies involving mixed cultures could be due to an unsuitable selection of the inoculum in which the required organisms are only present in low number, rather than a resistance to degradation due to molecular configuration.

(c) *Some general observations*

The above work provides a basis whereby some general relationships between biodegradability and chemical structure can be established but no hard and fast rules can be deduced. These studies have been concerned with the effect of certain molecules on whole cells of micro-organisms but more recently some work has appeared describing the effects of chemical structure on metabolism at the enzyme level. Kearney (1965, 1967) obtained an enzyme preparation from an organism capable of metabolizing the herbicide isopropyl *n*-(3-chlorophenyl)-carbamate (CIPC) which was able to hydrolyse this and other phenyl carbamates. The reaction catalysed by the enzyme is shown in Fig. 3, and the

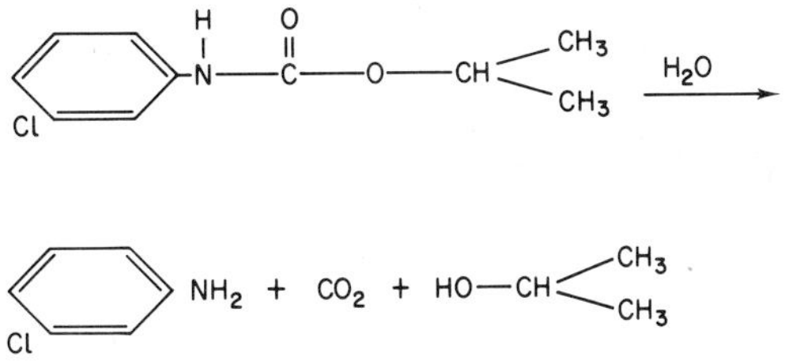

Fig. 3. Enzyme catalysed hydrolysis of CIPC

rates of hydrolysis with several *n*-phenylcarbamates with various ring substituents and alcohol groups were determined in an attempt to relate steric and electronic properties to breakdown. The hydrolysis rate was found to decrease as the size of the alcohol group was increased from ethyl to propyl to *iso*propyl. In addition, the rate decreased further when 1 or 2 chlorine atoms were introduced into the *iso*propyl alcohol moiety. These results can be interpreted in terms of the specificity of the enzyme; the more bulky the alcohol group, the less close the fit of the molecule to the reactive site of the enzyme. A different explanation is necessary for the alteration of the rates of hydrolysis caused by different substitutions at the *meta* position of the phenyl ring because these groups are probably too far away from the reactive site to exert a noticeable steric effect. The rate of reaction was shown to decrease in the order $NO_2 > CH_3 CO- > Cl > CH_3 O- > H$, an order which correlates with a decrease in the inductive or electron-withdrawing effect of the groups. Groups which exert an electron-withdrawing effect affect the acidity of the proton on the nitrogen atom so that a direct relationship between the acidity of this proton and the rate of enzymic hydrolysis is established. Although studies of this kind do not take into account any effects the compounds may have on the whole organisms, they may provide some explanation of their persistence in the soil. It is interesting, in this respect, that some of the groups which have previously been shown to be a hindrance to microbial attack (NO_2-, Cl-, $CH_3 O$-) can, in some circumstances, be an apparent aid to metabolism but it must be noted that only a single enzymic step is considered here and the products of the reaction, namely alcohols and substituted anilines, may be difficult to degrade.

Recent reports (Focht & Alexander, 1970, 1971) have demonstrated some of the structural features responsible for the persistence of DDT. A hydrogenomonad was isolated by enrichment culture using diphenylmethane as sole carbon source and was tested for its ability to metabolize various related compounds which were analogues or known metabolites of DDT. The results (Fig. 4) indicated that the organism could utilize diphenylethane, benzhydrol and *p*-chlorobenzhydrol as growth substrates but could not metabolize DDT or *p*-dichlorobenzophenone under any conditions. Benzophenone, *p*-chloro-

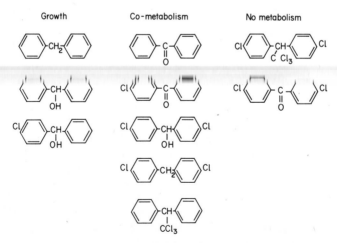

Fig. 4. Metabolic activities of a *Hydrogenomonas* sp. towards diphenylmethane derivatives (from Focht & Alexander, 1970).

benzophenone, p,p'-dichlorobenzhydrol, p,p'-dichlorophenylmethane, and 1,1-diphenyl-2,2,2-trichloroethane were co-metabolized by washed cell suspensions of the organism with the production of compounds which indicated that at least one of the phenyl rings had been cleaved. These results show that presence of a p-chloro substitution and of a carbonyl or trichloromethyl group at the carbon atom linking the two phenyl nuclei confers increased resistance to microbial attack. However, organisms have been shown to be able partially to degrade the DDT molecule and several metabolites have been identified (Wedemeyer, 1967a,b). Two of these metabolites, p,p'-dichlorophenylmethane and p,p'-dichlorophenylbenzhydrol, have now been shown to be further metabolized and it is possible to trace a metabolic pathway which is theoretically capable of effecting considerable degradation of the DDT molecule.

5. Conclusion

From the foregoing, one can deduce that certain molecular arrangements restrict degradation of a compound by micro-organisms. No rules which are generally applicable can be derived but trends can be identified one of which is that certain chemical groupings (e.g. multiple chlorine atoms or branched structures) can be expected to reduce biodegradability. From a practical viewpoint, these results might provide some help to the organic chemist, in his search for active pesticide materials, by indicating the type of molecule which could be expected to degrade readily. The problem is complicated, however, by the fact that it is often not pesticides themselves which are persistent but their breakdown products. However, organisms have been found that will degrade compounds which include some of the unfavourable features in their structures and this suggests that many more data are required before it will be possible to predict whether a compound is biodegradable solely by considerations of its structure.

6. References

Alexander, M. (1965). Biodegradation: Problems of molecular recalcitrance and microbial fallibility. *Adv. appl. Microbiol.* **7,** 35.

Alexander, M. & Aleem, M. I. H. (1961). Effect of chemical structure of microbial decomposition of aromatic herbicides. *J. agric. Fd Chem.* **9,** 44.

Alexander, M. & Lustigman, B. K. (1966). Effect of chemical structure on microbial degradation of substituted benzenes. *J. agric. Fd Chem.* **14,** 410.

Cain, R. B. & Farr, D. R. (1968). Metabolism of arylsulphonates by micro-organisms. *Biochem.J.* **106,** 859.

Cartwright, N. J. & Cain, R. B. (1959). Bacterial degradation of the nitrobenzoic acids. *Biochem. J.* **71,** 248.

Feist, C. F. & Hegeman, G. D. (1969). Phenol and benzoate metabolism by *Pseudomonas putida:* regulation of tangential pathways. *J. Bact.* **100,** 869.

Focht, D. D. & Alexander, M. (1970). DDT metabolites and analogs: ring fission by *Hydrogenomonas. Science, N. Y.* **170,** 91.

Focht, D. D. & Alexander, M. (1971). Aerobic co-metabolism of DDT analogs by *Hydrogenomonas* sp. *J. agric. Fd Chem.* **19,** 20.

Gale, E. F. (1952). *The Chemical Activities of Bacteria.* New York: Academic Press.

Hamdi, Y. A. & Tewfiks, M. S. (1970). Degradation of 3,5-dinitrocresol by *Rhizobium* and *Arthrobacter* spp. *Soil Biol. Biochem.* **2,** 163.

Hauck, R. D. & Stephenson, H. F. (1964). Nitrification of triazine nitrogen. *J. agric. Fd Chem.* **12,** 147.

Heukelekian, H. & Rand, M. C. (1955). Biochemical oxygen demand of pure organic compounds. *Sewage ind. Wastes* **27,** 1040.

Horvath, R. S. & Alexander, M. (1970). Co-metabolism of *m*-chlorobenzoate by an arthrobacter. *Appl. Microbiol.* **20,** 254.

Hughes, D. E. (1965). The metabolism of halogen-substituted benzoic acids by *Pseudomonas fluorescens. Biochem. J.* **96,** 181.

Jensen, H. L. & Abdel-Ghaffar, A. S. (1969). Cyanuric acid as nitrogen source for micro-organisms. *Arch. Mikrobiol.* **67,** 1.

Jensen, H. L. & Lautrup-Larsen, G. (1967). Micro-organisms that decompose nitro-aromatic compounds, with special reference to dinitro-ortho-cresol. *Acta Agric. scand.* **17,** 115.

Kameda, Y., Toyoura, E. & Kimura, Y. (1957). Metabolic activities of soil bacteria towards derivatives of benzoic acid, amino acids and acyl amino acids. *Kanazawa Diagaku Yakugakuba Nempo,* **7,** 37; via *Chem. Abstr.* **52,** 4081 (1958).

Kearney, P. C. (1965). Purification and properties of an enzyme responsible for hydrolysing phenylcarbamates. *J. agric. Fd Chem.* **13,** 561.

Kearney, P. C. (1967). Influence of physicochemical properties on biodegradability of phenylcarbamate herbicides. *J. agric. Fd Chem.* **15,** 568.

Leavitt, R. I. (1967). Microbial oxidation of hydrocarbons. Oxygen of *p*-isopropyltoluene by a *Pseudomonas* sp. *J. gen. Microbiol.* **49,** 411.

van der Linden, A. C. & Thijsse, G. J. E. (1965). The mechanisms of microbial oxidations of petroleum hydrocarbons. *Adv. Enzymol.* **27,** 469.

Ludzack, F. J. & Ettinger, M. B. (1960). Chemical structures resistent to aerobic biochemical stabilization. *J. Wat. Pollut. Control Fed.* **32,** 1173.

MacRae, I. C. & Alexander, M. (1965). Microbial degradation of selected herbicides in soil. *J. agric. Fd Chem.* **13,** 72.

Ribbons, D. W. (1965). The microbiological degradation of aromatic compounds. *Ann. Rep. Chem. Soc.* **62,** 445.

Swisher, R. D. (1963). Biodegradation rates of isomeric diheptylbenzene sulphonates. *Dev. ind. Microbiol.* **4,** 39.

Walker, N. & Harris, D. (1970). Metabolism of 3-chlorobenzoic acid by Arthrobacter species. *Soil Biol. Biochem.* **2,** 27.

Wedemeyer, G. (1967a). Dechlorination of 1, 1, 1-trichloro-2, 2-bis (*p*-chlorophenyl) ethane by *Aerobacter aerogenes. Appl. Microbiol.* **15,** 569.

Wedemeyer, G. (1967b). Biodegradation of dichlorodiphenyltrichloroethane intermediates in dichlorodiphenyl acetic acid metabolism by *Aerobacter aerogenes. Appl. Microbiol.* **15,** 1494.

Biodeterioration and Biodegradation
of Synthetic Polymers

H. O. W. Eggins and J. Mills

Biodeterioration Information Centre, Department of Biological Sciences
University of Aston in Birmingham
80 Coleshill Street, Birmingham, England

A. Holt and G. Scott

Department of Chemistry, University of Aston
in Birmingham, Birmingham, England

CONTENTS

1. Introduction

THE FACT that plastic materials are susceptible to biological breakdown became evident during the Second World War, plastics up to that time having their greatest usage in temperate zones with little evidence of attack. During the Second World War, however, large amounts of plastics-containing equipment were moved into tropical regions of the world, areas more conducive to large scale biological attack.

The discovery that plastics formulations could be attacked by micro-organisms under conditions of high (tropical) temperatures and high humidities led to intensive research programmes, and much of the earlier reported work tends to be confusing; no distinction is made between the susceptibility of the polymeric constituents and the plasticizers, fillers, etc., or whether, in fact, micro-organisms found on the surface of plastic were responsible for mechanical breakdown of the plastic. The principles behind the biological breakdown of materials were formulated by Hueck in 1959 and later published (Hueck, 1965). He defined biodeterioration as "any undesirable change in the properties of a material of economic importance caused by the vital activities of organisms".

267

The operative words are "undesirable changes in the properties", and strictly speaking much of the work published on the biodegradation of plastics is really on the biodeterioration of plastics since it centres mainly around trying to prevent biological attack on plastics. We reserve the term 'biodegradation' for the breakdown and/or conversion of waste materials into innocuous or useful products by micro-organisms. Examples of biodegradation of materials would be the composting of town waste into a saleable product. However, in recent years the term biodegradation has become applied to plastic materials, mainly packaging plastics, and the search has begun for biodegradable plastics which will be broken down completely by micro-organisms after their commercial life.

2. The Biodeterioration of Plastics

Using the system of Hueck (1965) we may classify the biodeterioration of plastics as being caused by: (1) mechanical processes; (2) soiling (or fouling) processes; and (3) chemical processes, (a) assimilatory, (b) dissimilatory, and then go on to ask "what is the cause of each phenomenon?" and "how can it be checked and prevented?"

With plastics, mechanical damage is caused by the gnawing activities of termites, insects and rodents and by the boring activities of molluscs. The larva of the False Cloth moth (*Hofmannophila pseudospretella*) is known to eat its way though polythene, polystyrene and nylon, and the larvae of aquatic moths have been found damaging polythene linings of ponds, eating their way through the polythene to the rougher surface underneath, to which they attach themselves before pupating (Whalley, 1965). The attack of p.v.c. cable coatings by termites is well known though the plastic itself does not seem to form a foodstuff for the termites. Rodents, particularly rats and squirrels, will gnaw through any type of plastic pipe or coating when they obstruct access to food; again there is no evidence that rats actually eat the plastic or that they gnaw it in preference to other substrates. These damaging mechanical processes, usually accompanied by failure of electricity or water supplies, happen simply because the plastics cable or pipe is in the way of burrowing or food searching activities. Plastics susceptible to this type of biodeterioration are usually protected, therefore with insect or rodent repellents.

Closely connected to this type of damage is that caused by fouling or soiling in which organisms use a plastic surface as a holdfast, e.g. algae attach themselves to polythene or p.v.c. liners in pools and irrigation ditches, and with their associated bacteria can cause clogging of pipelines. Fungal colonies are also frequently seen on plastic shower curtains. They utilize the soap particles on the surface for food and cause an aesthetic deterioration of the plastic.

By far the most important type of biodegradation is the type caused by chemical activities of living organisms, mainly by the fungi and the bacteria. These chemical processes can be divided into: (a) assimilatory processes in which constituents of the plastics serve as sources of nutrients to organisms and (b) dissimilatory processes, in which the plastic is not used as a source of carbon but may be chemically damaged by corrosive substances secreted by an organism living on detritus lying on the surface of the plastic. An example of a dissimilatory process of plastics is in its staining and discoloration, the best known example being the pink staining of nylon caused by *Penicillium janthinellum* (Dayal *et al.*, 1962). However, the most important process causing the biodeterioration of plastics is the assimilatory process in which constituents of the plastic are used as sources of carbon by micro-organisms.

Synthetic polymers, in general, are resistant to microbial enzymic attack and are not broken down, but cellulose nitrate, melamine-formaldehyde resins and polyvinyl acetate polymers are exceptions showing some signs of susceptibility (Brown, 1945; Abrams, 1948). Little information relating clearly to the susceptibility of polymers is to be found.

The plasticizers, as a group of low M.W. compounds, influence the susceptibility of a plastic to microbial attack more than any other group. Brown (1945) reported that almost half of 144 plasticizers tested could serve as fungal nutrients. Removal of plasticizers, usually in p.v.c. formulations, results in undesirable physical changes in the properties of the plastic, such as changes in tensile strength, loss of flexibility and changes in the electrical properties. Plasticizers known to be susceptible to microbial breakdown include derivatives of lauric, oleic, ricinoleic and stearic acids; phosphoric acid and phthalic acid derivatives are in general resistant. However, much of this work was done using questionably chosen pure cultures or a mixture of a few species of micro-organisms, and it has since been shown that several plasticizers previously labelled "funginert" could in fact be broken down by careful selection of organisms (Klausmeier & Jones, 1961) and the presence of additional essential microbial nutrients (Klausmeier, 1966). A variety of test methods has been used to determine the extent of plasticizer removal and its subsequent effects on test plastics. These test criteria can be summarized as: (1) visual examination of the extent and profuseness of microbial growth. This sometimes gives misleading results but can be valuable when assessing the protection of a plasticizer by a fungicide in an agar medium (Bomar, 1968); (2) weight of fungal mycelium produced; (3) loss in weight of the plastic after fungal growth has been removed from the surface (Burgess & Darby, 1965; Hazeu, 1967; Hitz & Zinkernagel, 1967; Booth & Robb, 1968). This test has recently been reviewed (Wendt *et al.*, 1970) and has been found to produce as sensitive or more sensitive results than other methods for estimating the biodeterioration of plasticized p.v.c. in a 1–3 weeks time scale; (4) changes in surface and volume resistivities (Dolezel, 1967, Allakhverdiev *et al.*, 1967*a*, *b*); (5) mechanical methods such as changes in tensile strength elongation, stress at elongation and flexibility (Berk, 1950; Hazeu & Waterman, 1965; Hazeu, 1967; Dolezel, 1967; Wälchli & Zinkernagel, 1966; Booth & Robb, 1968); (6) respirometric methods, where the uptake of oxygen by organisms breaking down plasticizers is measured and related to breakdown of that plasticizer (Burgess & Darby, 1965; Cavett & Woodrow, 1969).

As noted above, much of the work to date has been directed towards the goal of producing plastics which are resistant to biodeterioration, and many biocidal additives have been suggested. Ideally a biocide should possess these properties: (a) a broad spectrum of activity; (b) high toxicity to micro-organisms but low toxicity to humans and animals; (c) low cost; (d) stability in use; (e) ease of application and (f) compatibility with the other additives in the plastic.

Plastics for use in hospitals frequently contain hexachlorophene (Hopfenberg & Tulis, 1970; Taylor, 1970) whilst biocides for use in p.v.c. include quaternary ammonium compounds, pentachlorophenols (Baseman, 1966), halogenated dinitrobenzene compounds (Palei et al., 1965) copper 8-quinolinolate (Kaplan et al., 1970), organotin compounds (U.S. Patent 3,445,249, 1969; Wessel & Bejuki, 1959) and many others (see Heap, 1965).

3. The Biodegradation of Plastics

Because of their nature, plastics are eminently suitable for the production of packaging materials, and as such are usually rejected after a single use. Such plastics contribute to industrial and domestic waste and to a less extent to the problem of litter. In this country in 1969 plastics waste in domestic refuse amounted to 1·12% of the total collected refuse, occupying 3–5% of the refuse volume. Trends in the use of throwaway plastic packages and bottles will present local authorities with an ever increasing amount of waste plastics (a projected 5% by 1980 for disposal (Staudinger, 1970).

Many authorities, especially those of large cities, have adopted an incineration policy for disposal of their refuse, but smaller authorities resort to either controlled tipping or to composting of their refuse. There are many problems associated with the incineration of plastics wastes (Staudinger, 1970; Cheater, 1970) and furnaces are being designed to burn plastics more efficiently without release of atmospheric pollutants. Tipping of refuse for landfilling operation still has a high priority in long range waste management, but in many cases the refuse has to be transported sometimes quite considerable distances to the site (Abrahams, 1969; Clutterbuck, 1971), with associated increases in cost. An alternative to landfilling and incineration is the composting of refuse, a biological process in which natural high M.W. polymers such as cellulose are degraded by micro-organisms to a humus-based product which can be used directly for land reclamation (as opposed to burying untreated refuse in landfill operations), or which can be processed into a saleable product and used as a soil conditioner and fertilizer.

(a) Degraded plastics in fertilizers

Some of the problems related to using this type of compost are that it is low in nutrient value, difficult to transport and to apply, and it usually contains a certain amount of undegraded plastics waste. However, as a cheap form of soil conditioner and when used in conjunction with chemical fertilizers it could be a valuable asset in reclaiming and revitalizing poor types of soil. Recent experiments using composted town waste have shown very promising results with the growth of certain vegetables (Duckworth, 1970).

Plastics waste could still be picked out of the compost by hand, but here again the objectionable nature of the work would add to the cost of producing a product, and the picked out plastics would still have to be disposed. If such plastics waste could be produced in a biodegradable form this sorting operation would become unnecessary and a more acceptable product would be produced, still at a reasonably low cost.

Recalcitrant organic material such as humic acids, petroleum products and waxes are already known in nature. Apart from their polymeric form, probably one of the most important features of plastics in regard to biodegradation is their hydrophobic nature and their density and homogeneity as solids.

(b) Thermophilic microbial degradation systems

One line of attack has been through our investigation of the microbiological biodegradability of plasticizers and polymers in simulated biological refuse systems. If one considers the possibility of composting as a biological recycling system which in future could commend itself as an alternative to incineration, the effect of thermophilic systems should be studied, as it is these which would be central to such processes. To date, emphasis has been on mesophilic systems but we have concentrated on thermophilic activities as being more pertinent to possible biological refuse disposal systems.

Respirometric experiments at 48° using the thermophilic microbial population of soil have shown that of 12 plasticizers tested 9 were able to enhance the respiration of the soil population above its normal thermophilic endogenous rate (Fig. 1). In this process the plasticizers are either being used directly as microbial carbon sources or, more likely, the esters are being split initially by microbial enzymes or by interaction with other chemicals in the soil and the products are then being utilized by the microbial population.

Tri-n-butyl citrate which has been reported to support no fungal growth (Brown, 1945) in the mesophilic temperature range produced 1390% increase in

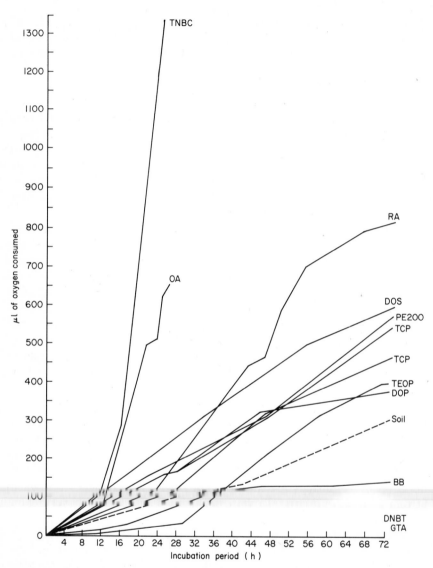

Fig. 1. The uptake of oxygen at 48° by the thermophilic microbial population of a pastureland soil enriched with various plasticizers.

the rate of soil respiration within 24 h and supported the growth of 7 thermophilic fungi in pure culture work when used as the sole C source in a mineral salts medium. Dioctyl sebacate, a plasticizer known to be degradable and much used in this work, increased the soil respiration by 100% in 72 h and supported the growth of 17 of 18 thermophilic fungi tested. Polyethylene glycol

200, tricresyl phosphate, tritolyl phosphate, triethyl-*o*-phosphate and dioctyl phthalate have all been reported to be unable to support fungal growth in the mesophilic temperature range (Brown, 1945) but all enhanced the respiration of the soil population in the thermophilic temperature range (Table 1) and supported the growth of individual thermophilic fungi (Table 2). Benzyl benzoate and glyceryl triacetate depressed respiration during the 72 h test period but supported individual thermophiles in pure culture. Di-*n*-butyl tartrate was actively biocidal and would only support the growth of 1 of 18 fungi tested. Oleic and ricinoleic acids, products of plasticizer breakdown, enhanced respiration and supported good growth when tested.

These results (Mills & Eggins, to be published) using pure plasticizers in the thermophilic temperature range, have shown that there is a need for assessing the biodeterioration of plasticized plastics with thermophilic micro-organisms when those plastics are to be exposed to high temperatures and humidities, e.g. in the tropics, or when they are to be wholly or partially buried in the soil in such environments.

Looking at the polymers themselves, our work has shown that polymers currently used in packaging plastics, e.g. polyethylene, are not susceptible to biodegradation in the thermophilic temperature range. We therefore investigated

Table 1

*Comparative cumulative oxygen uptakes by soil micro-organisms
at 48° in soil enriched with various plasticizers*

Plasticizer	Cumulative oxygen uptake (μl) at 72 h	% inhibition or enhancement of soil respiration due to plasticizer
Soil endogenous respiration	292·5	Control
Tri-*n*-butyl citrate (TNBC)	1338 (after 24 h)	+ 1390 (after 24 h)
Di-octyl sebacate (DOS)	586·4	+ 100
Polyethylene glycol 200 (PEG 200)	561·7	+91·5
Tri-cresyl phosphate (TCP)	539·9	+84·5
Tri-tolyl phosphate (TTP)	464·7	+59·0
Tri-ethyl orthophosphate (TEOP)	396·5	+35·0
Di-octyl phthalate (DOP)	377·9	+28·5
Benzyl benzoate (BB)	134·2	−54·3
Di-*n*-butyl tartrate (DNBT)	0	Completely
Glycerol triacetate (GTA)	0	inhibits respiration
Products of plasticizer breakdown		
Oleic acid	648·8*	+ 620%* (24 h)
Ricinoleic acid	805·0	+ 170%

Table 2

*The capacity of thermophilic fungi to grow on a mineral salts
agar medium with plasticizer as the sole carbon source*

| Plasticizer | No. of thermophilic fungi 18 | |
	capable of growth on plasticizer*	not capable of growth on plasticizer
Tri-*n*-butyl citrate	7	11
Dioctyl sebacate	17	1
Polyethylene glycol 200	10	8
Tricresyl phospate	3	15
Tritolyl phosphate	2	16
Triethyl orthophosphate	4	14
Dioctyl phthalate	4	14
Benzyl benzoate	6	12
Di-*n*-butyl tartrate	1	17
Glyceryl triacetate	5	13
Oleic acid	9	9
Ricinoleic acid	14	4

Incubation period, 14 days at 48°.
* As sole C source: amount added to medium, 1%.

the possibility of partially degrading the polyethylene polymer initially by physicochemical means and then using the breakdown (in this case oxidation) products for further breakdown by thermophilic fungi (Mills & Eggins, 1970). The polyethylene polymer was partially oxidized to low M.W. dicarboxylic acids which were then used as a carbon source for thermophilic fungi at 40 and 50°. Of the 10 thermophiles tested, all were able to grow and produce biomass to a greater or less extent, some growing better and some worse at 40° than at 50°.

(c) *Disposal of plastics waste*

These results show that this could be one method to dispose of unwanted plastic waste, i.e. the pretreatment of existing nondegradable polymers to partially break them down, and then feeding the lower M.W. entities to micro-organisms. for example in a refuse disposal system.

Plastics litter at the present time seems to present more of a nuisance problem than a serious environmental pollution problem. In the near future, however, it could become a serious pollution problem and here again there would seem to be a case for producing biodegradable packaging plastics which would decompose completely when they became litter.

If these assumptions prove to be valid (and there appears to be a need for biodegradable packaging plastics) we feel it is possible to indicate certain broad lines along which further investigations could proceed. The main problem arises from a few high-tonnage materials such as polyethylene, polypropylene and p.v.c. Few estimates of the tonnage of plastics litter in Britain have been made. However, it is known that 100,000 tons of high density polyethylene is used/year for making nonreturnable sacks and 45,000 tons for throwaway containers. The main objection to such litter is aesthetic, although it can present dangers to grazing animals. Unlike cellulosic wrappings, it does not easily compost and, as mentioned before, its hydrophobic nature contributes greatly to its nuisance value when litter.

It would be possible to separate the uses for which plastic materials are used, developing biodegradative polymers for uses likely to end up as litter. Considering polyethylene and polypropylene, we see that their basic chemical structure as saturated hydrocarbons is similar to that encountered in natural products such as fats and waxes and in paraffin wax. Fats and waxes are attacked by micro-organisms, although their susceptibilities vary with the physical state of the material, e.g. micro-organisms usually attack this type of material at a liquid to liquid or liquid to solid interface rather than by actual invasion of the fatty material. The lower M.W. paraffins also appear to be attacked.

From the fact that halogenated compounds are relatively rare in nature it might perhaps be anticipated that the p.v.c. polymer could resist attack. The considerations presented so far indicate the need for investigations on model compounds to determine relationships between M.W., physical form and biological attack. This line of research might possibly lead to the production of a new polymer containing susceptible groupings along its chain length. Biological attack and removal of these groupings would lead to a depolymerization of the molecule and a subsequent attack on the lower M.W. entities. However, most plastics are used because of their hydrophobic and biodegradation resistant nature, and it is difficult to devise systems which can "switch on" as soon as plastics become litter.

Suggestions have included reactions to changes in pH value, temperature and other factors, but to date the most attractive appears to be the encouragement of oxidative degradation due to UV absorption and subsequent microbial breakdown of the more susceptible end product.

(d) Oxidative degradation during environmental exposure

The chemical reactions occurring during environmental exposure of plastics involve a free radical chain reaction leading to cleavage of the chain and the introduction of functional groups such as carboxyl groups.

By far the most important environmental factor present under weathering conditions is electromagnetic radiation. Although the basic degradation sequence remains unchanged in the presence of UV light, the latter has a profound effect on the initiation reaction. This results from the sensitivity to UV light of many of the functional groups present in the polymers. They may be present either as a normal part of the polymer chain or as impurities introduced during polymer manufacture or processing.

By far the most important of these from a practical point of view is the hydroperoxide group because, under conditions where hydroperoxides are thermally stable, UV light causes their photolysis to free radicals which initiate the oxidation process:

$$ROOH \xrightarrow{\quad UV \quad} RO\cdot \quad + \quad OH$$
$$RO\cdot \ + \ 2\,RH \longrightarrow ROOH \ + \ ROH \ + \ R\cdot$$
$$HO\cdot \ + \ 2\,RH \longrightarrow ROOH \ + \ H_2O \ + \ R\cdot$$

Hydroperoxides are therefore powerful photoactivators for autoxidation and have the practical effect of reducing the induction period to rapid degradation. This process, like thermal oxidation, is normally autocatalytic and for the same reason, namely, that the rate of oxidation is proportional to the hydroperoxide concentration in both cases.

Chain interruptive antioxidants have much less effect in this situation because they do not prevent the formation of hydroperoxides which, under photochemical conditions, are initiators at room temperature, but which in the absence of UV light are stable to relatively high temperatures. The processing operation and environmental exposure are thus closely interdependent and in designing polymers for service life the total history of the polymer has to be considered.

The nature of the antioxidant added to stabilize the polymer during processing has a crucial effect on the environmental stability of the polymer. In general it may be concluded that if resistance to UV light is required, a peroxide-decomposing antioxidant should be used in addition to UV screening agents because they will clearly act in a complimentary way to the latter and synergize with them. If on the other hand a polymer is required which is short lived under environmental conditions clearly the more effective peroxide decomposers should normally be avoided since they will be antagonists for this process.

(e) *Accelerated environmental degradation*

The foregoing has given the scientific background to the problem of accelerated degradation processes. It is clearly possible in principle to accelerate the speed of environmental degradation of plastics by taking advantage of the electromagnetic energy of the sun's rays, but the problem is to do this with maximum efficiency at minimum cost. An important aspect of the economics of any such process must be that it does not appreciably increase the cost of the fabricated product. There are 2 major factors involved in this. The first is that the polymer should not be appreciably more expensive to the fabricator. The second is that there should be minimal interference with the processing stage.

The materials cost of the bulk polymer is likely always to be a dominating factor, and for this reason it seems unlikely that, except for very specialized uses, a new polymer can be contemplated. Even a co-polymer modification of one of the major polymers used in packaging would inevitably be more expensive than present materials due to the more limited scale of manufacture.

In principle it is possible to introduce hydroperoxides into a polymer during processing and, because hydroperoxides are important photo-initiators, it has already been seen that the environmental life of the polymer can be controlled in this way.

Ideally, what is required is an added chromophore which has no effect on the processing stability of the polymer or on the shelf life of the polymer indoors, but which begins to initiate polymer oxidation as soon as exposure to the outdoor environment occurs. This implies that the initiating process must be triggered only by light of shorter wavelength than 320 nm, which is that normally excluded from the indoor environment by window glass, and that the oxidation processes continues, albeit more slowly, even if the source of light is removed.

A search has revealed a variety of chemical chromophores which satisfy this requirement. A very important consequence of this photochemical oxidation process for the subsequent fate of the plastic is that during the degradation process not only is the surface area of the polymer increased enormously, so that micro-organisms can now attack the polymer more readily, but also chemical modification has occurred on the surface of the particles (by the introduction of hydrophilic groups such as carboxyl) which also favour attack by waterborne micro-organisms. We believe that we have here for the first time the possibility of a combined photo- and bio-degradation process in which the degradation commenced by the former can be completed by the latter, ultimately converting the plastic to useful humus, water and CO_2 which are the normal products of cellulose breakdown.

In conclusion, we should like to make a plea for ensuring that this problem is kept in perspective. Plastics have brought great benefits to man, one of which is the large scale use of a class of materials basically resistant to biodeterioration. However, they have now been used so widely that the existing litter laws are not sufficient to cope with the problems of plastic litter. We have suggested a possible approach which recognizes these aspects.

4. References

Abrahams, J. H. (1969). Packaging industry looks at waste utilisation. *Compost Sci.* **10**, 13.

Abrams, E. (1948). Microbiological deterioration of organic materials; its prevention and methods of test. U.S. Dept. of Commerce, National Bureau of Standards Misc. Publ. 188, Washington D.C.

Allakhverdiev, G. A., Martirosova, T. A. & Tariverdiev, R. D. (1967a). Change in the mechanical properties of polymeric films under the action of micro-organisms in soil. *Soviet Plastics* 2, 22.

Allakhverdiev, G. A., Tariverdiev, R. D., Kulieva, A. S., Kerimov, P. M., Drugii, E. D. & Khasaev, G. G. (1967b). Changes in the electrical resistivity of plastics films on ageing in soil. *Soviet Plastics* 3, 77.

Baseman, A. L. (1966). Antimicrobial agents for plastics. *Plastics Technol.* 12, 33.

Berk, S. (1950). American Society for Testing Materials Bulletin. (T.P. 181) p. 53.

Bomar, M. (1968). Some results of antifungal tests on combinations of plasticisers and fungicides. *Int. Biodetn Bull.* 4, 43.

Booth, G. H. & Robb, J. A. (1968). Bacterial degradation of plasticised p.v.c. – Effect of some physical properties. *J. appl. Chem.* 18, 194.

Brown, A. E. (1945). The problem of fungal growth on synthetic resins, plastics and plasticisers. O.S.R.D. Report No. 6067. *Modern Plastics* 23, 189.

Burgess, R. & Darby, A. E. (1965). Microbiological testing of plastics. *Br. Plastics* 38, 165.

Cavett, J. H. & Woodrow, M. N. (1969). A rapid method for determining degradation of plasticised p.v.c. by micro-organisms. In *Biodeterioration of Materials,* Part 4. London: Elsevier.

Cheater, G. (1970). Incineration of rubber and plastics waste. *Publ. Cleans.* 60, 187.

Clutterbuck, D. (1971). Wealth in waste. *New Scientist* 14 January.

Dayal, H. M., Makeshwary, K. L., Agarwal, P. N. & Nigan, S. S. (1962). Isolation of *Penicillium janthinellum* Biourge from parachute nylon fabric. *J. Sci. ind. Res.* 21-C, 356.

Dolezel, B. (1967). The resistance of plastics to micro-organisms. *Br. Plastics* 40, 105.

Duckworth, L. (1970). Shall we bury our rubbish–or be buried by it? *Birmingham Post* 12, October.

Hazeu, W. (1967). Results of the first interlaboratory tests on plastics. *Int. Biodetn Bull.* 3, 15.

Hazeu, W. & Waterman, H. A. (1965). Evalution of the microbial degradation of plastics by the vibrating reed method. Centraal Laboratorium Communication No. 229, p. 7.

Heap, W. M. (1965). Microbiol deterioration of rubbers and plastics. RAPRA Information Circular No. 476.

Hitz, H. R. & Zinkernagel, R. (1967). Test tube methods for evaluation of biodegradation of plasticised p.v.c. by *Pseudomonas aeruginosa. Int. Biodetn Bull.* 3, 21.

Hopfenberg, H. B. & Tulis, J. H. (1970). Advances in antibacterial plastics. *Mod. Plastics* 47, 110.

Hueck, H. J. (1965). The biodeterioration of materials as part of hylobiology. *Mat.* ~~Communication 1, 1~~

~~Kaplan, A. M., Greenberger, M. & Wendt, T. M. (1970). Evaluation of biocides for treatment~~ of polyvinyl chloride film. *Polymer Engng Sci.* 10, 241.

Klausmeier, R. E. (1966). The effect of extraneous nutrients on the biodeterioration of plastics. Society of Chemical Industry Monograph No. 23, p. 232. London: Society of Chemical Industry.

Klausmeier, R. E. & Jones, W. A. (1961). Microbial degradation of plasticisers. *Developments in Industrial Microbiology.* Vol. 2, p. 47. New York: Plenum Press.

Mills, J. & Eggins, H. O. W. (1970). Growth of thermophilic fungi on oxidation products of polyethylene. *Int. Biodetn Bull.* 6, 13.

Palei, M. I., Trepelkova, L. I., Akopdzhanyan, E. A. & Golodnaya, S. L. (1965). Fungus resistance of acoustic materials based on p.v.c. resin. *Soviet Plastics* p. 67.

Staudinger, J. J. P. (1970) Plastics waste and litter. Society of Chemical Industry Monograph No. 35. London: Society of Chemical Industry.

Taylor, G. F. (1970). Survival of Gram negative bacteria on plastic compounded with hexachlorophene. *Appl. Microbiol.* 19, 131.

Walchli, O. & Zinkernagel, R. (1966). Comparison of test methods for determining the resistance of plastics to microbial attack. *Mat. Organismen.* **1**, 161.
Wendt, T. M. Kaplan, A. M. & Greenberger, M. (1970). Weight loss as a method for estimating the microbial deterioration of p.v.c. film in soil burial. *Int. Biodetn Bull.* **6**, 137.
Wessel, C. J. & Bejuki, W. M. (1959). Industrial fungicides. *Ind. Engng Chem.* **51**, 25A.
Whalley, P. E. S. (1965). Damage to polythene by aquatic moths. *Nature. Lond.* **207**, 104.

Disposal of Infective Laboratory Materials

H. M. DARLOW

*Microbiological Research Establishment, Porton Down,
Salisbury, England*

CONTENTS

1. Introduction

DISPOSAL OF infective laboratory material involves somewhat more than the title would suggest; the subject must embrace not only everything from dead mice to their expiring breaths, but also materials that may be contaminated, in addition, with toxins or radio-isotopes. Furthermore, in its broadest sense it must surely include the disposal of wastes which, whilst no longer infective, still present problems. Such wastes include liquid effluents containing excess of germicides, solvents and other chemicals which can present hazards both to plumbing and sewage disposal, and 'throw-away' hypodermic needles and syringes which constitute a hazard to the 'totting' refuse operative and an attraction to drug addicts. The most important first step, however, is obviously taken in the field of disinfection, or, more properly, sterilization, and it is to this that this review is mainly devoted.

The history of disinfection, particularly 'in the cold', is as paved with good intentions as the road to Hell, but there are pitfalls between the stones waiting for the unwary to fall into, not infrequently accompanied by innocents who re-emerge as cases of laboratory infection or hospital sepsis. Disinfection is not synonymous with sterilization, and, whilst the latter is not always absolutely essential, it must surely be the preferable target in the laboratory, since without it microbiology as we know it would be impossible. There are almost as many methods of achieving this end as there are materials to be sterilized, and some of these are discussed below in the appropriate contexts, together with other relevant disposal problems.

2. Infected Air

Most microbiology laboratory procedures and accidents create aerosols of particles containing live organisms. These may present little more hazard than their capacity to contaminate cultures and ruin experiments, but analysis of accumulated data on human laboratory infection has shown that the inhalation of aerosols is likely to be by far the commonest cause of morbidity (Sulkin, 1961). This demands not only careful technique, awareness of the hazard and the use of safety devices at the bench (Darlow, 1969a) but also that effluent air from the laboratory and safety hoods be rendered safe before return to general circulation. There are many methods of achieving this with varying degrees of efficiency; heating, electrostatic precipitation, scrubbing, centrifugation, irradiation and mechnical filtration, or combination of 2 or more of these techniques. The most effective single method, however, is undoubtedly filtration by high efficiency fibrous filters, since this is not only economic, if correctly applied, but constitutes a barrier to particles alive or dead, independent of energy sources other than the ventilation fan, and prevents escape by diffusion or air currents should this also fail. The filter system must be installed as close as possible to the infected area to prevent contamination of ducts and exhaust fans, and means must be provided for *in situ* sterilization by fumigation. The fitting of coarse disposable prefilters is important as it greatly prolongs the life of the master units. These should be mounted actually within the toxic areas (rooms and cabinets) to permit easy inspection and servicing; indeed there is much to be said for mounting the master filters in the same situation. This eliminates the hazards of siting potentially dangerous equipment in clean crawl spaces, and facilitates simultaneous fumigation of the area and its filters together.

The charring or total combustion of airborne organisms, antigens and allergens by burning has attractions because it involves little resistance to air flow and ensures continuous sterility, provided that high temperatures are maintained. It is, however, uneconomic, unless the waste heat can be utilized, and it presents an explosive or toxic hazard where volatile solvents and reactive gases are involved. It must be remembered that the hydrogen which floated the ill-fated airship, R101, was made by passing steam over red-hot iron, and minor laboratory explosions have resulted from just this. On the whole the technique should be reserved for special circumstances, where no other method is permissible and where carefully designed equipment is available (Barbeito, Taylor & Seiders, 1968; Chatigny, Sarshad & Pike, 1970).

Very much the same remarks apply to subcharring temperatures, but here holding time becomes even more important; air is a poor conductor of heat and even the most sensitive organisms take time to heat up and die. Despite a literature too large to quote here, this fact is sometimes not appreciated. Relatively low temperatures, however, come into their own in conjunction with fibrous filters wherever there is a risk of the filter becoming wet, e.g. with fermenters and aspiration devices. For several reasons organisms can penetrate wet filters, which also have an increased resistance to air flow and, if made of glass fibre paper, a reduced bursting strength. The use of heated filters overcome this problem.

Electrostatic precipitation and UV radiation also have the attraction of low resistance to air flow, but they are prone to electrical failure and they present high voltage hazards. The former is of considerable value in producing dust-free air supplies, but should not be relied on to eliminate highly infective aerosols of small particle size; and due to many factors inherent in both the installation and micro-organisms, the latter cannot be relied on to perform efficiently in more than a few specialized contexts. Moreover, it can damage exposed plastics and human tissue. Cyclones and scrubbers, though they have application in dust removal, are best forgotten in the treatment of infective air.

3. Infected Dust and Surfaces

In well designed and regulated laboratories dust should not accumulate, though this is seldom the case. Spores, both bacterial and fungal, cocci, mycobacteria, some of the tougher viruses and, indeed, many other organisms can persist in laboratory dust or even constitute a major fraction thereof to the hazard of the worker and the detriment of his cultures. Accumulations of dust can of course be removed by washing down with a germicide though this is unsuitable in most types of laboratory. Vacuum cleaners can be used, though they can re-emit small particles unless fitted with high efficiency filters. Furthermore, as they concentrate the hazard, provision must be made for the sterilization of collected debris before removal from the machine, or piped vacuum system.

Sterilization of surfaces *in situ* prior to orthodox dusting methods provides the best answer. Again UV radiation is often employed, but it suffers from the defects already mentioned and, in addition, shadowing. Gaseous fumigation, on the other hand, with mists or vapours of formaldehyde, glutaraldehyde or β-propiolactone constitutes a simple and effective method of sterilizing all those inaccessible crannies in fittings, furniture and large pieces of laboratory apparatus into which infective dust particles or liquid splashes can penetrate. Various methods have been described and reviewed (Darlow, 1969b; Hoffman, 1971). Small, delicate and heat labile pieces of equipment can be treated similarly, though in this instance fumigation in ethylene oxide chambers (Hoffman, 1971) is generally to be preferred, as it avoids the risk of damage from moisture and polymer residues.

4. Hollow-ware and Cultures

The term 'hollow-ware' is used here to embrace traditional bacteriological glassware and the disposable plastics equipment which is now replacing it in a large measure. Though the ultimate disposal of the 2 materials is different, the initial treatment is the same, namely, the sterilization of contained cultures, culture residues and other infective material such as blood, pus, faeces, sputum or live vaccines.

There is a tendency in some laboratories to discard contaminated hollow-ware into tanks of disinfectant, and even to rely on this as the sole method of sterilization prior to wash-up or consignment to the dustbin. This procedure has much to recommend it in at least some contexts, but also has sufficient limitations to warrant its use only as a prelude to autoclaving. The disinfectant must be freshly made up, known to be effective against the organisms being handled and sufficiently concentrated to compensate for inactivation by organic and chemical residues. Even then, organisms can escape destruction owing to airlocks, blood clots, etc, or the failure of liquid residues to mix adequately with the germicide. Some disinfectants, too, can be adsorbed on glass or rubber liners in sufficient quantity to interfere with their subsequent bacteriological use. Heat sterilization, therefore, should be resorted to whenever possible. It can be argued that if hypochlorites are used as an initial 'dunk', as they very often are, corrosive damage may be done to the autoclave, if, for instance, they are introduced in pipette discard pots. This is undoubtedly true, but corrosion can be avoided readily by prior neutralization with thiosulphate.

The disposal of plastic items is of course a big talking point, and presents a problem in instances where incineration or burial cannot be carried out within the limits of the establishment. Refuse grinders capable of reducing not only plastics but also glass, hypodermic needles and even tin cans to fine fragments are in common use in other contexts and their extension to laboratories would clearly save in respect to bulk and the problems raised in the opening paragraph.

There are 2 plastics, albeit not often involved, which carry a hazard: nitrocellulose, usually encountered in the form of centrifuge tubes, not only constitutes a fire risk, but can explode in the autoclave (Silver, 1963), and PTFE which can emit highly toxic fluorine-containing pyrolysis products at temperatures $>400°$. Both, however, can be disposed of in an efficient incinerator, if fed in as small loads.

5. Liquid Effluents

There are many instances on record of the pollution of water supplies by pathogens escaping from leaking sewers and inefficient sewage disposal plants. It is therefore incumbent on the laboratory worker to ensure that this situation is not aggravated by any fault of his own.

Live cultures of pathogens should never be decanted down the drains of laboratory sinks. Trivial though such an act may sound, it can contribute a far higher concentration of dangerous organisms to a district sewer than a natural excreter of, say, typhoid, turbercle or polio. Not only this, but the act itself can generate an aerosol which escapes directly into the laboratory and into other areas via puff pipes. In short, laboratory cultures should always be autoclaved before disposal.

Larger volumes of toxic effluent from the drains of animal houses, fermenters or vaccine plants, for example, demand proportionately more complex measures. Chemical methods have proved unsatisfactory on account of the cost of a truly germicidal (as opposed to bacteriostatic) concentration and the need to inactivate it before final discharge to the sewers. Furthermore, the problems of ensuring adequate mixing of the chemicals and effluent, and of storage until sterility has been achieved, present additional difficulties.

Heat treatment is the only truly reliable method. This is achieved either by continuous flow or by batch treatment types of plant. The former is rarely necessary, is prone to fail for a variety of reasons, and, as a result, may demand a monitoring procedure before discharge, which again necessitates the use of holding tanks until sterility has been checked. Batch treatment in pressure vessels at $128°$ (30 lb/in^2 steam pressure) for 30 min followed by cooling and direct discharge to the sewers is by far the simplest method. The plant is relatively simple to operate, but should be equipped with the means to sterilize air displaced during filling, internal steam jets to prevent settling of muds and instrumentation to indicate temperatures, pressures and fluid levels.

Drains of all kinds of course require traps, but they are useless unless kept regularly topped up, a frequently neglected precaution in the case of forgotten sinks in dark corners. The drains of each floor of a building should have separate fall pipes to sewer level to prevent up-welling on lower floors in the event of blockage of a common fall pipe. Cases have occurred in which blockage coupled with the decanting of live cultures in an upper floor laboratory have caused widespread infection in the staff. Puff pipes should vent above roof level and, where infective effluent is unavoidable, should be equipped with air filters to prevent escape of infectious agents or migration along the puff pipe system to other parts of the building via dried out drains. Finally, it is obviously desirable in some cases at least to provide means of sterilizing drains by another method than a brief drench of disinfectant, and, whilst steam pressure sterilization is not generally practicable, the use of free-steaming, with or without the addition of formalin, is a useful technique for drains presenting a major hazard, such as those serving animal houses.

6. Laboratory Clothing

Style in protective clothing varies from laboratory to laboratory, at least as much from the vagaries of human whim as at the dictates of safety, and insufficient attention is often paid to the problems of safe removal of this clothing, when contaminated, and its subsequent decontamination. Three facts must be borne in mind: (1) laboratory clothing inevitably becomes contaminated; (2) the front is more vulnerable to splashes and spills than the back; (3) the garment presents not only a contact hazard but is a potent shedder of infective particles.

The results of failure to appreciate these points are best illustrated by an all too common example. The laboratory worker, if he does not commit the folly of consuming tea in his laboratory, makes his way to the canteen wearing his clinical-looking white coat, probably oblivious of the fact that the front and cuffs have become contaminated. Material may even have penetrated between the buttons and reached his shirt. He leaves a trail of airborne organisms behind him. Later, he removes his coat, undoing the buttons with hands which, if not already contaminated, certainly will be as a result. The coat is subsequently collected by a cleaner and sent, via the supply department, to the laundry. The potential extent of the consequent contamination and the number of persons put at risk is great, and has been known to result in outbreaks of infectious disease in laundry staff. To include a paragraph on the disposal of this messy worker himself would doubtless risk an accusation of facetiousness and almost depopulate some laboratories. Nevertheless, one must only remember the appalling cross infection spread by the clothing and hands of man-midwives in the middle of the last century to realize that the principle is still with us.

This is not the moment to discuss clothing design, but the ideal garment for bench work, giving total protection to vulnerable areas, must be readily removable aseptically, e.g. by invertion as with surgical gowns, and preferably, should be insufficiently attractive to be used as a status symbol. It must not be allowed to leave the laboratory unless safely encapsulated in a plastic bag for safe transit to the autoclave or, if of paper disposable type, the incinerator. Much the same applies to towels and gloves, and also to aprons, gum boots, caps and specialized undergarments that high risk techniques demand. There are of course items, such as ventilated suits or hoods and gas masks, which do not take kindly to heat treatment and require resort to decontamination by chemical means. Such means exist, but the important thing is to ensure that an efficient drill is worked out in detail in advance of embarking on projects involving this type of sophisticated equipment.

7. Animals and Cage Litter

Many ways of disposing of cage contents are in current use. Some are entirely unsuitable for infective materials, and the remainder present difficulties which, whilst not insuperable, leave room for doubts as to their efficiency. Clearly, when an infectious hazard exists, composting or conversion to liquid effluent by processing in a garbage disposal unit should not be permitted, and there is even a case for avoiding these measures in supposedly hazard-free circumstances, since highly infectious diseases, e.g. rabbit-pox, pseudotuberculosis or salmonellosis, may break out in well run experimental animal colonies. Incineration is the only reliable means of disposal, but the incinerator must be of good quality and not, as is so often the case, a mere funeral pyre. Unless efficient combustion is obtained smoke, offensive odours and viable bacteria can be emitted to pollute the neighbourhood (Barbeito & Gremillion, 1968).

Transport of animals, bedding and dung from cage to incinerator also presents problems. In some establishments this is achieved by highly sophisticated vacuum extraction techniques which transfer refuse and even small disposable cages, dead mouse and all, to a hopper in the crematorium. This is highly convenient, but creates the further problems of how to decontaminate the interior of such a system and how to dispose of the exhaust air, loaded as it is with moist infective debris. The latter can of course be injected direct into the incinerator, provided that it is burning at the time and requires more air for combustion than the full output of the vacuum system, but it is doubtful whether reliable decontamination of the interior of suction pipes, vacuum pump, hopper, etc; can really be achieved quickly and conveniently by chemical means. On the whole this system is not recommended for high risk conditions (Runkle & Phillips, 1969).

Once again the solution lies in simplicity. Cage contents or entire cages can be enclosed in plastic bags (tough, leak proof and red for danger) or preferably metal bins, which are then externally chemically decontaminated and transferred to the autoclave for cooking. After this the sterilized contents can be safely disposed of in whatever way is appropriate to the local circumstances, e.g. the municipal refuse destructor. Care, however, must be taken to ensure that adequate steam penetration of loads is achieved; 50 kg of dead rabbits in a dust bin is a severe challenge for any autoclave.

8. Miscellaneous Problems

Though the primary objective in disposal throughout this discussion has been sterilization, each type of material presents special problems. The list of materials is almost inexhaustible, but it seems fitting here to mention a few more, if only to awaken awareness to the fact that there is always another problem waiting round the corner, largely because the development of methods and materials almost invariably occurs without full consideration of the consequences.

Vacuum pump oil is liable to contamination in freeze drying machines and high speed centrifuges. Many bacteria and viruses can survive for weeks and even months in this material and may be re-aerosolized in the oil mist. The problem can again be solved by installing air filters on the negative pressure side of the pump which, contrary to popular belief, does not interfere unduly with the pulling of an effective vacuum or with dehydration. Once the pump is contaminated there is little that can be done. Pumps do not take kindly to autoclaving, though they are not irreparable, but oil films in any case protect some organisms from steam. Modest heating, however, is effective, except in the case of spores, when the only nondestructive course is to drain off the oil and wash out the pump with a suitable solvent. Ethylene oxide can then be used in the final sterilizing process.

The disposal of radio-isotope labelled cultures also presents difficulties, though much depends on the level of activity and the half life. Clearly it is undesirable to run the risk of contaminating an autoclave or other sterilizer as a result of a vessel bursting therein or emitting active volatile substances. Fortunately many of the more commonly used isotopes are short lived and so permit of economic storage until a satisfactory degree of decay has occurred. This also allows time for autolysis of the culture or its inactivation by chemicals to take place.

On the subject of autoclaves it must be stressed that the bursting or overflow of a culture vessel during the pulling of the first vacuum can contaminate the drain line and sewers. When cultures are placed in an autoclave requiring vacuum, provision must be made to prevent this by placing the hollow-ware in a second vessel or tray of adequate volume to retain the contents.

In conclusion it must be admitted that many questions remain undiscussed, covering the disposal of materials ranging from waste paper to tissues and chemicals or antibiotics likely to interfere with sewage disposal. In many cases the answers can be found in the literature quoted above, but a very important factor is the maintenance of friendly liaison with local Public Health Authorities.

9. References

Barbeito, M. S., Taylor, L. A. & Seiders, R. W. (1968). Microbiological evaluation of a large volume air incinerator. *Appl. Microbiol.* **16**, 490.

Barbeito, M. S. & Gremillion, G. G. (1968). Microbiological safety evaluation of an industrial refuse incinerator. *Appl. Microbiol.* **16**, 291.

Chatigny, M. A., Sarshad, A. A. & Pike, G. F. (1970). Design and evaluation of a system for thermal decontamination of process air. *Biotech. Bioeng.* **12**, 483.

Darlow, H. M. (1969a). Safety in the microbiological laboratory. In *Methods in Microbiology*, Vol. I (Eds J. N. Norris & D. W. Ribbons). London: Academic Press.

Darlow, H. M. (1969b). Disinfection in fermentation laboratories. *Proc. Biochem.* 4, 15.

Hoffman, R. K. (1971). Toxic gases. In *Destruction of the Microbial Cell.* (Ed. W. B. Hugo). London: Academic Press.

Runkle, R. S. & Phillips, G. B. (1969). *Microbial Contamination Control Facilities.* New York: van Nostrand Reinhold.

Silver, I. H. (1963). Explosion in an autoclave caused by cellulose nitrate tubes. *Nature, Lond.* 199, 102.

Sulkin, S. E. (1961). Laboratory acquired infections. *Bact. Rev.* 25, 203.

Evolution of Desert Biota

Evolution of
Desert Biota

Edited by David W. Goodall

University of Texas Press Austin & London

Publication of this book was financed in part by
the Desert Biome and the Structure of Ecosystems programs of
the U.S. participation in the International
Biological Program.

Library of Congress Cataloging in Publication Data
Main entry under title:

Evolution of desert biota.

   Proceedings of a symposium held during the First
International Congress of Systematic and Evolution-
ary Biology which took place in Boulder, Colo.,
during August, 1973.
   Bibliography: p.
   Includes index.
   1. Desert biology—Congresses. 2. Evolution—
Congresses. I. Goodall, David W., 1914–
II. International Congress of Systematic and Evolu-
tionary Biology, 1st, Boulder, Colo., 1973.
QH88.E95   575'.00915'4   75-16071
ISBN 0-292-72015-7

Printed in the United States of America

Contents

Evolution of Desert Biota

1. Introduction   David W. Goodall

In the broad sense, "deserts" include all those areas of the earth's surface whose biological potentialities are severely limited by lack of
water. If one takes them as coextensive with the arid and semiarid
zones of Meigs's classification, they occupy almost one-quarter of the
terrestrial surface of the globe. Though the largest arid areas are to be
found in Africa and Asia, Australia has the largest proportion of its
area in this category. Smaller desert areas occur in North and South
America; Antarctica has cold deserts; and the only continent virtually
without deserts is Europe.

When life emerged in the waters of the primeval world, it could hardly have been predicted that the progeny of these first organisms
would extend their occupancy even to the deserts. Regions more different in character from the origin and natural home of life would be
hard to imagine. Protoplasm is based on water, rooted in water. Some
three-quarters of the mass of active protoplasm is water; the biochemical reactions underlying all its activities take place in water and depend on the special properties of water for the complex mechanisms
of enzymatic and ionic controls which integrate the activity of cell
and organisms into a cybernetic whole. It is, accordingly, remarkable
that organisms were able to adapt themselves to environments in
which water supplies were usually scanty, often almost nonexistent,
and always unpredictable.

The first inhabitants of the deserts were presumably opportunistic.
On the margins of larger bodies of water were areas which were alter-
nately wetted and dried for longer or shorter periods. Organisms living there acquired the possibility of surviving the dry periods by drying
out and becoming inactive until rewetted, at which time their activity
resumed where it had left off. While in the dry state, these organisms

—initially, doubtless, Protista—were easily moved by air currents and thus could colonize other bodies of water. Among them were the very temporary pools formed by the occasional rainstorms in desert areas. Thus the deserts came to be inhabited by organisms whose ability to dry and remoisten without loss of vitality enabled them to take advantage of the short periods during which limited areas of the deserts deviate from their normally arid state.

Yet other organisms doubtless—the blue green algae among them —similarly took advantage of the much shorter periods, amounting perhaps to an hour at a time, during which the surface of the desert was moistened by dew, and photosynthesis was possible a few minutes before and after sunrise to an organism which could readily change its state of hydration.

In the main, though, colonization of the deserts had to wait until colonization of other terrestrial environments was well advanced. For most groups of organisms, the humid environments on land presented less of a challenge in the transition from aquatic life than did the deserts. By the time arthropods and annelids, mollusks and vertebrates, fungi and higher plants had adapted to the humid terrestrial environments, they were poised on the springboard where they could collect themselves for the ultimate leap into the deserts. And this leap was made successfully and repeatedly. Few of the major groups of organisms that were able to adapt to life on land did not also contrive to colonize the deserts.

Some, like the arthropods and annual plants, had an adaptational mechanism—an inactive stage of the life cycle highly resistant to desiccation—almost made to order to match opportunistically the episodic character of the desert environment. For others the transition was more difficult. for mammals, whose circulatory mechanism assumes the availability of liquid water; for perennial plants, whose photosynthetic mechanism normally carries the penalty of water loss concurrent with carbon dioxide intake. But the evolutionary process surmounted these difficulties; and the deserts are now inhabited by a range of organisms which, though somewhat inferior to that of more favored environments, bears testimony to the inventiveness and success of evolution in filling niches and in creating diversity.

The most important modifications and adaptations needed for life in the deserts are concerned with the dryness of the environment there.

But an important feature of most desert environments is also their unpredictability. Precipitation has a higher coefficient of variability, on any time scale, than in other climatic types, with the consequence that desert organisms may have to face floods as well as long and highly variable periods of drought. The range of temperatures may also be extreme—both diurnal and seasonal. Under the high radiation of the subtropical deserts, the soil surface may reach a temperature which few organisms can survive; and, in the cold deserts of the great Asian land mass, extremely low winter temperatures are recorded. Sand and dust storms made possible by the poor stability of the surface soil are also among the environmental hazards to which desert organisms must become adapted.

Like other climatic zones, the deserts have not been stable in relation to the land masses of the world. Continental drift, tectonic movements, and changes in the earth's rotation and in the extent of the polar icecaps have led to secular changes in the area and distribution of arid land surfaces. But, unlike other climatic zones, the arid lands have probably always been fragmented—constituting a number of discrete areas separated from one another by zones of quite different climate. The evolutionary process has gone on largely independently in these separate areas, often starting from different initial material, with the consequence that the desert biota is highly regional. Elements in common between the different main desert areas are few, and, as between continents or subcontinents, there is a high degree of endemism. The smaller desert areas of the world are the equivalent of islands in an ocean of more humid environments.

These are among the problems to be considered in the present volume. It reports the proceedings of a symposium which was held on August 10, 1973, at Boulder, Colorado, as part of the First International Congress of Systematic and Evolutionary Biology.

2. The Origin and Floristic Affinities of the South American Temperate Desert and Semidesert Regions   Otto T. Solbrig

Introduction

In this paper I will attempt to summarize the existent evidence regarding the floristic relations of the desert and semidesert regions of temperate South America and to explain how these affinities came to exist.

More than half of the surface of South America south of the Tropic of Capricorn can be classed as semidesert or desert. In this area lie some of the richest mineral deposits of the continent. These regions consequently are important from the standpoint of human economy. From a more theoretical point, desert environments are credited with stimulating rapid evolution (Stebbins, 1952; Axelrod, 1967) and, further, present some of the most interesting and easy-to-study adaptations in plants and animals.

Although, at present, direct evidence regarding the evolution of desert vegetation in South America is still meager, enough data have accumulated to make some hypotheses. It is hoped this will stimulate more research in the field of plant micropaleontology in temperate South America. Such research in northern South America has advanced our knowledge immensely (Van der Hammen, 1966), and high rewards await the investigator who searches this area in the temperate regions of the continent.

The Problem

If a climatic map of temperate South America is compared with a phytogeographic map of the same region drawn on a physiognomic basis and with one drawn on floristic lines, it will be seen that they do not coincide. Furthermore, if the premise (not proven but generally held to be true) is accepted that the physical environment is the determinant of the structure of the ecosystem and that, as the physical environment (be it climate, physiography, or both) changes, the structure of the vegetation will also change, then an explanation for the discrepancy between climatic and phytogeographic maps has to be provided. Alternative explanations to solve the paradox are (1) the premise on which they are based is entirely or partly wrong; (2) our knowledge is incomplete; or (3) the discrepancies can be explained on the basis of the historical events of the past. It is undoubtedly true that floristic and paleobotanical knowledge of South American deserts is incomplete and that much more work is needed. However, I will proceed under the assumption that a sufficient minimum of information is available. I also feel that our present insights are sufficient to accept the premise that the ecosystem is the result of the interaction between the physical environment and the biota. I shall therefore try to find in the events of the past the answer for the discrepancy.

I shall first describe the semidesert regions of South America and their vegetation, followed by a brief discussion of Tertiary and Pleistocene events. I shall then look at the floristic connections between the regions and the distributional patterns of the dominant elements of the area under study. From this composite picture I shall try to provide a coherent working hypothesis to explain the origin and floristic affinities of the desert and semidesert regions of temperate South America.

Theory

Biogeographical hypotheses such as the ones that will be made further on in this paper are based on certain theoretical assumptions. In most cases, however, these assumptions are not made explicit; consequently, the reader who disagrees with the author is not always certain whether he disagrees with the interpretation of the evidence or with the assumptions made. This has led to many futile controversies. The fundamental assumptions that will be made here follow from the general theory of evolution by natural selection, the theory of speciation, and the theory of geological uniformitarianism.

The first assumption is that a continuous distributional range reflects an environment favorable to the plant, that is, an environment where it can compete successfully. Since the set of conditions (physical, climatical, and biological) where the plant can compete successfully (the realized niche) bounds a limited portion of ecological space, it will be further assumed that the range of a species indicates that conditions over that range do not differ greatly in comparison with the set of all possible conditions that can be given. It will be further assumed that each species is unique in its fundamental and realized niche (defined as the hyperspace bounded by all the ecological parameters to which the species is adapted or over which it is able to compete successfully). Consequently, no species will occupy exactly the same geographical range, and, as a corollary, some species will be able to grow over a wide array of conditions and others over a very limited one.

When the vegetation of a large region, such as a continent, is mapped, it is found that the distributional ranges of species are not independent but that ranges of certain species tend to group themselves even though identical ranges are not necessarily encountered. This allows the phytogeographer to classify the vegetation. It will be assumed that, when neighboring geographical areas do not differ greatly in their present physical environment or in their climate but differ in their flora, the reason for the difference is a historical one reflecting different evolutionary histories in these floras and requiring an explanation.

Disjunctions are common occurrences in the ranges of species. In a strict sense, all ranges are disjunct since a continuous cover of a species over an extensive area is seldom encountered. However, when similar major disjunctions are found in the ranges of many species whose ranges are correlated, the disjunction has biogeographical significance. Unless there is evidence to the contrary, an ancient continuous range will be assumed in such instances, one that was disrupted at a later date by some identifiable event, either geological or climatological.

It will also be assumed that the atmospheric circulation and the basic meteorological phenomena in the past were essentially similar to those encountered today, unless there is positive evidence to the contrary. Further, it will be assumed that the climatic tolerances of a living species were the same in the past as they are today. Finally, it will be assumed that the spectrum of life forms that today signify a rain forest, a subtropical forest, a semidesert, and so on, had the same meaning in the past too, implying with it that the basic processes of carbon gain and water economy have been essentially identical at least since the origin of the angiosperms.

From these assumptions a coherent theory can be developed to reconstruct the past (Good, 1953; Darlington, 1957, 1965). No general assumptions about age and area will be made, however, because they are inconsistent with speciation theory (Stebbins, 1950; Mayr, 1963). In special cases when there is some evidence that a particular group is phylogenetically primitive, the assumption will be made that it is also geologically old. Such an assumption is not very strong and will be used only to support more robust evidence.

The Semidesert Regions of South America

In temperate South America we can recognize five broad phytogeographical regions that can be classed as "desert" or "semidesert" regions. They are the Monte (Haumann, 1947; Morello, 1958), the Patagonian Steppe (Cabrera, 1947), the Prepuna (Cabrera, 1971), and the Puna (Cabrera, 1958) in Argentina, and the Pacific Coastal Desert in Chile and Peru (Goodspeed, 1945; Ferreyra, 1960). In addition, three other regions—the Matorral or "Mediterranean" region in Chile (Mooney and Dunn, 1970) and the Chaco and the Espinal in Argentina (Fiebrig, 1933; Cabrera, 1953, 1971), although not semideserts, are characterized by an extensive dry season. Finally, the high mountain vegetation of the Andes shows adaptations to drought tolerance (fig. 2-1).

The Monte

The Monte (Lorentz, 1876; Haumann, 1947; Cabrera, 1953; Morello, 1958; Solbrig, 1972, 1973) is a phytogeographical province that extends from lat. 24°35′ S to lat. 44°20′ S and from long. 62°54′ W on the Atlantic coast to long. 69°50′ W at the foothills of the Andes (fig. 2-1).

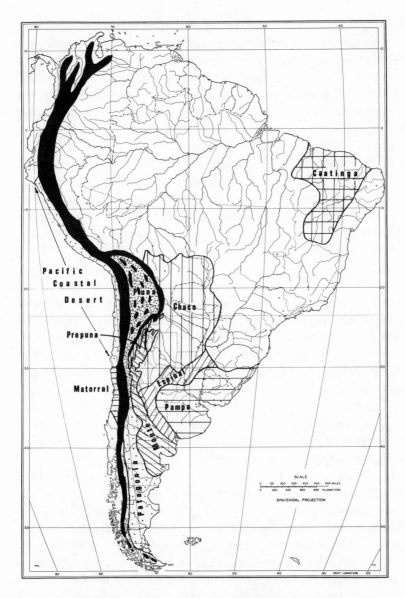

Fig. 2-1. Geographical limits of the phytogeographical provinces of the Andean Dominion (stippled) and of the Chaco Dominion (various hatchings) according to Cabrera (1971). The high cordillera vegetation is indicated in solid black. Goode Base Map, copyright by The University of Chicago, Department of Geography.

Rains average less than 200 mm a year in most localities and never exceed 600 mm; evaporation exceeds rainfall throughout the region. The rain falls in spring and summer. The area is bordered on the west by the Cordillera de los Andes, which varies in height between 5,000 and 7,000 m in this area. On the north the region is bordered by the high Bolivian plateau (3,000–5,000 m high) and on the east by a series of mountain chains (Sierras Pampeanas) that vary in height from 3,000 to 5,000 m in the north (Aconquija, Famatina, and Velazco) to less than 1,000 m (Sierra de Hauca Mahuida) in the south. Physiographically, the northern part is formed by a continuous barrier of high mountains which becomes less important farther south as well as lower in height. The Monte vegetation occupies the valleys between these mountains as a discontinuous phase in the northern region and a more or less continuous phase from approximately lat. 32° S southward.

The predominant vegetation of the Monte is a xerophytic scrubland with small forests along the rivers or in areas where the water table is quite superficial. The predominant community is dominated by the species of the creosote bush or *jarilla* (*Larrea divaricata*, *L. cuneifolia*, and *L. nitida* [Zygophyllaceae]) associated with a number of other xerophytic or aphyllous shrubs: *Condalia microphylla* (Rhamnaceae), *Monttea aphylla* (Scrophulariaceae), *Bougainvillea spinosa* (Nyctaginaceae), *Geoffroea decorticans* (Leguminosae), *Cassia aphylla* (Leguminosae), *Bulnesia schickendanzii* (Zygophyllaceae), *B. retama*, *Atamisquea emarginata* (Capparidaceae), *Zuccagnia punctata* (Leguminosae), *Gochnatia glutinosa* (Compositae), *Proustia cuneifolia* (Compositae), *Flourensia polyclada* (Compositae), and *Chuquiraga erinacea* (Compositae).

Along water courses or in areas with a superficial water table, forests of *algarrobos* (mesquite in the United States) are observed, that is, various species of *Prosopis* (Leguminosae), particularly *P. flexuosa*, *P. nigra*, *P. alba*, and *P. chilensis*. Other phreatophytic or semiphreatophytic species of small trees or small shrubs are *Cercidium praecox* (Leguminosae), *Acacia aroma* (Leguminosae), and *Salix humboldtiana* (Salicaceae).

Herbaceous elements are not common. There is a flora of summer annuals formed principally by grasses.

The Patagonian Steppe

The Patagonian Steppe (Cabrera, 1947, 1953, 1971; Soriano, 1950, 1956) is limited on its eastern and southern borders by the Atlantic Ocean and the Strait of Magellan. On the west it borders quite abruptly with the *Nothofagus* forest; the exact limits, although easy to determine, have not yet been mapped precisely (Dimitri, 1972). On the north it borders with the Monte along an irregular line that goes from Chos Malal in the state of Neuquen in the west to a point on the Atlantic coast near Rawson in the state of Chubut (Soriano, 1949). In addition, a tongue of Patagonian Steppe extends north from Chubut to Mendoza (Cabrera, 1947; Böcher, Hjerting, and Rahn, 1963). Physiognomically the region consists of a series of broad tablelands of increasing altitude as one moves from east to west, reaching to about 1,500 m at the foot of the cordillera. The soil is sandy or rocky, formed by a mixture of windblown cordilleran detritus as well as *in situ* eroded basaltic rocks, the result of ancient volcanism.

The climate is cold temperate with cold summers and relatively mild winters. Summer means vary from 21°C in the north to 12°C in the south (summer mean maxima vary from 30°C to 18°C) with winter means from 8°C in the north to 0°C in the south (winter mean minima 1.5°C to −3°C). Rainfall is very low, averaging less than 200 mm in all the Patagonian territory with the exception of the south and west borders where the effect of the cordilleran rainfall is felt. The little rainfall is fairly well distributed throughout the year with a slight increase during winter months.

The Patagonian Steppe is the result of the rain-shadow effect of the southern cordillera in elevating and drying the moist westerly winds from the Pacific. Consequently the region not only is devoid of rains but also is subjected to a steady westerly wind of fair intensity that has a tremendous drying effect. The few rains that occur are the result of occasional eruptions of the Antarctic polar air mass from the south interrupting the steady flow of the westerlies.

The dominant vegetation is a low scrubland or else a vegetation of low cushion plants. In some areas xerophytic bunch grasses are also common. Among the low (less than 1 m) xerophytic shrubs and cushion plants, the *neneo*, *Mulinum spinosum* (Umbelliferae), is the dominant form in the northwestern part, while *Chuquiraga avellanedae* (Compositae) and *Nassauvia glomerulosa* (Compositae) are dominant over extensive areas in central Patagonia. Other important shrubs are *Trevoa patagonica* (Rhamnaceae), *Adesmia campestris* (Compositae), *Colliguaja integerrima* (Euphorbiaceae), *Nardophyllum obtusifolium* (Compositae), and *Nassauvia axillaris*. Among the grasses are *Stipa humilis*, *S. neaei*, *S. speciosa*, *Poa huecu*, *P. ligularis*, *Festuca argentina*, *F. gracillima*, *Bromus macranthus*, *Hordeum comosus*, and *Agropyron fuegianum*.

The Puna

The Puna (Weberbauer, 1945; Cabrera, 1953, 1958, 1971) is situated in the northwestern part of Argentina, western and central Bolivia, and southern Peru. It is a very high plateau, the result of the uplift of an enormous block of an old peneplane, which started to lift in the Miocene but mainly rose during the Pliocene and the Pleistocene to a mean elevation of 3,400–3,800 m. The Puna is bordered on the east by the Cordillera Real and on the west by the Cordillera de los Andes that rises to 5,000–6,000 m; the plateau is peppered by a number of volcanoes that rise 1,000–1,500 m over the surface of the Puna.

The soils of the Puna are in general immature, sandy to rocky, and very poor in organic matter (Cabrera, 1958). The area has a number of closed basins, and high mountain lakes and marshes are frequent.

The climate of the Puna is cold and dry with values for minimum and maximum temperatures not too different from Patagonia but with the very significant difference that the daily temperature amplitude is very great (values of over 20°C are common) and the difference between summer and winter very slight. The precipitation is very irregular over the area of the Puna, varying from a high of 800 mm in the northeast corner of Bolivia to 100 mm/year on the southwest border in Argentina. The southern Puna is undoubtedly a semidesert region, but the northern part is more of a high alpine plateau, where the limitations to plant growth are given more by temperature than by rainfall.

The typical vegetation of the Puna is a low, xerophytic scrubland formed by shrubs one-half to one meter tall. In some areas a grassy steppe community is found, and in low areas communities of high mountain marshes are found.

Among the shrubby species we find *Fabiana densa* (Solanaceae), *Psila boliviensis* (Compositae), *Adesmia horridiuscula* (Leguminosae), *A. spinossisima, Junellia seriphioides* (Verbenaceae), *Nardophyllum armatum* (Compositae), and *Acantholippia hastatula* (Verbenaceae). Only one tree, *Polylepis tomentella* (Rosaceae), grows in the Puna, strangely enough only at altitudes of over 4,000 m. Another woody element is *Prosopis ferox*, a small tree or large shrub. Among the grasses are *Bouteloua simplex, Muhlenbergia fastigiata, Stipa leptostachya, Pennisetum chilense*, and *Festuca scirpifolia*. Cactaceae are not very frequent in general, but we find locally abundant *Opuntia atacamensis, Oreocerus trollii, Parodia schroebsia*, and *Trichocereus poco*.

Although physically the Puna ends at about lat. 30° S, Puna vegetation extends on the eastern slope of the Andes to lat. 35° S, where it merges into Patagonian Steppe vegetation.

The Prepuna

The Prepuna (Czajka and Vervoorst, 1956; Cabrera, 1971) extends along the dry mountain slopes of northwestern Argentina from the state of Jujuy to La Rioja, approximately between 2,000 and 3,400 m. It is characterized by a dry and warm climate with summer rains; it is warmer than the Puna, colder than the Monte; and it is a special formation strongly influenced by the exposure of the mountains in the region.

The vegetation is mainly formed by xerophytic shrubs and cacti. Among the shrubs, the most abundant are *Gochnatia glutinosa* (Compositae), *Cassia crassiramea* (Leguminosae), *Aphyllocladus spartioides, Caesalpinia trichocarpa* (Leguminosae), *Proustia cuneifolia* (Compositae), *Chuquiraga erinacea* (Compositae), *Zuccagnia punctata* (Leguminosae), *Adesmia inflexa* (Leguminosae), and *Psila boliviensis* (Compositae). The most conspicuous member of the Cactaceae is the cardon, *Trichocereus pasacana*; there are also present *T. poco* and species of *Opuntia, Cylindropuntia, Tephrocactus, Parodia*, and *Lobivia*. Among the grasses are *Digitaria californica, Stipa leptostachya, Monroa argentina*, and *Agrostis nana*.

The Pacific Coastal Desert

Along the Peruvian and Chilean coast from lat. 5° S to approximately lat. 30° S, we find the region denominated "La Costa" in Peru (Weberbauer, 1945; Ferreyra, 1960) and "Northern Desert," "Coastal Desert," or "Atacama Desert" in Chile (Johnston, 1929; Reiche, 1934; Goodspeed, 1945). This very dry region is under the influence of the combined rain shadow of the high cordillera to the east and the cold Humboldt Current and the coastal upwelling along the Peruvian coast. Although physically continuous, the vegetation is not uniform, as a result of the combination of temperature and rainfall variations in such an extended territory. Temperature decreases from north to south as can be expected, going from a yearly mean to close to 25°C in northern Peru (Ferreyra, 1960) to a low of 15°C at its southern border. Rainfall is very irregular and very meager. Although some localities in Peru (Zorritos, Lomas de Lachay; cf. Ferreyra, 1960) have averages of 200 mm, the average yearly rainfall is below 50 mm in most places. This has created an extreme xerophytic vegetation often with special adaptations to make use of the coastal fog.

Behind the coastal area are a number of dry valleys, some in Peru but mostly in northern Chile, with the same kind of extreme dry conditions as the coastal area.

The flora is characterized by plants with extreme xerophytic adaptations, especially succulents, such as *Cereus spinibaris* and *C. coquimbanus*, various species of *Echinocactus*, and *Euphorbia lactifolia*. The most interesting associations occur in the so-called *lomas*, or low hills (less than 1,500 m), along the coast that intercept the coastal fog and produce very localized conditions favorable for some plant growth. Almost all of these formations from the Equadorian border to central Chile constitutes a unique community. Over 40 percent of the plants in the Peruvian coastal community are annuals (Ferreyra, 1960), although annuals apparently are less common in Chile (Johnston, 1929); of the perennials, a large number are root perennials or succulents. Only about 5 percent are shrubs or trees in the northern sites (Ferreyra, 1960), while shrubs and semishrubs constitute a higher proportion in the Chilean region. From the Chilean region should be mentioned *Oxalis gigantea* (Oxalidaceae), *Heliotropium philippianum* (Boraginaceae), *Salvia gilliesii* (Labiatae), and

Proustia tipia (Compositae) among the shrubs; species of *Poa, Eragrostis, Elymus, Stipa,* and *Nasella* among the grasses; and *Alstroemeria violacea* (Amaryllidaceae), a conspicuous and relatively common root perennial. In southern Peru *Nolana inflata, N. spathulata* (Nolanaceae), and other species of this widespread genus; *Tropaeolum majus* (Tropaeolaceae), *Loasa urens* (Loasaceae), and *Arcythophyllum thymifolium* (Rubiaceae); in the *lomas* of central Peru the *amancay, Hymenocallis amancaes* (Amaryllidaceae), *Alstroemeria recumbens* (Amaryllidaceae), *Peperomia atocongona* (Piperaceae), *Vicia lomensis* (Leguminosae), *Carica candicans* (Caricaceae), *Lobelia decurrens* (Lobeliaceae), *Drymaria weberbaueri* (Caryophyllaceae), *Capparis prisca* (Capparidaceae), *Caesalpinia tinctoria* (Leguminosae), *Pitcairnia lopezii* (Bromeliaceae), and *Haageocereus lachayensis* and *Armatocereus* sp. (Cactaceae). Finally, in the north we find *Tillandsia recurvata, Fourcroya occidentalis, Apralanthera ferreyra, Solanum multinterruptum,* and so on.

Of great phytogeographic interest is the existence of a less-xerophytic element in the very northern extreme of the Pacific Coastal Desert, from Trujillo to the border with Ecuador (Ferreyra, 1960), known as *algarrobal.* Principal elements of this vegetation are two species of *Prosopis, P. limensis* and *P. chilensis*; others are *Cercidium praecox, Caesalpinia paipai, Acacia huarango, Bursera graveolens* (Burseraceae), *Celtis iguanea* (Ulmaceae), *Bougainvillea peruviana* (Nyctaginaceae), *Cordia rotundifolia* (Boraginaceae), and *Grabowskia boerhaviifolia* (Solanaceae).

Geological History

The present desert and subdesert regions of temperate South America result from the existence of belts of high atmospheric pressure around lat. 30° S, high mountain chains that impede the transport of moisture from the oceans to the continents, and cold water currents along the coast, which by cooling and drying the air that flows over them act like the high mountains.

The Pacific Coastal Desert of Chile and Peru is principally the result of the effect of the cold Humboldt Current that flows from south to north; the Patagonian Steppe is produced by the Cordillera de los Andes that traps the moisture in the prevailing westerly winds; while the Monte and the Puna result from a combination of the cordilleran rain shadow in the west and the Sierras Pampeanas in the east and the existence of the belt of high pressure.

The high-pressure belt of mid-latitudes is a result of the global flow of air (Flohn, 1969) and most likely has existed with little modification throughout the Mesozoic and Cenozoic (however, for a different view, see Schwarzenbach, 1968, and Volkheimer, 1971). The mountain chains and the cold currents, on the other hand, are relatively recent phenomena. The latter's low temperature is largely the result of Antarctic ice. But aridity results from the interaction of temperature and humidity. In effect, when ambient temperatures are high, a greater percentage of the incident rainfall is lost as evaporation and, in addition, plants will transpire more water. Consequently, in order to reconstruct the history of the desert and semidesert regions of South America, we also have to have an idea of the temperature and pluvial regimes of the past.

In this presentation I will use two types of evidence: (1) the purely geological evidence regarding continental drift, times of uplifting of mountain chains, marine transgressions, and existence of paleosoils and pedemonts; and (2) paleontological evidence regarding the ecological types and phylogenetical stock of the organisms that inhabited the area in the past. With this evidence I will try to reconstruct the most likely climate for temperate South America since the Cretaceous and deduce the kind of vegetation that must have existed.

Cretaceous

This account will start from the Cretaceous because it is the oldest period from which we have fossil records of angiosperms, which today constitute more than 90 percent of the vascular flora of the regions under consideration. At the beginning of the Cretaceous, South America and Africa were probably still connected (Dietz and Holden, 1970), since the rift that created the South Atlantic and separated the two continents apparently had its origin during the Lower Cretaceous. The position of South America at this time was slightly south (approximate-

ly lat. 5°–10° S) of its present position and with its southern extremity tilted eastward. There were no significant mountain chains at
that time.

Northern and western South America are characterized in the Cretaceous by extensive marine transgressions in Colombia, Venezuela,
Ecuador, and Peru (Harrington, 1962). In Chile, during the middle Cretaceous, orogeny and uplift of the Chilean Andes began (Kummel,
1961). This general zone of uplift, which was accompanied by active
volcanism and which extended to central Peru, marks the beginning of
the formation of the Andean cordillera, a phenomenon that will have its
maximum expression during the upper Pliocene and Pleistocene and
that is not over yet.

Although the first records of angiosperms date from the Cretaceous
(Maestrichtian), the known fossil floras from the Cretaceous of South
America are formed predominantly by Pteridophytes, Bennettitales,
and Conifers (Menéndez, 1969). Likewise, the fossil faunas are
formed by dinosaurs and other reptilian groups. Toward the end of the
Cretaceous (or beginning of Paleocene) appear the first mammals
(Patterson and Pascual, 1972).

Climatologically, the record points to a much warmer and possibly
wetter climate than today, although there is evidence of some aridity,
particularly in the Lower Cretaceous.

All in all, the Cretaceous period offers little conclusive evidence of
extensive dry conditions in South America. Nevertheless, during the
Lower Cretaceous before the formation of an extensive South Atlantic
Ocean, conditions in the central portion of the combined continent
must have been drier than today. In effect, the high rainfall in the
present Amazonian region is the result of the condensation of moisture from rising tropical air that is cooling adiabatically. This air is
brought in by the trade winds and acquires its moisture over the North
and South Atlantic. Before the breakup of Pangea, trade winds must
have been considerably drier on the western edge of the continent
after blowing over several thousand miles of hot land. It is interesting
that some characteristic genera of semidesert regions, such as *Prosopis* and *Acacia*, are represented in both eastern Africa and South
America. This disjunct distribution can be interpreted by assuming
Cretaceous origin for these genera, with a more or less continuous

Cretaceous distribution that was disrupted when the continents separated (Thorne, 1973). This is in accordance with their presumed primitive position within the Leguminosae (L. I. Nevling, 1970, personal communication). There is some geomorphological evidence also for at least local aridity in the deposits of the Lower Cretaceous of Córdoba and San Luis in Argentina, which are of a "typical desert phase" according to Gordillo and Lencinas (1972).

Cenozoic

Paleocene. The marine intrusions of northern South America still persisted at the beginning of the Paleocene but had become much less extensive (Haffer, 1970). The Venezuelan Andes and part of the Caribbean range of Venezuela began to rise above sea level (Liddle, 1946; Harrington, 1962). In eastern Colombia, Ecuador, and Peru continental deposits were laid down to the east of the rising mountains, which at this stage were still rather low. The sea retreated from southern Chile, but there was a marine transgression in central eastern Patagonia.

At the beginning of the Paleocene the South American flora acquired a character of its own, very distinct from contemporaneous European floras, although there are resemblances to the African flora (Van der Hammen, 1966). The first record of Bombacaceae is from this period (Van der Hammen, 1966).

There are remains of crocodiles from the Paleocene of Chubut in Argentina, indicating a probable mean temperature of 10°C or higher for the coldest month (Volkheimer, 1971), some fifteen to twenty degrees warmer than today. The early Tertiary mammalian fossil faunas consist of marsupials, edentates of the suborder Xenarthra, and a variety of ungulates (Patterson and Pascual, 1972). These forms appear to have lived in a forested environment, confirming the paleobotanical evidence (Menéndez, 1969, 1972; Petriella, 1972).

The climate of South America during the Paleocene was clearly warmer and more humid than today. With the South Atlantic now fairly large and with no very great mountain range in existence, probably no extensive dry-land floras could have existed.

Eocene. During the Eocene the general features of the northern Andes were little changed from the preceding Paleocene. The northern extremity of the eastern cordillera began to be uplifted. In western Colombia and Ecuador the Bolívar geosyncline was opened (Schuchert, 1935; Harrington, 1962). Thick continental beds were deposited in eastern Colombia-Peru, mainly derived from the erosion of the rising mountains to the east. In the south the slow rising of the cordillera continued. There was an extensive marine intrusion in eastern Patagonia.

The flora was predominantly subtropical (Romero, 1973). It was during the Eocene that the tropical elements ranged farthest south, which can be seen very well in the fossil flora of Río Turbio in Argentine Patagonia. Here the lowermost beds containing *Nothofagus* fossils are replaced by a rich flora of tropical elements with species of *Myrica*, *Persea*, *Psidium*, and others, which is then again replaced in still higher beds by a *Nothofagus* flora of more mesic character (Hünicken, 1966; Menéndez, 1972).

However, in the Eocene we also find the first evidence of elements belonging to a more open, drier vegetation, particularly grasses (Menéndez, 1972; Van der Hammen, 1966).

The Eocene was also a time of radiation of several mammalian phyletic lines, particularly marsupials, xenarthrans, ungulates, and notoungulates (Patterson and Pascual, 1972). Of particular interest for our purpose is the appearance of several groups of large native herbivores (Patterson and Pascual, 1972). More interesting still is "the precocity shown by certain ungulates in the acquisition of high-crowned, or hyposodont, and rootless, or hypselodont, teeth" (Patterson and Pascual, 1972). By the lower Oligocene such teeth had been acquired by no fewer than six groups of ungulates. Such animals must have thrived in the evolving pampas areas. True pampas are probably younger, but by the Eocene it seems reasonable to propose the existence of open savanna woodlands, somewhat like the llanos of Venezuela today.

The climate appears to have been fairly wet and warm until a peak was reached in middle Eocene, after which time a very gradual drying and cooling seems to have occurred.

Oligocene. The geological history of South America during the Oligocene followed the events of the earlier periods. There were further uplifts of the Caribbean and Venezuelan mountains and also the Cordillera Principal of Peru. In Patagonia the cordillera was uplifted and the coastal cordillera also began to rise. At the same time, erosion of these mountains was taking place with deposition to the east of them.

In Patagonia elements of the Eocene flora retreated northward and the temperate elements of the *Nothofagus* flora advanced. In northern South America all the evidence points to a continuation of a tropical forest landscape, although with a great deal of phyletic evolution (Van der Hammen, 1966).

The paleontological record of mammals shows the continuing radiation and gradual evolution of the stock of ancient inhabitants of South America. The Oligocene also records the appearance of caviomorph rodents and platyrrhine primates, which probably arrived from North America via a sweepstakes route (Simpson, 1950; Patterson and Pascual, 1972), although an African origin has also been proposed (Hoffstetter, 1972).

Miocene. During Miocene times a number of important geological events took place. In the north the eastern cordillera of Colombia, which had been rising slowly since the beginning of the Tertiary, suffered its first strong uplift (Harrington, 1962). The large deposition of continental deposits in eastern Colombia and Peru continued, and by the end of the period the present altiplano of Peru and Bolivia had been eroded almost to sea level (Kummel, 1961). In the southern part of the continent one sees volcanic activity in Chubut and Santa Cruz as well as continued uplifting of the cordillera. By the end of the Miocene we begin to see the rise of the eastern and central cordilleras of Bolivia and the Puna and Pampean ranges of northern Argentina (Harrington, 1962).

During the Miocene the southern *Nothofagus* forest reached an all tension similar to that of today. By the end of this time the pampa, large grassy extensions in central Argentina, became quite widespread (Patterson and Pascual, 1972). We also see the appearance and radiation of Compositae, a typical element in nonforested areas today (Van der Hammen, 1966). Among the fauna no major changes took place.

The climate continued to deteriorate from its peak of wet-warm in the middle Eocene. It still was more humid than today, as the presence of thick paleosoils in Patagonia seem to indicate (Volkheimer, 1971).

Nevertheless, the southern part of the continent, other than locally, was no longer occupied by forest but most certainly by either grassland or a parkland. The reduced rainfall, together with the ever-increasing rain-shadow effect of the rising Andes, must have led to long dry seasons in the middle latitudes. Indirect evidence from the evolutionary history of some bird and frog groups appears to indicate that the *Nothofagus* forest was not surrounded by forest vegetation at this time (Hecht, 1963; Vuilleumier, 1967). It is also very likely that semidesert regions existed in intermountain valleys and in the lee of the rising mountains in the western part of the continent from Patagonia northward.

Pliocene. From the Pliocene we have the first unmistakable evidence for the existence of more or less extensive areas of semidesert. Geologically it was a very active period. In the north we see the elevation of the Bolívar geosyncline and the development of the Colombian Andes in their present form, leading to the connection of South and North America toward the end of the period (Haffer, 1970). In Peru we see the rising of the cordillera and the bodily uplift of the altiplano to its present level, followed by some rifting. In Chile and Argentina we see the beginning of the final rise of the Cordillera Central as well as the uplift of the Sierras Pampeanas and the precordillera. All this increased orogenic activity was accompanied by extensive erosion and the deposition of continental sediments to the east in the Amazonian and Paraná-Paraguay basins (Harrington, 1962).

The lowland flora of northern South America, particularly that of the Amazonas and Orinoco basins, was not too different from today's flora in physiognomy or probably in floristic composition. However, because of the rise of the cordillera we find in the Pliocene the first indications of the existence of a high mountain flora (Van der Hammen, 1966) as well as the first clear indication of the existence of desert vegetation (Simpson Vuilleumier, 1967; Van der Hammen, 1966).

With the disappearance of the Bolívar geosyncline in late Pliocene, South America ceased to be an island and became connected to North America. This had a very marked influence on the fauna of the continent (Simpson, 1950; Patterson and Pascual, 1972). In effect, extensive faunistic interchanges took place during the Pliocene and Pleistocene between the two continents.

By the end of the Pliocene the landscape of South America was essentially identical to its present form. The rise of the Peruvian and Bolivian areas that we know as the Puna had taken place creating the dry highlands; the uplift of the Cordillera Central of Chile and the Sierras Pampeanas of Argentina had produced the rain shadows that make the area between them the dry land it is; and, finally, the rise of the southern cordillera of Chile must have produced dry, steppelike conditions in Patagonia. Geomorphological evidence shows this to be true (Simpson Vuilleumier, 1967; Volkheimer, 1971). The coastal region of Chile and Peru was probably more humid than today, since the cold Humboldt Current probably did not exist yet in its present form (Raven, 1971, 1973). However, although the stage is set, the actors are not quite ready. In effect, the Pleistocene, although very short in duration compared to the Tertiary events just described, had profound effects on species formation and distribution by drastically affecting the climate. Furthermore, because of its recency we also have a much better geological and paleontological record and therefore knowledge of the events of the Pleistocene.

Pleistocene

The deterioration of the Cenozoic climate culminated in the Pleistocene, when temperatures in the higher latitudes were lowered sufficiently to allow the accumulation and spread of immense ice sheets in the northern continents and on the highlands of the southern continents. Four major glacial periods are usually recognized in the Northern Hemisphere (Europe and North America), with three milder interglacial periods between them, and a fourth starting about 10,000 B.P. (Holocene), in which we are presently living. It is generally agreed (Charlesworth, 1957; Wright and Frey, 1965; Frenzel, 1968) that the Pleistocene has been a time of great variations in climate, both in temperature and in humidity, associated with rather significant changes in sea level (Emiliani, 1966). In general, glacial maxima correspond to colder and wetter climates than exist today; interglacials to warmer and often drier periods. But the march of events was more complex, and the temperature and humidity changes were not necessarily correlated (Charlesworth, 1957). Neither the exact series of events nor their ultimate causes are entirely clear.

Simpson Vuilleumier (1971), Van der Hammen and González (1960), and Van der Hammen (1961, 1966) have reviewed the Pleistocene events in South America. In northern South America (Venezuela, Colombia, and Ecuador) one to three glaciations took place, corresponding to the last three events in the Northern Hemisphere (Würm, Riss, and Mindel). In Peru, Bolivia, northern Chile, and Argentina there were at least three, in some areas possibly four. In Patagonia there were three to four glaciation events (table 2-1). All these glaciations, with the possible exception of Patagonia (Auer, 1960; Czajka, 1966), were the result of mountain glaciers.

The alternation of cold, wet periods with warm-dry and warm-wet periods had drastic effects on the biota. During glacial periods snow lines were lowered with an expansion of the areas suitable for a high mountain vegetation (Van der Hammen, 1966; Simpson Vuilleumier, 1971). At the same time glaciers moving along valleys created barriers to gene flow in some cases. During interglacials the snow line moved up again, and the areas occupied by high mountain vegetation no doubt were interrupted by low-lying valleys, which were occupied by more mesic-type plants. On the other hand, particularly at the beginning of interglacials, large mountain lakes were produced, and later on, with the rise of sea level, marine intrusions appeared. These events also broke up the ranges of species and created barriers to gene flow. To these happenings have to be added the effects of varying patterns of aridity and humidity. Let us then briefly review the events and their possible effects on the semidesert areas of temperate South America.

Patagonia. Glacial phenomena are best known from Patagonia (Caldenius, 1932; Feruglio, 1949; Frenguelli, 1957; Auer, 1960; Czajka, 1966; Flint and Fidalgo, 1968). Three or four glacial events are recorded. Along the cordillera the *Nothofagus* forest retreated north. The ice in its maximum extent covered probably most of Tierra del Fuego, all the area west of the cordillera, and some 100 km east of the mountains. Furthermore, during the glacial maxima, as a result of the lowering of the sea level, the Patagonian coastline was situated almost 300 km east of its present position. The climate was definitely colder and more humid. Studies by Auer (1958, 1960) indicate, however, that the *Nothofagus* forest did not expand eastward to any con-

Table 2-1. *Summary of Glacial Events in South America*

Localities	Glaciations (no.)	Age of Glaciations Relative to Europe	Present Snow Line (m)
Venezuela Mérida, Perija	1 or 2	Würm or Riss & Würm	4,800–4,900
Colombia Santa Marta and Cordillera C.	Variable, 1 to 3	Mindel to Würm	4,200–4,500
Peru All high Andean peaks	3	Mindel to Würm	5,800(W); 5,000(E)
Bolivia All NE ranges; high peaks in SE; few peaks in SW	3 or 4	Günz or Mindel to Würm	5,900(W); ca. 5,300(E)
Argentina and Chile Peaks between lat. 30° and 42°S; all land to the west of main Andean chain; to the east only to the base of the cordillera	3 or 4	Günz or Mindel to Würm	Variable: above 5,900 m in north to 800 m in south
Paraguay, Brazil, and Argentina Paraná basin		no glaciations	
Brazil Mt. Itatiaia	1 or 2	Würm or Riss & Würm	none
Brazil Amazonas basin		no glaciations	

Source: Modified from Simpson Vuilleumier, 1971.

Glacial Snow Line (m)	Glacial Climate	Interglacial Climate
2,700–3,300		
4,500(W); 4,200(E)	wet, temp. 4° to 11° lower than present	dry, temp. 2° to 3° higher than present
4,500(W); 4,200(E)	wet, temp. 7° lower than present	
5,000–5,300(W); 4,600–5,000(E)	wet, temp. 6° lower than present	
500 m at Santiago, Chile; sea level south of lat. 42°	wet	more genial than present
	cool, dry	humid, warm
2,300		
	cool, dry	humid, warm

siderable extent. It must be remembered that, even though the climate was more humid, the prevailing winds still would have been westerlies and they still would have discharged most of their humidity when they collided with the cordillera as is the case today. The drastically lowered snow line and the cold-dry conditions of Patagonia, on the other hand, must have had the effect of allowing the expansion of the high mountain flora that began to evolve as a result of the uplift of the cordillera in the Pliocene and earlier.

Monte. The essential semidesert nature of the Monte region was probably not affected by the events of the Pleistocene, but the extent of the area must have fluctuated considerably during this time. In effect, during glacial maxima not only did some regions become covered with ice, such as the valley of Santa María in Catamarca, but they also became colder. On the other hand, during interglacials there is evidence for a moister climatic regime, as the existence of fossil woodlands of *Prosopis* and *Aspidosperma* indicates (Groeber, 1936; Castellanos, 1956). Also, the present patterns of distribution of many mesophytic (but not wet-tropical) species or pairs of species, with populations in southern Brazil and the eastern Andes, could probably only have been established during a wetter period (Smith, 1962; Simpson Vuilleumier, 1971). On the other hand, geomorphological evidence from the loess strata of the Paraná-Paraguay basin (Padula, 1972) shows that there were at least two periods when the basin was a cool, dry steppe. During these periods the semidesert Monte vegetation must have expanded northward and to the east of its present range.

Puna. It has already been noted that during glacial maxima the snow line was lowered and the area open for colonization by the high mountain elements was considerably extended. Nowhere did that become more significant than in the Puna area (Simpson Vuilleumier, 1971). During glacial periods a number of extensive glaciers were formed in the mountains surrounding it, particularly the Cordillera Real near La Paz (Ahlfeld and Branisa, 1960). Numerous and extensive glacial lakes were also formed (Steinmann, 1930; Ahlfeld and Branisa, 1960; Simpson Vuilleumier, 1971). However, the basic nature of the Puna vegetation was probably not affected by these events. They must, however, have produced extensive shifts in ranges and isolation of populations, events that must have increased the rate of evolution and speciation.

Pacific Coastal Desert. The Pacific Coastal Desert is the result of the double rain shadow produced by the Andes to the east and the cold Humboldt Current to the west. The Andes did not reach their present size until the end of the Pliocene or later. The cold Humboldt Current did not become the barrier it is until its waters cooled considerably as a result of being fed by melt waters of Antarctic ice. The coastal cordillera, however, was higher in the Pleistocene than it is today (Cecioni, 1970). It is not possible to state categorically when the conditions that account for the Pacific Coastal Desert developed, but it was almost certainly not before the first interglacial. Consequently, it is safe to say that the Pacific Coastal Desert is a Pleistocene phenomenon, as is the area of Mediterranean climate farther south (Axelrod, 1973; Raven, 1973).

During the Pleistocene the snow line in the cordillera was considerably depressed and may have been as low as 1,300 m in some places (Simpson Vuilleumier, 1971). Estimates of temperature depressions are in the order of 7°C (Ahlfeld and Branisa, 1960). Although the ice did not reach the coast, the lowered temperature probably resulted in a much lowered timber line and expansion of Andean elements into the Pacific Coastal Desert. There is also evidence for dry and humid cycles during interglacial periods (Simpson Vuilleumier, 1971). The cold glacial followed by the dry interglacial periods probably decimated the tropical and subtropical elements that occupied the area in the Tertiary and allowed the invasion and adaptive radiation of cold- and dry-adapted Andean elements.

Holocene

We finally must consider the events of the last twelve thousand years, which set the stage for today's flora and vegetation. Evidence from Colombia, Brazil, Guyana, and Panama (Van der Hammen, 1966; Wijmstra and Van der Hammen, 1966; Bartlett and Barghoorn, 1973) indicates that the period started with a wet-warm period that lasted for two to four thousand years, followed by a period of colder and drier weather that reached approximately to 4,000 B.P. when the forest retreated, after which present conditions gradually became established. The wet-humid periods were times of expansion of the tropical vegetation, while the dry period was one of retreat and expansion of savannalike vegetation which appears to have occupied extensive areas of what is today the Amazonian basin (Van der Hammen, 1966; Haffer, 1969; Vanzolini and Williams, 1970; Simpson Vuilleumier, 1971). Unfortunately, no such detailed observations exist for the temperate regions of South America, but it is likely that the same alternations of wet, dry, and wet took place there, too.

The Floristic Affinities

Cabrera (1971) divides the vegetation of the earth into seven major regions, two of which, the *Neotropical* and *Antarctic* regions, include the vegetation of South America. The latter region comprises in South America only the area of the *Nothofagus* forest along both sides of the Andes from approximately lat. 35° S to Antarctica and the subantarctic islands (fig. 2-1). The Neotropical region, which occupies the rest of South America, is divided further into three dominions comprising, broadly speaking, the tropical flora (Amazonian Dominion), the subtropical vegetation (Chaco Dominion), and the vegetation of the Andes (Andean-Patagonian Dominion). The Chaco Dominion is further subdivided into seven phytogeographical provinces. Two of these are semidesert regions: the Monte province and the Prepuna province. The remaining five provinces of the Chaco Dominion are the Matorral or central Chilean province, the Chaco province, the Argentine Espinal (not to be confused with the Chilean Espinal), the region of the Pampa, and the region of the Caatinga in northeastern Brazil. With exception of the Matorral and the Caatinga, the other provinces of the Chaco Dominion are contiguous and reflect a different set of temperature, rainfall, and soil conditions in each case. The other dominion of the Neotropical flora that concerns us here is the Andean-Patagonian one, with three provinces: Patagonia, the Puna, and the vegetation of the high mountains. We see then that, of the five subdesert temperate provinces, two have a flora that is subtropical in origin and three a flora that is related to the high mountain vegetation. We will now briefly discuss the floristic affinities of each of these regions.

The Monte

The vegetation, flora, and floristic affinities of the Monte are the best known of all temperate semidesert regions (Vervoorst, 1945, 1973; Czajka and Vervoorst, 1956; Morello, 1958; Sarmiento, 1972; Solbrig, 1972, 1973). There is unanimous agreement that the flora of the Monte is related to that of the Chaco province (Cabrera, 1953, 1971; Sarmiento, 1972; Vervoorst, 1973).

Sarmiento (1972) and Vervoorst (1973) have made statistical comparisons between the Chaco and the Monte. Sarmiento, using a number of indices, shows that the Monte scrub is most closely related, both floristically and ecologically, to the contiguous dry Chaco woodland. Vervoorst, using a slightly different approach, shows that certain Monte communities, particularly on mountain slopes, have a greater number of Chaco species than other more xerophytic communities, particularly the *Larrea* flats and the vegetation of the sand dunes. Altogether, better than 60 percent of the species and more than 80 percent of the genera of the Monte are also found in the Chaco.

The most important element in the Monte vegetation is the genus *Larrea* with four species. Three of these—*L. divaricata*, *L. nitida*, and *L. cuneifolia*—constitute the dominant element over most of the surface of the Monte, either singly or in association (Barbour and Díaz, 1972; Hunziker et al., 1973). The fourth species, *L. ameghinoi*, a low-creeping shrub, is found in depressions on the southern border of the Monte and over extensive areas of northern Patagonia. Of the three remaining species, *L. cuneifolia* is found in Chile in the area between the Matorral and the beginning of the Pacific Coastal Desert, known locally as Espinal (not to be confused with the Argentine Espinal). *Larrea divaricata* has the widest distribution of the species in the genus. In Argentina it is found throughout the Monte as well as in the dry parts of the Argentine Espinal and Chaco up to the 600-mm isohyet (Morello, 1971, personal communication). However, there is some question whether the present distribution of *L. divaricata* in the Chaco is natural or the result of the destruction of the natural vegetation by man since *L. divaricata* is known to be invasive. This species is also found in Chile in the central provinces and in two isolated localities in Bolivia and Peru: the valley of Chuquibamba in Peru and the region of Tarija in Bolivia (Morello, 1958; Hunziker et al., 1973). Finally, *L. divaricata* is found in the semidesert regions of North America from Mexico to California (Yang, 1970; Hunziker et al., 1973).

The second most important genus in the Monte is *Prosopis*. Of the species of *Prosopis* found there, two of the most important ones (*P. alba*, *P. nigra*) are characteristic species in the Chaco and Argentine Espinal where they are widespread and abundant. A third very characteristic species of *Prosopis*, *P. chilensis*, is found in central and northern Chile, in the Matorral where it is fairly common and in some interior localities of the Pacific Coastal Desert, as well as in northern Peru. The records of *P. chilensis* from farther north in Ecuador and Colombia, and even from Mexico, correspond to the closely related species *P. juliflora*, considered at one time conspecific with *P. chilensis* (Burkart, 1940, 1952). *Prosopis alpataco* is found in the Monte and in Patagonia. Most other species of the genus have more limited distributions.

Another conspicuous element in the Monte is *Cercidium*. The genus is distributed from the semidesert regions in the United States where it is an important element of the flora, south along the Cordillera de los Andes, with a rather large distributional gap in the tropical region from Mexico to Ecuador. *Cercidium* is found in dry valleys of the Pacific Coastal Desert and in the cordillera in Peru and Chile. In Argentina it is found, in addition to the Monte, in the western edge of the Chaco, in the Prepuna, and also in the Puna (Johnston, 1924).

Bulnesia is represented in the Monte by two species, *B. retama* and *B. schickendanzii*. The first of these species is found also in the Pacific Coastal Desert in the region of Ica and Nazca; *B. schickendanzii*, however, is a characteristic element of the Prepuna province. Other interesting distributions among characteristic Monte species are the presence of *Bougainvillea spinosa* in the department of Moquegua in Peru (where it grows with *Cercidium praecox*). The highly specialized *Monttea aphylla* is endemic to the Monte, but a very closely related species, *Monttea chilensis*, is found in northern Chile. *Geoffroea decorticans*, the *chañar*, which is an important element both in the Chaco and in the Monte, is also found in northern Chile where it is common. These are but a few of the more important examples of Monte species that range into other semidesert phytogeographical provinces, particularly the Pacific Coastal Desert.

In summary, the Monte has its primary floristic connection with the Chaco but also has species belonging to an Andean stock. In addition, a number of important Monte elements are found in isolated dry pockets in southern Bolivia (Tarija), northern Chile, and coastal Peru and are hard to classify.

The Prepuna

There are no precise studies on the flora or the floristic affinities of the Prepuna. However, a look at the common species indicates a clear affinity with the Chaco and the Monte, such as *Zuccagnia punctata* (Monte), *Bulnesia schickendanzii* (Monte), *Bougainvillea spinosa* (Monte), *Trichocereus tertscheckii* (Monte), and *Cercidium praecox* (Monte and Chaco). Other elements are clearly Puna elements: *Psila boliviensis*, *Junellia juniperina*, and *Stipa leptostachya*. Although the Prepuna province has a physiognomy and floral mixture of its own, it undoubtedly has a certain ecotone nature, and its limits and its individuality are most probably Holocene events.

The Puna and Patagonia

Although the floristic affinities of the Puna and Patagonia have not been studied in as much detail as those of the Monte, they do not present any special problem. The flora of both regions is clearly part of the Andean flora. This important South American floristic element is relatively new (since it cannot be older than the Andes). This is further shown by the paucity of endemic families (only two small families, Nolanaceae [also found in the Galápagos Islands] and Malesherbiaceae, are endemic to the Andean Dominion) and by the large number of taxa belonging to such families as Compositae, Gramineae, Verbenaceae, Solanaceae, and Cruciferae, considered usually to be relatively specialized and geologically recent. The Leguminosae, represented in the Chaco Dominion mostly by Mimosoideae (among them some primitive genera), are chiefly represented in the Andean Dominion by more advanced and specialized genera of the Papilionoideae.

The Patagonian Steppe is characterized by a very large number of endemic genera, but particularly of endemic species (over 50%, cf. Cabrera, 1947). Of the species whose range extends beyond Patagonia, the great majority grow in the cordillera, a few extend into the Nothofagus forest, and a very small number are shared with the Monte. This is surprising in view of some similarities in soil and water stress between the two regions and also in view of the lack of any obvious physical barrier between the two phytogeographical provinces.

The Pacific Coastal Desert

The flora of the Pacific Coastal Desert is the least known. The relative lack of communications in this region, the almost uninhabited nature of large parts of the territory, and the harshness of the climate and the physical habitat have made exploration very difficult. Furthermore, a large number of species in this region are ephemerals, growing and blooming only in rainy years. Our knowledge is based largely on the works of Weberbauer and Ferreyra in Peru and those of Philippi, Johnston, and Reiche in Chile.

One of the characteristics of the region is the large number of endemic taxa. The only two endemic families of the Andean Dominion, the Malesherbiaceae and the Nolanaceae, are found here; many of the genera and most of the species are also endemic.

The majority of the species and genera are clearly related to the Andean flora. The common families are Compositae, Umbelliferae, Cruciferae, Caryophyllaceae, Gramineae, and Boraginaceae, all families that are considered advanced and geologically recent. In this it is similar to Patagonia. However, the region does not share many taxa with Patagonia, indicating an independent history from Andean ancestral stock, as is to be expected from its geographical position.

On the other hand, contrary to Patagonia, the Pacific Coastal Desert has elements that are clearly from the Chaco Dominion. Among them are Geoffroea decorticans, Prosopis chilensis, Acacia caven, Zuccagnia punctata, and pairs of vicarious species in Monttea (M. aphylla, M. chilensis), Bulnesia (B. retama, B. chilensis), Goch-

natia, and *Proustia*. In addition there are isolated populations of *Bulnesia retama*, *Bougainvillea spinosa*, and *Larrea divaricata* in Peru. Because the Monte and the Pacific Coastal Desert are separated today by the great expanse of the Cordillera de los Andes that reaches to over 5,000 m and by a minimum distance of 200 km, these isolated populations of Monte and Chaco plants are very significant.

Discussion and Conclusions

In the preceding pages a brief description of the desert and semidesert regions of temperate South America was presented, as well as a short history of the known major geological and biological events of the Tertiary and Quaternary and the present-day floristic affinities of the regions under consideration. An attempt will now be made to relate these facts into a coherent theory from which some verifiable predictions can be made.

The paleobotanical evidence shows that the Neotropical flora and the Antarctic flora were distinct entities already in Cretaceous times (Menéndez, 1972) and that they have maintained that distinctness throughout the Tertiary and Quaternary in spite of changes in their ranges (mainly an expansion of the Antarctic flora). The record further indicates that the Antarctic flora in South America was always a geographical and floristic unit, being restricted in its range to the cold, humid slopes of the southern Andes. The origin of this flora is a separate problem (Pantin, 1960; Darlington, 1965) and will not be considered here. Some specialized elements of this flora expanded their range at the time of the lifting of the Andes (*Drimys*, *Lagenophora*, etc.), but the contribution of the Antarctic flora to the desert and semidesert regions is negligible. The discussion will be concerned, therefore, exclusively with the Neotropical flora from here on.

The data suggest that at the Cretaceous-Tertiary boundary (between Maestrichtian and Paleocene) the Neotropical angiosperm flora covered all of South America with the exception of the very southern tip. The evidence for this assertion is that the known fossil floras of that time coming from southern Patagonia (Menéndez, 1972) indicate the existence then of a tropical, rain-forest-type flora in a region that today supports xerophytic, cold-adapted scrub and cushion-plant vegetation. The reasoning is that if at that time it was hot and humid enough in the southernmost part of the continent for a rain forest, undoubtedly such conditions would be more prevalent farther north. Such reasoning, although largely correct, does not take into account all the factors.

If we accept that the global flow of air and the pattern of insolation of the earth were essentially the same throughout the time under consideration (see "Theory"), it is reasonable to assume that a gradient of increasing temperature from the poles to the equator was in existence. But it is not necessarily true that a similar gradient of humidity existed. In effect, on a perfect globe (one where the specific heat of water and land is not a factor) the equatorial region and the middle high latitudes (around 40°–60°) would be zones of high rainfall while the middle latitudes (25°–30°) and the polar regions would be regions of low rainfall. This is the consequence of the global movements of air (rising at the poles and middle high latitudes and consequently cooling adiabatically and discharging their humidity, falling in the middle latitudes and the poles and consequently heating and absorbing humidity). But the earth is not a perfect globe, and, consequently, the effects of distribution of land masses and oceans have to be taken into account. When air flows over water, it picks up humidity; when it flows over land, it tends to discharge humidity; when it encounters mountains, it rises, cools, and discharges humidity; behind a mountain it falls and heats and absorbs humidity (which is the reason why Patagonia is a semidesert today).

As far as can be ascertained, at the beginning of the Tertiary there were no large mountain chains in South America. Therefore the expected air flow probably was closer to the ideal, that is, humid in the tropics and in the middle low latitudes, relatively dry in mid-latitudes. I would like to propose, therefore, that at the beginning of the Tertiary South America was not covered by a blanket of rain forest, but that at middle latitudes, particularly in the western part of the continent, there existed a tropical (since the temperature was high) flora adapted to a seasonally dry climate. This was not a semidesert flora but most likely a deciduous or semideciduous forest with some xerophytic adaptations. I would further hypothesize that this flora persisted with extensive modification into our time and is what we today call the flora of the Chaco Dominion. I will call this flora "the Tertiary-Chaco paleo-

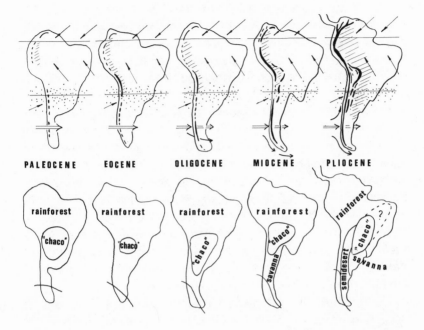

Fig. 2-2. Reconstruction of the outline of South America, mountain chains, and probable vegetation during the Tertiary. Solid line in Patagonia indicates the extent of the Nothofagus *forest. Arrows indicate main global wind patterns.*

flora" (fig. 2-2). I would like to hypothesize further that *Prosopis*, *Acacia*, and other Mimosoid legumes were elements of that flora as well as taxa in the Anacardiaceae and Zygophyllaceae or their ancestral stock. My justification for this claim is the prominence of these elements in the Chaco Dominion. Another indication of their primitiveness is their present distributional ranges, particularly the fact that many are found in Africa, which was supposedly considerably closer to South America at the beginning of the Tertiary than it is now. Furthermore, fossil remains of *Prosopis*, *Schinopsis*, *Schinus*, *Zygophyllum* (=*Guaiacum* [?], cf. Porter, 1974), and *Aspidosperma* have been reported (Berry, 1930; MacGinitie, 1953, 1969; Kruse, 1954; Axelrod, 1970) from North American floras (Colorado, Utah, and Wyoming) of Eocene or Eocene-Oligocene age (Florissant and Green River). This would indicate a wide distribution for these elements already at the beginning of the Tertiary. Verification of this hypothesis can be obtained from the study of the geology of Maestrichtian and Paleocene deposits of central Argentina and Chile as well as from the study of microfossils (and even megafossils) of this area.

Axelrod (1970) has proposed that some of the genera of the Monte and Chaco that have no close relatives in their respective families, such as *Donatia*, *Puya*, *Grabowskia*, *Monttea*, *Bredemeyera*, *Bulnesia*, and *Zuccagnia*, are modified relicts from the original upland angiosperm stock, which he feels is of pre-Cretaceous origin. The question of a pre-Cretaceous origin for the angiosperms is a speculative point due, to large measure, to the lack of corroborating fossil evidence (Axelrod, 1970). Assuming Axelrod's thesis were correct, not only would the Tertiary-Chaco paleoflora be the ancient nucleus of the subtropical elements adapted to a dry season, but presumably it would also represent the oldest angiosperm stock on the continent. However, some of the genera cited (*Bulnesia*, *Monttea*, *Zuccagnia*) are not primitive but are highly specialized.

The Tertiary in South America is characterized by the gradual lifting of the Andean chain. The process became much accelerated after the Eocene. The Tertiary is also characterized by the gradual cooling and drying of the climate after the Eocene, apparently a world-wide event (Wolfe and Barghoorn, 1960; Axelrod and Bailey, 1969; Wolfe, 1971). As a result, during the Paleocene and Eocene the Tertiary-Chaco paleoflora must have been fairly restricted in its distribution. However, after the Eocene it started to expand and differentiate. During the Eocene the first development of the steppe elements of the pampa region occurred. (The present pampa flora is part of the Chaco Dominion.) Evidence is found in the evolution of mammals adapted to eating grass and living in open habitats (Patterson and Pascual, 1972) and in the first fossil evidence of grasses (Van der Hammen, 1966). Since Tertiary deposits exist in the pampa region that have yielded animal microfossils (Padula, 1972), a study of these cores for plant microfossils may produce uncontroversial direct evidence for the evolution of the pampas. Less certain is the evolution of a dry Chaco or Monte vegetation from the Tertiary-Chaco paleoflora during the latter part of the Tertiary. The present-day distribution of *Larrea*, *Bulnesia*, *Monttea*, *Geoffroea*, *Cercidium*, and other typical dry Chaco or Monte elements makes me think that by the Pliocene a semidesert-type vegetation, not just one adapted to a dry season, was in existence in western middle South America. In effect, all these elements, so characteristic of extensive areas of semidesert in Argentina east of the Andes, are represented by small or, in some cases (*Geoffroea*, *Prosopis*), fairly abundant populations west of the Andes in a region (the Pacific Coastal Desert) dominated today by Andean elements that originated at a later date. Furthermore, the Mediterranean region of Chile is formed in part by Chaco elements, and we know that a Mediterranean climate probably did not evolve until the Pleistocene (Raven, 1971, 1973; Axelrod, 1973). It is, consequently, probable that toward the end of the Tertiary a more xerophytic flora was evolving from the Tertiary-Chaco paleoflora, perhaps as a result of xeric local conditions in the lee of the rising mountains. It should be pointed out, too, that the coastal cordillera in Chile rose first and was bigger in the Tertiary than it is today (Cecioni, 1970).

The Pliocene is characterized by a great increase in orogenic activity that created the Cordillera de los Andes in its present form (Kummel, 1961; Harrington, 1962; Haffer, 1970). In a sea of essentially tropical vegetation an alpine environment was created, ready to be colonized by plants that could withstand not only the cold but also the great daily (in low latitudes) or seasonal climate variation (in high latitudes). The rise of the cordillera also interfered with the free flow of winds and produced changes in local climate, creating the Patagonian and Monte semideserts as we know them today.

Most of the elements that populate the high cordillera were drawn from the Neotropical flora, although some Antarctic elements invaded the open Andean regions (*Lagenophora*) as well as some North American elements, such as *Alnus*, *Sambucus*, *Viburnum*, *Erigeron*, *Aster*, and members of the Ericaceae and Cruciferae (Van der Hammen, 1966). Particularly, elements in the Compositae, Caryophyllaceae, Umbelliferae, Cruciferae, and Gramineae radiated and became dominant in the newly opened habitats. The Puna is an integral part of the Andes and consequently must have become populated by the Andean elements as it became uplifted in the Pliocene, displacing the original tropical Amazonian elements present there in Miocene times (Berry, 1917) which were ill-adapted to the new climatic conditions. A similar but less-evident pattern must have taken place in Patagonia. The tropical flora present in Patagonia in the Paleocene and Eocene and, on mammalian evidence, up to the Miocene (Patterson and Pascual, 1972) had migrated north in response to the cooling climate and had been apparently replaced by an open steppe presumably of Chaco origin. This flora and that of the evolving dry Chaco-Monte vegetation should have been able to adapt to the increasing xerophytic environment of Patagonia in the Pliocene and perhaps did. Only better fossil evidence can tell. But it is the events of the Pleistocene that account for the present dry flora of Patagonia. The same can be said for the Pacific Coastal Desert. The lifting of the cordillera created a disjunct area of Chaco vegetation on the Pacific coast. Before the development of the cold Humboldt Current, presumably a Pleistocene event (in its present condition), the environment must have been more mesic, especially in its southern regions, and the Chaco-Monte flora should have been able to adapt to those conditions. Again it is the Pleistocene events that explain the present flora.

The Pleistocene is marked by a series (up to four) of very drastic fluctuations in temperature as well as some (presumably also up to four) extreme periods of aridity. During the cold periods permanent snow lines dropped, mountain glaciers formed, and the high mountain vegetation expanded. During the last glacial event, the Würm, which seems to have been the most drastic in South America, ice fields extended in Patagonia from the Pacific Coast to some 100 km or more east of the high mountain line, and the Patagonian climate was that of an arctic steppe with extremely cold winters. During those periods any existing subtropical elements disappeared and were replaced by cold-adapted Andean taxa. As conditions gradually improved, phyletic evolution of that cold-adapted flora of Andean origin produced today's Patagonian flora. The same is true in the Pacific Coastal Desert. Only here some elements of the old Chaco flora managed to survive the Pleistocene and account for the range disjunctions of such species as *Larrea divaricata*, *Bougainvillea spinosa*, or *Bulnesia retama*. Some elements of the old Chaco flora of the Pacific coast moved north and are found today in the *algarrobal* formation of northern Peru and Ecuador and in dry inter-Andean valleys of Colombia and Venezuela. Others moved south and are part of the Espinal and Matorral formations of Chile.

The Pleistocene also affected the Monte region. The northern inter-

Andean valleys and *bolsones* became colder, and the mountain slope vegetation was replaced by a vegetation of high-mountain type, so that the Monte vegetation was compressed into a smaller area farther south and east, principally in Mendoza, La Pampa, San Luis, and La Rioja. Cold periods were followed by warmer and wetter periods, characterized by more mesic subtropical elements which expanded their range from the east slope of the Andes to Brazil (Smith, 1962; Simpson Vuilleumier, 1971). After these mesic periods came very extensive dry periods, marked by expansion of the Monte flora and the breakup of the ranges of the subtropical elements, many of which have now disjunct distributions in Brazil and the eastern Andes.

Acknowledgments

This paper is the result of my long-standing interest in the flora and vegetation of the Monte in Argentina and of temperate semidesert and desert regions in general. Too many people to name individually have aided my interest, stimulated my curiosity, and satisfied my knowledge for facts. I would like, however, to acknowledge my particular indebtedness to Professor Angel L. Cabrera at the University of La Plata in Argentina, who first initiated me into floristic studies and who, over the years, has continuously stimulated me through personal conversations and letters, and through his writings. Other people whose help I would like to acknowledge are Drs. Humberto Fabris, Juan Hunziker, Harold Mooney, Jorge Morello, Arturo Ragonese, Beryl Simpson, and Federico Vervoorst. With all of them and many others I have discussed the ideas in this paper, and no doubt these ideas became modified and were changed to the point where it is hard for me to state now exactly what was originally my own. I further would like to thank the Milton Fund of Harvard University, the University of Michigan, and the National Science Foundation, which made possible yearly trips to South America over the last ten years. I particularly would like to acknowledge two NSF grants for studies in the structure of ecosystems that have supported my active research in the Monte and Sonoran desert ecosystem for the last three years. Sergio Archangelsky, Angel L. Cabrera, Philip Cantino, Carlos Menéndez,

Bryan Patterson, Duncan Porter, Beryl Simpson, and Rolla Tryon read the manuscript and made valuable suggestions for which I am grateful.

Summary

The existent evidence regarding the floristic relations of the semi-desert regions of South America and how they came to exist is reviewed.

The regions under consideration are the phytogeographical provinces of Patagonia and the Monte in Argentina; the Puna in Argentina, Bolivia, and Peru; the Espinal in Chile; and the Pacific Coastal Desert in Chile and Peru. It is shown that the flora of Patagonia, the Puna, and the Pacific Coastal Desert are basically of Andean affinities, while the flora of the Monte has affinities with the flora of the subtropical Chaco. However, Chaco elements are also found in Chile and Peru. From these considerations and those of a geological and geoclimatological nature, it is postulated that there might have existed an early (Late Cretaceous or early Tertiary) flora adapted to living in more arid —although not desert—environments in and around lat. 30° S.

The present flora of the desert and semidesert temperate regions of South America is largely a reflection of Pleistocene events. The flora of the Andean Dominion that originated the flora that today populates Patagonia and the Pacific Coastal Desert, however, evolved largely in the Pliocene, while the Chaco flora that gave origin to the Monte and Prepuna flora had its beginning probably as far back as the Cretaceous.

References

Ahlfeld, F., and Branisa, L. 1960. *Geología de Bolivia*. La Paz: Instituto Boliviano de Petroleo.
Auer, V. 1958. The Pleistocene of Fuego Patagonia. II. The history of the flora and vegetation. *Suomal. Tiedeakat. Toim. Ser. A 3*. 50:1–239.

————. 1960. The Quaternary history of Fuego-Patagonia. *Proc. R. Soc. Ser. B.* 152:507–516.

Axelrod, D. I. 1967. Drought, diastrophism and quantum evolution. *Evolution* 21:201–209.

————. 1970. Mesozoic paleogeography and early angiosperm history. *Bot. Rev.* 36:277–319.

————. 1973. History of the Mediterranean ecosystem in California. In *The convergence in structure of ecosystems in Mediterranean climates*, ed. H. Mooney and F. di Castri, pp. 225–284. Berlin: Springer.

Axelrod, D. I., and Bailey, H. P. 1969. Paleotemperature analysis of Tertiary floras. *Paleogeography, Paleoclimatol. Paleoecol.* 6:163–195.

Barbour, M. G., and Díaz, D. V. 1972. *Larrea* plant communities on bajada and moisture gradients in the United States and Argentina. *U.S./Intern. biol. Progr.: Origin and Structure of Ecosystems Tech. Rep.* 72–6:1–27.

Bartlett, A. S., and Barghoorn, E. S. 1973. Phytogeographic history of the Isthmus of Panama during the past 12,000 years. In *Vegetation and vegetational history of northern Latin America*, ed. A. Graham, pp. 203–300. Amsterdam: Elsevier.

Berry, E. W. 1917. Fossil plants from Bolivia and their bearing upon the age of the uplift of the eastern Andes. *Proc. U.S. natn. Mus.* 54:103–164.

————. 1930. Revision of the lower Eocene Wilcox flora of the southeastern United States. *Prof. Pap. U.S. geol. Surv.* 156:1–196.

Böcher, T.; Hjerting, J. P.; and Rahn, K. 1963. Botanical studies in the Atuel Valley area, Mendoza Province, Argentina. *Dansk bot. Ark.* 22:7–115.

Burkart, A. 1940. Materiales para una monografía del género *Prosopis*. *Darwiniana* 4:57–128.

————. 1952. *Las leguminosas argentinas silvestres y cultivadas*. Buenos Aires: Acme Agency.

Cabrera, A. L. 1947. La Estepa Patagónica. In *Geografía de la República Argentina*, ed. GAEA, 8:249–273. Buenos Aires: GAEA.

————. 1953. Esquema fitogeográfico de la República Argentina. *Revta Mus. La Plata (nueva Serie), Bot.* 8:87–168.

————. 1958. La vegetación de la Puna Argentina. *Revta Invest. agríc., B. Aires* 11:317–412.

————. 1971. Fitogeografía de la República Argentina. *Boln Soc. argent. Bot.* 14:1–42.

Caldenius, C. C. 1932. Las glaciaciones cuaternarias en la Patagonia y Tierra del Fuego. *Geogr. Annlr* 14:1–164.

Castellanos, A. 1956. Caracteres del pleistoceno en la Argentina. *Proc.IV Conf. int. Ass. quatern. Res.* 2:942–948.

Cecioni, G. 1970. *Esquema de paleogeografía chilena*. Santiago: Editorial Universitaria.

Charlesworth, J. K. 1957. *The Quaternary era*. 2 vols. London: Arnold.

Czajka, W. 1966. Tehuelche pebbles and extra-Andean glaciation in east Patagonia. *Quaternaria* 8:245–252.

Czajka, W., and Vervoorst, F. 1956. Die naturräumliche Gliederung Nordwest-Argentiniens. *Petermanns geogr. Mitt.* 100:89–102, 196–208.

Darlington, P. J. 1957. *Zoogeography: The geographical distribution of animals*. New York: Wiley.

————. 1965. *Biogeography of the southern end of the world*. Cambridge, Mass.: Harvard Univ. Press.

Dietz, R. S., and Holden, J. C. 1970. Reconstruction of Pangea: Breakup and dispersion of continents, Permian to present. *J. geophys. Res.* 75:4939–4956.

Dimitri, M. J. 1972. Consideraciones sobre la determinación de la superficie y los limites naturales de la región andino-patagónica. In *La región de los Bosques Andino-Patagonicos*, ed. M. J. Dimitri, 10:59–80. Buenos Aires: Col. Cient. del INTA.

Emiliani, C. 1966. Isotopic paleotemperatures. *Science, N.Y.* 154:851.

Ferreyra, R. 1960. Algunos aspectos fitogeográficos del Perú. *Publnes Inst. Geogr. Univ. San Marcos (Lima)* 1(3):41–88.

Feruglio, E. 1949. *Descripción geológica de la Patagonia*. 2 vols. Buenos Aires: Dir. Gen. de Y.P. F.

Fiebrig, C. 1933. Ensayo fitogeográfico sobre el Chaco Boreal. *Revta Jard. bot. Mus. Hist. nat. Parag.* 3:1–87.

Flint, R. F., and Fidalgo, F. 1968. Glacial geology of the east flank of the Argentine Andes between latitude 39°-10′ S and latitude 41°-20′ S. *Bull. geol. Soc. Am.* 75:335–352.

Flohn, H. 1969. *Climate and weather*. New York: McGraw-Hill Book Co.

Frenguelli, J. 1957. El hielo austral extraandino. In *Geografía de la República Argentina*, ed. GAEA, 2:168–196. Buenos Aires: GAEA.

Frenzel, B. 1968. The Pleistocene vegetation of northern Eurasia. *Science, N.Y.* 161:637.

Good, R. 1953. *The geography of the flowering plants*. London: Longmans, Green & Co.

Goodspeed, T. 1945. The vegetation and plant resources of Chile. In *Plants and plant science in Latin America*, ed. F. Verdoorn, pp. 147–149. Waltham, Mass.: Chronica Botanica.

Gordillo, C. E., and Lencinas, A. N. 1972. Sierras pampeanas de Córdoba y San Luis. In *Geología regional Argentina*, ed. A. F. Leanza, pp. 1–39. Córdoba: Acad. Nac. de Ciencias.

Groeber, P. 1936. Oscilaciones del clima en la Argentina desde el Plioceno. *Revta Cent. Estud. Doct. Cienc. nat., B. Aires* 1(2):71–84.

Haffer, J. 1969. Speciation in Amazonian forest birds. *Science, N.Y.* 165:131–137.

————. 1970. Geologic-climatic history and zoogeographic significance of the Uraba region in northwestern Colombia. *Caldasia* 10: 603–636.

Harrington, H. J. 1962. Paleogeographic development of South America. *Bull. Am. Ass. Petrol. Geol.* 46:1773–1814.

Haumann, L. 1947. Provincia del Monte. In *Geografía de la República Argentina*, ed. GAEA, 8:208–248. Buenos Aires: GAEA.

Hecht, M. K. 1963. A reevaluation of the early history of the frogs. Pt. II. *Syst. Zool.* 12:20–35.

Hoffstetter, R. 1972. Relationships, origins and history of the Ceboid monkeys and caviomorph rodents: A modern reinterpretation. *Evol. Biol.* 6:323–347.

Hünicken, M. 1966. Flora terciaria de los estratos del río Turbio, Santa Cruz. *Revta Fac. Cienc. exact. fís. nat. Univ. Córdoba, Ser. Cienc. nat.* 27:139–227.

Hunziker, J. H.; Palacios, R. A.; de Valesi, A. G.; and Poggio, L. 1973. Species disjunctions in *Larrea*: Evidence from morphology, cytogenetics, phenolic compounds, and seed albumins. *Ann. Mo. bot. Gdn* 59:224–233.

Johnston, I. 1924. Taxonomic records concerning American spermatophytes. 1. Parkinsonia and Cercidium. *Contr. Gray Herb. Harv.* 70:61–68.

———. 1929. Papers on the flora of northern Chile. *Contr. Gray Herb. Harv.* 85:1–171.

Kruse, H. O. 1954. Some Eocene dicotyledoneous woods from Eden Valley, Wyoming. *Ohio J. Sci.* 54:243–267.

Kummel, B. 1961. *History of the earth*. San Francisco: W. H. Freeman & Co.

Liddle, R. A. 1946. *The geology of Venezuela and Trinidad*. 2d ed. Ithaca: Pal. Res. Inst.

Lorentz, P. 1876. Cuadro de la vegetación de la República Argentina. In *La República Argentina*, ed. R. Napp, pp. 77–136. Buenos Aires: Currier de la Plata.

MacGinitie, H. D. 1953. Fossil plants of the Florissant beds, Colorado. *Publs Carnegie Instn* 599:1–180.

———. 1969. The Eocene Green River flora of northwestern Colorado and northeastern Utah. *Univ. Calif. Publs geol. Sci.* 83:1–140.

Mayr, E. 1963. *Animal species and evolution*. Cambridge, Mass: Harvard Univ. Press.

Menéndez, C. A. 1969. Die fossilen floren Südamerikas. In *Biogeography and ecology in South America*, ed. E. J. Fittkau, J. Illies, H. Klinge, G. H. Schwabe, and H. Sioli, 2:519–561. The Hague: Dr. W. Junk.

———. 1972. Paleofloras de la Patagonia. In *La región de los Bosques Andino-Patagonicos*, ed. M. J. Dimitri, 10:129–184. Col. Cient. Buenos Aires: del INTA.

Mooney, H., and Dunn, E. L. 1970. Convergent evolution of Mediterranean climate evergreen colorophyll chrubc. *Evolution* 24:202 UUU.

Morello, J. 1958. La provincia fitogeográfica del Monte. *Op. lilloana* 2:1–155.

Padula, E. L. 1972. Subsuelo de la mesopotamia y regiones adyacentes. In *Geología regional Argentina*, ed. A. F. Leanza, pp. 213–236. Córdoba: Acad. Nac. de Ciencias.

Pantin, C. F. A. 1960. A discussion on the biology of the southern cold temperate zone. *Proc. R. Soc. Ser. B.* 152:431–682.

Patterson, B., and Pascual, R. 1972. The fossil mammal fauna of South America. In *Evolution, mammals, and southern continents*, ed. A. Keast, F. C. Erk, and B. Glass, pp. 247–309. Albany: State Univ. of N.Y.

Petriella, B. 1972. Estudio de maderas petrificadas del Terciario inferior del área de Chubut Central. *Revta Mus. La Plata (Nueva Serie), Pal.* 6:159–254.

Porter, D. M. 1974. Disjunct distributions in the New World Zygophyllaceae. *Taxon* 23:339–346.

Raven, P. H. 1971. The relationships between "Mediterranean" floras. In *Plant life of South-West Asia*, ed. P. H. Davis, P. C. Harper, and I. C. Hedge, pp. 119–134. Edinburgh: Bot. Soc.

———. 1973. The evolution of Mediterranean floras. In *The convergence in structure of ecosystems in Mediterranean climates*, ed. H. Mooney and F. di Castri, pp. 213–224. Berlin: Springer.

Reiche, K. 1934. *Geografía botánica de Chile*. 2 vols. Santiago: Imprenta Universitaria.

Romero, E. 1973. Ph.D. dissertation, Museo La Plata Argentina.

Sarmiento, G. 1972. Ecological and floristic convergences between seasonal plant formations of tropical and subtropical South America. *J. Ecol.* 60:367–410.

Schuchert, C. 1935. *Historical geology of the Antillean-Caribbean region*. New York: Wiley.

Schwarzenbach, M. 1968. Das Klima des rheinischen Tertiärs. *Z. dt. geol. Ges.* 118:33–68.

Simpson, G. G. 1950. History of the fauna of Latin America. *Am. Scient.* 1950:361–389.

Simpson Vuilleumier, B. 1967. The systematics of Perezia, section Perezia (Compositae). Ph.D. thesis, Harvard University.

———. 1971. Pleistocene changes in the fauna and flora of South America. *Science, N.Y.* 173:771–780.

Smith, L. B. 1962. Origins of the flora of southern Brazil. *Contr. U.S. natn. Herb.* 35:215–250.

Solbrig, O. T. 1972. New approaches to the study of disjunctions with special emphasis on the American amphitropical desert disjunctions. In *Taxonomy, phytogeography and evolution*, ed. D. D. Valentine, pp. 85–100. London and New York: Academic Press.

———. 1973. The floristic disjunctions between the "Monte" in Argentina and the "Sonoran Desert" in Mexico and the United States. *Ann. Mo. bot. Gdn* 59:218–223.

Soriano, A. 1949. El limite entre las provincias botánicas Patagónica y Central en el territorio del Chubut. *Revta argent. Agron.* 17:30–66.

———. 1950. La vegetación del Chubut. *Revta argent. Agron.* 17:30–66.

———. 1956. Los distritos floristicos de la Provincia Patagónica. *Revta Invest. agríc., B. Aires* 10:323–347.

Stebbins, G. L. 1950. *Variation and evolution in plants*. New York: Columbia Univ. Press.

———. 1952. Aridity as a stimulus to evolution. *Am. Nat.* 86:33–44.

Steinmann, G. 1930. *Geología del Perú*. Heidelberg: Winters.

Thorne, R. F. 1973. Floristic relationships between tropical Africa and tropical America. In *Tropical forest ecosystems in Africa and South America: A comparative review*, ed. B. J. Meggers, E. S. Ayensu, and D. Duckworth, pp. 27–40. Washington, D.C.: Smithsonian Instn. Press.

Van der Hammen, T. 1961. The Quaternary climatic changes of northern South America. *Ann. N.Y. Acad. Sci.* 95:676–683.

———. 1966. Historia de la vegetación y el medio ambiente del norte sudamericano. In *1° Congr. Sud. de Botánica, Memorias de Symposio*, pp. 119–134. Mexico City: Sociedad Botánica de Mexico.

Van der Hammen, T., and González, E. 1960. Upper Pleistocene and Holocene climate and vegetation of the "Sabana de Bogotá." *Leid. geol. Meded.* 25:262–315.

Vanzolini, P. E., and Williams, E. E. 1970. South American anoles: The geographic differentiation and evolution of the *Anolis chrysolepis* species group (Sauria, Iguanidae). *Archos Zool. Est. S Paulo* 19:1–298.

Vervoorst, F. 1945. *El Bosque de algarrobos de Pipanaco (Catamarca)*. Ph.D. dissertation, Universidad de Buenos Aires.

———. 1973. Plant communities in the bolsón de Pipanaco. *U.S./ Intern. biol. Progr.: Origin and Structure of Ecosystems Prog. Rep.* 73-3:3–17.

Volkheimer, W. 1971. Aspectos paleoclimatológicos del Terciario Argentina. *Revta Mus. Cienc. nat. B. Rivadavia Paleontol.* 1:243–262.

Vuilleumier, F. 1967. Phyletic evolution in modern birds of the Patagonian forests. *Nature, Lond.* 215:247–248.

Weberbauer, A. 1945. *El mundo vegetal de los Andes Peruanos*. Lima: Est. Exp. La Molina.

Wijmstra, T. A., and Van der Hammen, T. 1966. Palynological data on the history of tropical savannas in northern South America. *Leid. geol. Meded.* 38:71–90.

Wolfe, J. A. 1971. Tertiary climatic fluctuations and methods of analysis of Tertiary floras. *Paleogeography, Paleoclimatol. Paleoecol.* 9:27–57.

Wolfe, J. A., and Barghoorn, E. S. 1960. Generic change in Tertiary floras in relation to age. *Am. J. Sci.* 258A:388–399.

Wright, H. E., and Frey, D. G. 1965. *The Quaternary of the United States*. Princeton: Princeton Univ. Press.

Yang, T. W. 1970. Major chromosome races of *Larrea divaricata* in North America. *J. Ariz. Acad. Sci.* 6:41–45.

3. The Evolution of Australian Desert Plants   John S. Beard

Introduction

As an opening to this subject it may be well to outline briefly the where-abouts of the Australian desert, its climate and vegetation. The desert consists, of course, of the famous "dead heart" of Australia, covering the interior of the continent; and it has been defined on a map together with its component natural regions by Pianka (1969*a*). An important characteristic of this area is that, while certainly arid and classifiable as desert by most, if not all, of the better-known bioclimatic classifi-cations and indices, it is not as rainless as some of the world's deserts and is correspondingly better vegetated. The most arid portion of the Australian interior, the Simpson Desert, receives an average rainfall of 100 mm, while most of the rest of the desert receives around 200 mm. The desert is usually taken to begin, in the south, at the 10-inch, or 250-mm, isohyet. In the north, in the tropics under higher temper-atures, desert vegetation reaches the 20-inch, or 500-mm, isohyet.

Plant Formations in Australian Deserts

As a result of the rainfall in the Australian desert, it always possesses a plant cover of some kind, and we have no bare and mobile sand dunes and few sheets of barren rock. There are two principal plant formations: a low woodland of *Acacia* trees colloquially known as mul-ga, which covers roughly the southern half of the desert south of the tropic, and the "hummock grassland" (Beadle and Costin, 1952) col-loquially known as spinifex, which covers the northern half within the tropics. Broadly the two formations are climatically separated, although the preference of each of them for certain soils tends to obscure this relationship; thus, the hummock grassland appears on sand even in the southern half. The *Acacia* woodland is to be compared with those of other continents, but few Australian species of *Acacia* have thorns and few have bipinnate leaves. The hummock grassland, on the other hand, is, I think, a unique product of evolution in Australia. It is comparable with the grass steppe vegetation of other continents, but the life form of the grasses is different. Two genera are represented, *Triodia* and *Plectrachne*. Each plant branches repeatedly into a great number of culms which intertwine to form a hummock and bear rigid, terete, pungent leaves presenting a serried phalanx to the exterior. When flowering takes place in the second half of summer, given adequate rains, upright rigid inflorescences are produced above the crown of the hummock, rising 0.5 to 1 m above it. The flowers quickly set seed, which is shed within two months, although this is then the beginning of the dry season. The size of the hummock varies considerably according to the site from 30 cm in height and diameter on the poorest, stoniest sites up to about 1 m in height and 2 m in diameter on some deep sands. Old hummocks, if unburnt, tend to die out in the center or on one side, leading to ring or crescentic growth. At this stage the original root has died and the outer culms have rooted themselves adventitiously in the soil. Individual hummocks do not touch, and there is much bare ground between them.

The hummock grassland normally contains a number of scattered shrubs or scattered trees in less-arid areas where ground water is available. All of these must be resistant to fire, by which the grassland is regularly swept. After burning, the grasses regenerate from the root or from seed.

The *Acacia* woodlands, in which *A. aneura* is frequently the sole species in the upper stratum, contain a sparse lower layer of shrubs most frequently of the genera *Eremophila* and *Cassia*, 1–2 m tall, and an even sparser ground layer mainly of ephemerals and only locally of grasses.

These Australian desert formations are given distinctive character by the physiognomy of their commonest plants, that is:

Trees. Evergreen, sclerophyll. Leaves pendent in *Eucalyptus*; linear, erect, and glaucescent in *Acacia aneura*; vestigial in *Casuarina decaisneana*. Bark white in most species of *Eucalyptus*.

Shrubs. The larger shrubs are sclerophyll, typically phyllodal species of *Acacia*; the smaller shrubs, ericoid (*Thryptomene*).

Subshrubs. Many soft perennial subshrubs typically with densely pubescent or silver-tomentose stems and leaves, e.g., *Crotalaria cunninghamii*, and numerous Verbenaceae (*Dicrastyles, Newcastelia, Pityrodia* spp.). Also, suffrutices with underground rootstocks and ephemeral or more or less perennial shoots, often also densely pubescent or silver-tomentose, e.g., *Brachysema chambersii*, many *Ptilotus* spp., *Leschenaultia helmsii*, and *L. striata*. Some are viscid—*Goodenia azurea* and *G. stapfiana*.

Ephemerals. Many species of Compositae, *Ptilotus*, and *Goodenia* appear as brilliant-flowering annuals in season. Colors are predominantly yellow and mauve, with some white and pink. Red is absent.

Grasses. Grasses of the "short bunch-grass" type in the sense of Bews (1929) occur only on alluvial flats close to creeks or on plains of limited extent developed on or close to basic rocks. In these cases there is a fine soil with a relatively high water-holding capacity and probably also high-nutrient status. On sand, laterite, and rock in the desert, grasses belong almost entirely to the genera *Triodia* and *Plectrachne*, which adopt the hummock-grass form as previously described. This growth form appears to be peculiar to Australia and to be the only unique form evolved in the Australian desert.

It will therefore be seen that the Australian desert possesses special vegetative characters of its own which can be supposed to be of some adaptive significance, particularly *glaucescence* of bark and leaves, *pubescence* frequently in association with glaucescence, *suffrutescence*, the presence of vernicose and viscid leaf surfaces, and the *spinifex* habit in grasses. Other characters, such as tree and shrub growth forms and sclerophylly, are not peculiar to the desert Eremaea but are shared with other Australian vegetation.

Growth Forms

In most of the world's deserts special and peculiar growth forms have evolved which confer advantage in the arid environment. In North and

Central America the family Cactaceae has produced the well-known range of forms based on stem succulence, closely replicated by the Euphorbiaceae in Africa. In southern African deserts leaf succulence is a dominant feature that has been developed in many families, notably the Aizoaceae and Liliaceae. Leaf-succulent rosette plants in the Bromeliaceae are a feature of both arid northwest Brazil and the cold Andean Puna. In all cases we are accustomed to look also for deciduous, thorny trees and plants with underground perennating organs, especially bulbs and corms. In Australia there is an extraordinary lack of all these forms; where some of them exist they are confined to certain areas.

Leaf- and stem-succulent plants belonging to the family Chenopodiaceae in fact characterize two other important plant formations, less widespread than the principal formations described above and confined to certain soils. These I have named "succulent steppe" (Beard, 1969) following the usage of African ecologists; they comprise, first, saltbush and bluebush steppe dominated by species of *Atriplex* and *Kochia* respectively, and, second, samphire communities with *Arthrocnemum*, *Tecticornia*, and related genera. The former are small soft shrubs whose leaves are fleshy or semisucculent, associated with annual grasses and herbs, and sometimes with a sclerophyll tree layer of *Acacia* or *Eucalyptus*. The formation is confined to the southern half of the desert region and occupies alkaline soils, most commonly on limestone or calcareous clays. In the northern half such soils normally carry hummock grassland on limestone and bunch grassland on clays. The samphire communities, however, range throughout the region on very saline soils in depressions, usually in the beds of playa lakes or peripheral to them. The samphires are subshrubs with succulent-jointed stems. These formations are the only ones with a genuinely succulent character and are essentially halophytes.

On the siliceous soils of the desert, sclerophylly is the dominant characteristic, and stem succulence is represented in only a handful of species of no prominence, such as *Sarcostemma australe* (Asclepiadaceae), a divaricate, leafless plant found occasionally in rocky places. Others are *Spartothamnella teucriiflora* (Verbenaceae) and *Calycopeplus helmsii* (Euphorbiaceae). Likewise, leaf succulence is found in a variety of groups but is often weakly developed and never a conspicuous feature. *Gyrostemon ramulosus* (Phytolaccaceae) has somewhat fleshy foliage, which the explorers noted as a favorite feed of camels. The Aizoaceae in Australia are mostly tropical herbs, and the most genuinely succulent member, *Carpobrotus*, is not Eremaean. The Portulacaceae are a substantial group with twenty-seven species in *Calandrinia*, of which about twelve are Eremaean, and eight in *Portulaca*, which belong to the Northern Province. *Calandrinia* is herbaceous and leaf succulent, and several species are not uncommon, but it will be noted that they are not essentially desert plants. A weak leaf succulence can be seen in *Kallstroemia, Tribulus*, and *Zygophyllum* of the Zygophyllaceae and in *Euphorbia* and *Phyllanthus* of the Euphorbiaceae. Few of these are plants of any ecological importance.

Evolutionary History

The evolutionary significance of these different growth forms must now be discussed. Our view of the past history of biota has been transformed by the development of the theory of plate tectonics in quite recent years, with sanction given to the previously heretical ideas of continental drift. As long ago as 1856, in his famous preface to the *Flora Tasmaniae*, J. D. Hooker suggested that the modern Australian flora was compounded of three elements—an Indo-Malaysian element derived from southeast Asia, an autochthonous element evolved within Australia itself, and an Antarctic element comprising forms common to the southern continents which in some way should be presumed to have been transmitted via Antarctica. The trouble was that, while the reality of this Antarctic element could not be doubted, no means or mechanism save that of long-range dispersal could be used to account for it—unless one were very daring and, after Wegener and du Toit, were prepared to invoke continental drift. The thinking of those years of fixed-positional geology is typified by Darlington's book *Biogeography of the Southern End of the World* (1965), in which the southern continents are seen as refuges where throughout time odd forms from the Northern Hemisphere have established themselves and survived. Our Antarctic element would then become only a random selection of forms long extinct in the other hemisphere. This view is now discredited.

Although the breakup of Gondwanaland is dated rather earlier than the origin of the angiosperms, many of the continents do seem to have remained sufficiently close or, in some cases, in actual contact in such a way that explanations of the distribution of plant forms are materially assisted. Where Australia is concerned in this discussion of desert biota we need only go back to Eocene times, some 40 to 60 million years ago, when our continent was joined to Antarctica along the southern edge of its continental shelf and lay some 15° of latitude farther south than now (Griffiths, 1971). In middle Eocene times a rift occurred in the position of the present mid-oceanic ridge separating Australia and Antarctica; the two continents broke apart and drifted in opposite directions: Antarctica to have its biota largely extinguished by a polar icecap, Australia to move toward and into the tropics, passing in the process through an arid zone in which much of it still lies. The evolution of the desert flora of Australia has therefore occurred since the Eocene *pari passu* with this movement.

In discussions of Tertiary paleoclimates it is commonly assumed that the circulation of the atmosphere has always been much the same as it is today, so that the positions of major latitudinal climatic belts have also been fairly constant, even though there may have been cyclic variations in temperature and in quantity of rainfall. At the time, therefore, when Australia was situated 15° farther south, it would have lain squarely in the roaring forties; and it seems likely that a copious and well-distributed rainfall would have been received more or less throughout the continent. This is borne out by the fossil record which predominantly suggests a cover of rain forest of a character and composition similar to that found today in the North Island of New Zealand (Raven, 1972).

Paleontological evidence suggests rather warmer temperatures prevailing at that time and in those latitudes than exist there today. When the break from Antarctica took place, the southern coastline of Australia slumped and thin deposits of Eocene and Miocene sediments were laid down upon the continental margin. Fossils indicate deposition in seas of tropical temperature, continuing as late as Miocene times (Dorman, 1966; Cockbain, 1967; Lowry, 1970). This is consistent with the evidence of tropical flora extending to lat. 50° N in North America in the Eocene (Chaney, 1947) and to Chile and Patagonia (Skottsberg, 1956).

Evidence from the soil supports the concept of both high temperature and high rainfall. In the Canning and Officer sedimentary basins in Western Australia, the parts of the country now occupied by the Great Sandy and Gibson deserts, an outcrop of rocks of Cretaceous age has been very deeply weathered and thickly encrusted with laterite. Farther south than this an outcrop of Miocene limestone in the Eucla basin exhibits relatively little weathering or development of typical karst features and is considered to have been exposed to a climate not substantially wetter than the present since its uplift from the sea at the end of Miocene times (Jennings, 1967).

The laterization would indicate subjection for a long period to a warm, wet climate, which must therefore be early Tertiary in date. The present surface features of all of these sedimentary basins are in accord with presumed climatic history based on known latitudinal movement of the continent.

From Eocene times, therefore, Australian flora had to adapt itself to progressive desiccation. It is frequently assumed that it also had to adapt to warmer temperatures in moving northward, but I believe that this is a mistake. We have fossil evidence for warmer temperatures already in the Eocene, followed by a progressive cooling of the earth through the later Tertiary; and the northward movement of Australasia largely provided, I think, a compensation for the latter process. I do not concur with Raven and Axelrod (1972), for example, that we have to assume a developed adaptation to tropical conditions in those elements in the flora of New Caledonia which are of southern origin. Australasian flora, however, had to adapt to the greater extremes of temperature which accompany aridity, even though mean temperatures may not have greatly altered.

From my own consideration of the paleolatitudes and an attempt to map the probable paleoclimates (which I cannot now go into in detail), I believe that the first appearance of aridity may have been in the northwest in the Kimberley district of Western Australia in later Eocene times, expanding steadily to the southeast. The first Mediterranean climate with its winter-wet, summer-dry regime seems likely to have become established in the Pilbara district of Western Australia in the Oligocene and to have been progressively displaced south-ward.

The Roles of Fires and Soil

In addition to the climatic adaptations required, Australian flora also had to adapt itself to changes in soil which have accompanied the desiccation and to withstand fire. In the early Tertiary rain forests fire was probably unknown or a rarity. Such forests are able to grow even on a highly leached and impoverished substratum in the absence of fire, as a cycle of accumulation and decomposition of organic matter is built up and the forest is living on the products of its own decay. It has been shown that intense weathering and laterization occurred in the early Tertiary in some areas of Western Australia, and this may be observed elsewhere in the continent.

This process would have occurred initially under the forest without provoking significant changes, but with desiccation two things happen: fire ruptures the nutrient cycle leading to a collapse of the eco-system, and the laterites are indurated to duricrust. After burning and rapid removal of mineralized nutrients by the wind and the rain, a depauperate scrub community with a low-nutrient demand replaces the rain forest. In the absence of fire a slow succession back to the rain forest will ensue, but further fires stabilize the disclimax. This process may be seen in operation today in western Tasmania. It is intensified where laterite is present since induration of laterite by desiccation is irreversible and produces an inhospitable hardpan in the soil, usually followed by deflation of the leached sandy topsoil to leave a surface duricrust which is even more inhospitable.

Arid Australia is situated in those central and western parts of the continent where there has been little or no tectonic movement during the Tertiary to regenerate systems of erosion, so that after desiccation set in there was mostly no widespread removal of ancient weathered soil material or the rejuvenation of the soils. Great expanses of inert sand or surface laterite clothe the higher ground and offer an inhospit-able substratum to plants, poor in nutrients and in water-holding capacity. Leaching has continued, and its products have been deposited in the lower ground by evaporation where soils have been zonally accumulating calcium carbonate, gypsum, and chlorides.

Biogeographical Elements

Evolutionary adaptation to these changed conditions during the later Tertiary produced the autochthonous element in the Australian flora mostly by adaptation of forms present in the previously dominant Antarctic element. The Indo-Malaysian element is a relatively recent arrival and, as may be expected, has colonized mainly the moister tropical habitats. It has not contributed very significantly to the desert biota, but there are a few species whose very names betray their origin in that direction: *Trichodesma zeylanicum*, an annual herb in the Boraginaceae; *Crinum asiaticum*, a bulbous Amaryllid, bringing a life form (the perennating bulb) which is almost unknown in Australian desert biota in spite of its apparent evolutionary advantages.

Herbert (1950) pointed out that the autochthonous element is essentially one adapted to subhumid, semiarid, and desert conditions which has been evolved within Australia from forms whose relatives are of world-wide distribution. Evolution, said Herbert, took place in three ways: from ancestors already adapted to these drier climates, by survival of hardier types when increasing aridity drove back the more mesic vegetation, and by recolonization of drier areas by the more xerophytic members of mesic communities.

Burbidge (1960) examined the question more closely and acknowledged a suggestion made to her by Professor Smith-White of the University of Sydney that many of the elements in the desert flora may have developed from species associated with coastal habitats. Burbidge considered that such an opinion was supported by the number of genera in the desert flora of Australia which elsewhere in the world are associated with coastal areas, sand dunes, and habitats of saline type. It is certainly a very reasonable assumption that, in a well-watered early Tertiary continent, source material for future desert plants should lie in the flora of the littoral already adapted to drying winds, sand or rock as a substratum, or salt-marsh conditions. Burbidge went on to say that it is not until the late Pleistocene or early

Recent that there is any real evidence in the fossil record for the existence of a desert flora. However, this does not prove it was not there, and the evidence for the northward movement of Australia into the arid zone suggests strongly that it must have begun its evolution at least as early as the Miocene. Pianka (1969b) in discussing Australian desert lizards found that the species density was too great to permit evolution proceeding only from the sub-Recent. An identical argument is bound to apply to flora also. Speciation is too great and too diversified to have originated so recently.

Morphological Evolution

In addition to the systematic evolution of the desert flora, we may usefully discuss also its morphological evolution. It has been shown that some of the life forms considered most typical of desert biota in other continents are inconspicuous or lacking in Australia, for example, deciduousness, spinescence, and underground perennating organs. Other life forms, especially succulence, are limited to particular areas. Morphologically, there is a dualism in Australian desert flora. The typical plant forms of poor, leached siliceous soils are radically different from those of the base-rich alkaline and saline soils. The former are essentially sclerophyllous in the particular manner of so many Australian plant forms from all over the continent which are not confined to the desert. There has even been the evolution of a unique form of sclerophyll grass, the spinifex or hummock-grass form. On the other hand, succulent and semisucculent leaves replace the sclerophyll on base-rich soils. It is evident that aridity alone is not responsible for sclerophylly in Australian plants as has so often been thought. This evidence seems strongly to support the views of Professor N. C. W. Beadle, expressed in numerous papers (e.g., Beadle, 1954, 1966). Beadle has argued for a relationship between sclerophylly and nutrient deficiency, especially lack of soil phosphate. It certainly seems true to say that the plant forms of nutrient-deficient soils in the Australian desert have had the directions of their evolution dictated not only by aridity but by soil conditions as well, soil conditions largely peculiar to Australia as a continent so that this section of the Australian desert flora has acquired a unique character. It has evolved, we may say, within a straitjacket of sclerophylly. This limitation, however, has not been imposed on the ion-accumulating bottom-land soils where plant forms more similar to those of deserts in other continents have evolved.

To look back to what has been said about the taxonomic evolution of the desert flora, limitations are also imposed by the nature of the genetic source material. A subtropical and warm temperate rain forest is not a very promising source area for forms which will have the necessary genetic plasticity for adaptation to great extremes of temperature and aridity, as well as to extremes of soil deficiency. Certain Australian plant families have possessed this faculty, especially the Proteaceae, and this has resulted in a proliferation of highly specialized and adapted species in a relatively limited number of genera. This phenomenon is remarked especially on the soils which have the most extreme nutrient deficiencies or imbalances under widely differing climatic conditions, notably on the Western Australian sand plains, the Hawkesbury sandstone of New South Wales, and the serpentine outcrops in the mountains of New Caledonia, in all of which different species belonging to the same or related Australian genera can be seen forming a similar maquis or sclerophyll scrub. The sclerophyll desert flora has drawn heavily upon this source material, while the nonsclerophyll flora has been influenced particularly by the ability of the family Chenopodiaceae to produce forms adaptable to the particular conditions.

Summary

The Australian desert, covering the interior of the continent, receives an average rainfall of 100 to 250 mm annually and is well vegetated. There are two principal plant formations, *Acacia* low woodland and *Triodia-Plectrachne* hummock grassland, characteristic broadly of the sectors south and north of the Tropic of Capricorn. Component species are typically sclerophyll in form, even the grasses. Nonsclerophyll vegetation of succulent and semisucculent subshrubs locally occupies alkaline soils, in depressions or on limestone and calcareous clays. There is otherwise a notable absence of such xerophytic life forms as stem and leaf succulents, rosette plants, deciduous thorny trees, and plants with bulbs and corms.

Australian desert flora evolved gradually from the end of Eocene times as the continent moved northward into arid latitudes. As the previous vegetation was mainly a subtropical rain forest, it has been suggested that the source material for this evolution came largely from the littoral and seashore. Species had to adapt not only to aridity but also to soils deeply impoverished by weathering under previous humid conditions and not rejuvenated. It is believed that the siliceous, nutrient-deficient soils have been responsible for the predominantly sclerophyllous pattern of evolution; succulence has only developed on base-rich soils.

References

Beadle, N. C. W. 1954. Soil phosphate and the delimitation of plant communities in eastern Australia. *Ecology* 25:370–374.

————. 1966. Soil phosphate and its role in moulding segments of the Australian flora and vegetation with special reference to xeromorphy and sclerophylly. *Ecology* 47:991–1007.

Beadle, N. C. W., and Costin, A. B. 1952. Ecological classification and nomenclature. *Proc. Linn. Soc. N.S.W.* 77:61–82.

Beard, J. S. 1969. The natural regions of the deserts of Western Australia. *J. Ecol.* 57:677–711.

Bews, J. W. 1929. *The world's grasses*. London: Longmans, Green & Co.

Burbidge, N. T. 1960. The phytogeography of the Australian region. *Aust. J. Bot.* 8:75–211.

Chaney, R. W. 1947. Tertiary centres and migration routes. *Ecol. Monogr.* 17:141–148.

Cockbain, A. E. 1967. Asterocyclina from the Plantagenet beds near Esperance, W.A. *Aust. J. Sci.* 30:68.

Darlington, P. J. 1965. *Biogeography of the southern end of the world*. Cambridge, Mass.: Harvard Univ. Press.

Dorman, F. H. 1966. Australian Tertiary paleotemperatures. *J. Geol.* 74:49–61.

Griffiths, J. R. 1971. Reconstruction of the south-west Pacific margin of Gondwanaland. *Nature, Lond.* 234:203–207.

Herbert, D. A. 1950. Present day distribution and the geological past. *Victorian Nat.* 66:227–232.

Hooker, J. D. 1856. Introductory Essay. In *Botany of the Antarctic Expedition, vol. III flora Tasmaniae*, pp. xxvii–cxii.

Jennings, J. N. 1967. Some karst areas of Australia. In *Land form studies from Australia and New Guinea*, ed. J. N. Jennings and J. A. Mabbutt. Canberra: Aust. Nat. Univ. Press.

Lowry, D. C. 1970. Geology of the Western Australian part of the Eucla Basin. *Bull. geol. Surv. West. Aust.* 122:1–200.

Pianka, E. R. 1969*a*. Sympatry of desert lizards (*Ctenotus*) in Western Australia. *Ecology* 50:1012–1013.

———. 1969*b*. Habitat specificity, speciation and species density in Australian desert lizards. *Ecology* 50:498–502.

Raven, P. H. 1972. An introduction to continental drift. *Aust. nat. Hist.* 17:245–248.

Raven, P. H., and Axelrod, D. I. 1972. Plate tectonics and Australasian palaeobiogeography. *Science, N.Y.* 176:1379–1386.

Skottsberg, C. 1956. *The natural history of Juan Fernández and Easter Island. I(ii) Derivation of the flora and fauna of Easter Island.* Uppsala: Almqvist & Wiksell.

4. Evolution of Arid Vegetation in Tropical America

Guillermo Sarmiento

Introduction

More or less continuous arid regions cover extensive areas in the middle latitudes of both South and North America, forming a complex pattern of subtropical, temperate, and cold deserts on the western side of the two American continents. They appear somewhat intermingled with wetter ecosystems wherever more favorable habitats occur. These two arid zones are widely separated from each other, leaving a huge gap extending over almost the whole intertropical region (see fig. 4-1). South American arid zones, however, penetrate deeply into intertropical latitudes from northwestern Argentina through Chile, Bolivia, and Peru to southern Ecuador. But they occur either as high-altitude deserts, such as the Puna (high Andean plateaus over 3,000 m), or as coastal fog deserts, such as the Atacama Desert in Chile and Peru, the driest American area. This coastal region, in spite of its latitudinal position and low elevation, cannot be considered as a tropical warm desert, because its cool maritime climate is determined by almost permanent fog. In fact, in most of tropical America, either in the lowlands or in the high mountain chains, from southern Ecuador to southern Mexico, more humid climates and ecosystems prevail. In sharp contrast with the range areas of western North America and the high cordilleras and plateaus of western South America, the tropical American mountains lie in regions of wet climates from their piedmonts to the highest summits. The same is true for the lower ranges located in the interior of the Guianan and Brazilian plateaus.

Upon closer examination, however, it is apparent that, although warm, tropical rain forests and mountain forests, as well as savannas, are characteristic of most of the tropical American landscape, the arid

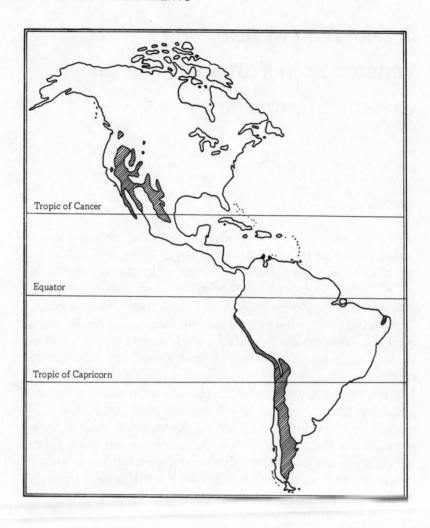

Fig. 4-1. American arid lands (after Meigs, 1953, modified).

ecosystems are far from being completely absent. If we look at a generalized map of arid-land distribution, such as that of Meigs (1953), we will notice two arid zones in tropical South America: one in northeastern Brazil and the other forming a narrow belt along the Caribbean coast of northern South America, including various small nearby islands. These two tropical areas share some common geographical features:

1. They are quite isolated from each other and from the two principal desert areas in North and South America. The actual distance between the northeast Brazilian arid Caatinga and the nearest desert in the Andean plateaus is about 2,500 km, while its distance from the Caribbean arid region is over 3,000 km. The distance from the Caribbean arid zone to the nearest South American continuous desert, in southern Ecuador, and to the closest North American continuous desert, in central Mexico, is in both cases around 1,700 km.

2. They appear completely encircled by tropical wet climates and plant formations.

3. The two areas are more or less disconnected from the spinal cord of the continent (the Andes cordillera), particularly in the case of the Brazilian region. This fact surely has had major biogeographical consequences.

Recently, interest in ecological research in American arid regions has been renewed, mainly through the wide scope and interdisciplinary research programs of the International Biological Program (Lowe et al., 1973). These studies give strong emphasis to a thorough comparison of temperate deserts in the middle latitudes of North and South America, with the purposes of disclosing the precise nature of their ecological and biogeographical relationships and also of assessing the degree of evolutionary convergence and divergence between corresponding ecosystems and between species of similar ecological behavior. Within this context, a deeper knowledge of tropical American arid ecosystems would provide additional valuable information to clarify some of the previous points, besides having a research interest per se, as a particular case of evolution of arid and semiarid ecosystems of Neotropical origin under the peculiar environmental conditions of the lowland tropics.

The aim of this paper is to present certain available data concerning tropical American arid and semiarid ecosystems, with particular reference to their flora, environment, and vegetation structure. The geographical scope will be restricted to the Caribbean dry region, of which I have direct field knowledge; there will be only occasional further reference to the Brazilian dry vegetation. The Caribbean dry region is still scarcely known outside the countries involved; a review book on arid lands, such as that of McGinnies, Goldman, and Paylore (1968), does not provide a single datum about this region.

In order to delimit more precisely the region I am talking about, a climatic and a vegetational criterion will be used. My field experience suggests that most dry ecosystems in this part of the world lie inside the 800-mm annual rainfall line, with the most arid types occurring below the 500-mm rainfall line. Figure 4-2 shows the course of these two climatic lines through the Caribbean area. Though some wetter ecosystems are included within this limit, particularly at high altitudes, few arid types appear outside this area except localized edaphic types on saline soils, beaches, coral reefs, dunes, or rock outcrops. Only in the Lesser Antilles does a coastal arid vegetation appear under higher rainfall figures, up to 1,200 mm, and this only on very permeable and dry soils near the sea (Stehlé, 1945).

This climatically dry region extends over northern Colombia and Venezuela and covers most of the small islands of the Netherlands Antilles—Aruba, Curaçao, and Bonaire—reaching a total area of about 50,000 km². The nearest isolated dry region toward the northwest is in Guatemala, 1,600 km away; in the north, a dry region is in Jamaica and Hispaniola, 800 km across the Caribbean Sea; while southward the nearest dry region is in Ecuador, 1,700 km away.

From the point of view of vegetation, only the extremes of the Seasonal Evergreen Formation Series and the Dry Evergreen Formation Series of Beard (1944, 1955) will be considered here, including the following four formations: Thorn Woodland, Thorn Scrub, Desert, and Dry Evergreen Bushland. Several papers have dealt with the vegetation of this dry area, but they analyze either only a restricted zone inside this whole region, as those of Dugand (1941, 1970), Tamayo (1941), Marcuzzi (1956), Stoffers (1956), and several others, or they are generalized accounts of plant cover for a whole country that include a short description of the arid types, like those of Cuatrecasas (1958) or Pittier (1926). The aim of this paper is to go one step further than previous investigations—first, considering the entire

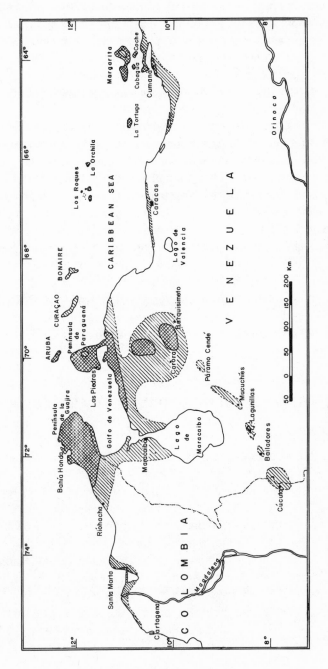

Fig. 4-2. Caribbean arid lands. Semiarid (500–800 mm rainfall) and arid zones (less than 500 mm rainfall) have been distinguished.

Caribbean dry region and, second, comparing it to the rest of American dry lands. My previous paper (1972) had a similar approach. The thorn forests and thorn scrub of tropical America were included in a floristic and ecological comparison between tropical and subtropical seasonal plant formations of South America. I will follow that approach here, but will restrict my scope to the dry extreme of the tropical American vegetation gradient.

To avoid a possible terminological misunderstanding as to certain concepts I am employing, it is necessary to point out that the words *arid* and *semiarid* refer to climatological concepts and will be applied both to climates and to plant formations and ecosystems occurring under these climates. *Dry* will refer to every type of xeromorphic vegetation, either climatically or edaphically determined, such as those of sand beaches, dunes, and rock pavements. *Desert* will be used in its wide geographical sense, that is, a region of dry climate where several types of dry plant formations occur, among them semidesert and desert formations. In this way, for instance, the Sonora and the Monte deserts have mainly a semidesert vegetation, while the Chile-Peru coastal desert shows mainly a desert plant formation. Each time I refer to a *desert vegetation* in contrast to a *desert region*, I shall clarify the point.

The Environment in the Caribbean Dry Lands

Geography

The Caribbean dry region, as its name suggests, is closely linked with the Caribbean coast of northern South America, stretching almost continuously from the Araya Peninsula in Venezuela, at long. 64° W, to a few kilometers north of Cartagena in Colombia, at long. 75° W. Along most of this coast the dry zone constitutes only a narrow fringe between the sea and the forest formation beginning on the lower slopes of the contiguous mountains: the Caribbean or Coast Range in Venezuela and the Sierra Nevada of Santa Marta in Colombia. In many places this arid fringe is no more than a few hundred meters wide. But in the two northernmost outgrowths of the South American continent, the Guajira and Paraguaná peninsulas, the dry region widens to cover these two territories almost completely (see fig. 4-2).

Besides these strictly coastal areas, dry vegetation penetrates deeper inside the hinterland around the northern part of the Maracaibo basin as well as in the neighboring region of low mountains and inner depressions known as the Lara-Falcón dry area of Venezuela. In this zone the aridity reaches more than 200 km from the coast.

Besides this almost continuous dry area in continental South America, the Caribbean dry region extends over the nearby islands along the Venezuelan coast, from Aruba through Curaçao, Bonaire, Los Roques, La Orchila, and other minor islands to Margarita, Cubagua, and Coche. The islands farthest from the continental coast lie 140 km off the Venezuelan coast. Dry vegetation entirely covers these islands, except for a few summits with an altitude over 500 m. The Lesser Antilles somehow connect this dry area with the dry regions of Hispaniola, Cuba, and Jamaica, because almost all of them show restricted zones of dry vegetation (Stehlé, 1945).

Both on the continents and in the islands dry plant formations occupy the lowlands, ranging in altitude from sea level to no more than 600–700 m, covering in this low climatic belt all sorts of land forms, rock substrata, and geomorphological units, such as coastal plains, alluvial and lacustrine plains, early and middle Quaternary terraces, rocky slopes, and broken hilly country of different ages. In the islands dry vegetation also occurs on coral reefs, banks, and on the less-extended occurrences of loose volcanic materials.

Apart from the nearly continuous coastal region and its southward extensions, I should point out that a whole series of small patches or "islands" of dry vegetation and climate occurs along the Andes from western Venezuela across Colombia and Ecuador to Peru. These small and isolated arid patches may be divided into two ecologically divergent types according to their thermal climate determined by altitude: those occurring below 1,500–1,800 m that have a warm or megathermal climate and those appearing above that altitude and belonging then to the meso- or microthermal climatic belts. The latter, such as the small dry islands in the Páramo Cendé and the upper Chama and upper Mocoties valleys of the Venezuelan Andes, even though they have low rainfall, have a less-unfavorable water budget because of their comparatively constant low temperature. Therefore, their vegetation has few features in common with the remaining dry Caribbean areas. On the other hand, the lower-altitude dry patches, like the middle Chama valley, the Tachira-Pamplonita depression, and the lower Chicamocha valley, are quite similar to the dry coastal regions in ecology, flora, and vegetation and will be considered in this study as part of the Caribbean dry lands. I shall point out further the biogeographical significance of this archipelago of Andean dry islands connecting the Caribbean dry region with the southern South American deserts.

Throughout the dry area of northern South America, dry plant formations appear bordered by one or other of three different types of vegetation units: tropical drought-deciduous forest, dry evergreen woodland, or littoral formations (mangroves, littoral woodlands, etc.). In the lower Magdalena valley, as well as in certain other partially flooded areas, marshes and other hydrophytic formations are also common, intermingled with thorn woodland or thorn scrub.

Climate

I propose to analyze the prevailing climatic features of the region enclosed within the 800-mm rainfall line, with particular reference to the main climatic factor affecting plant life, that is, the amount of rainfall and its seasonal distribution, but without disregarding other climatic elements that sharply differentiate tropical and extratropical climates, like minimal temperatures, annual cycle of insolation, and thermo- and photoperiodicity. Lahey (1958) provided a detailed discussion about the causes of the dry climates around the Caribbean Sea, and I shall refer to that paper for pertinent meteorological and climatological considerations on this topic. Porras, Andressen, and Pérez (1966) presented a detailed study of the climate of the islands of Margarita, Coche, and Cubagua, some of the driest areas of the Caribbean; some of the climatic data I will discuss have been taken from that paper.

As pointed out before, a major part of the region with annual rainfall figures below 800 mm is located in the megathermal belt, below 600–700 m, and has an annual mean temperature above 24°C. A few small patches along the Andes reach higher elevations, up to 1,500–1,800 m, and their annual mean temperatures go down to 20°C, fitting within what has been considered as the mesothermal belt. However, this temperature difference does not seem to introduce significant changes in vegetation physiognomy or ecology.

Mean annual temperatures in coastal and lowland localities range from a regional maximum of 28.7°C in Las Piedras, at sea level, to 24.2°C in Barquisimeto, a hinterland locality at 566-m elevation. Mean temperatures show very slight month-to-month variation (1° to 3.5°C), as is typical for low latitudes. The annual range of extreme temperatures in this ever-warm region is not so wide as in subtropical or temperate dry regions. The recorded absolute regional maximum does not reach 40°C, while the absolute minima are everywhere above 17°C. As we can see, then, in sharp contrast with the case in extratropical conditions, in the dry Caribbean region low temperatures never constitute an ecological limitation to plant life and natural vegetation.

I have already pointed out that, using natural vegetation as a guideline for our definition of aridity, an annual rainfall of 800 mm roughly separates semiarid and arid from humid regions in this part of the world. Excluding edaphically determined vegetation, the most open and sparse vegetation types appear where rainfall figures do not reach 500 mm. The lowest rainfall in the whole area has been recorded in the northern Guajira Peninsula (Bahía Honda: 183 mm) and in the island of La Orchila, which has the absolute minimum rainfall for the region, 150 mm. Rainfall figures below 300 mm also characterize the small islands of Coche and Cubagua and the central and driest part of Margarita. As we can see, these figures are really very low, fully comparable to many desert localities in temperate South and North America, but in our case these rainfall totals occur under constantly high temperatures and, therefore, represent a less-favorable water balance and a greater drought stress upon plant and animal life.

Concerning rainfall patterns, figure 4-3 shows the rainfall regime at eight localities, arranged in an east-to-west sequence from Cumaná at long. 64°11′ W to Pueblo Viejo at long. 74° 16′ W. The rainfall pattern varies somewhat among the localities appearing in the figure; some places show a unimodal distribution, with the yearly maximum slightly preceding the winter solstice (October to December), while other localities show a bimodal distribution, with a secondary maximum during the high sun period (May to June). It is clear, nevertheless, that all localities have a continuous drought throughout the year, with ten to twelve successive months when rainfall does not reach 100 mm and five to eight months with monthly rainfall figures below 50

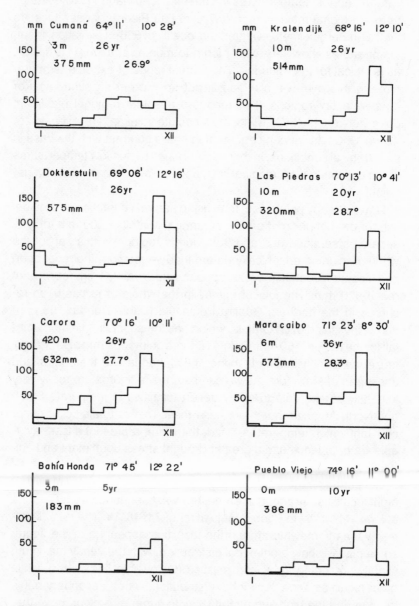

Fig. 4-3. Rainfall regimen for eight stations in the dry Caribbean region. Each climadiagram shows longitude, latitude, altitude, years of recording, mean annual rainfall, and mean temperature.

mm. The total number of rain days ranges over the whole area from forty to sixty.

As is typical for dry climates, rainfall variability is very high, reaching values of 40 percent and more where rainfall is less than 500 mm. This high interannual variability maintains the dryness of the climate and the drought stress upon perennial organisms.

In contrast to most other dry regions in temperate America, relative humidity in the Caribbean dry region is not as low, showing average monthly figures of 70 to 80 percent throughout the year and minimal monthly values of around 55 to 60 percent. Annual pan evaporation, however, is very high, generally exceeding 2,000 mm, with many areas having values as high as 2,500–2,800 mm. Potential evapotranspiration, calculated according to the Thornthwaite formula, reaches 1,600–1,800 mm.

According to its rainfall and temperature, this Caribbean dry region falls within the BSh and BWh climatic types of Koeppen's classification, that is, *hot steppe* and *hot desert* climates. We should remember that in this tropical region the rainfall value setting apart dry and humid climates in the Koeppen system is around 800 mm, which is precisely the limit I have taken according to natural vegetation. In turn, the 400-to-450-mm rainfall line separates BS, or semiarid, from BW, or arid, climates. Following the second system of climatic classification of Thornthwaite, this region comes within the DA' and EA' types, that is, *semiarid megathermal* and *arid megathermal* climates respectively. It is interesting that in both systems the climates corresponding to the dry Caribbean area are the same as those found in the dry subtropical regions of America, such as the Chaco and Monte regions of South America and the Sonoran and Chihuahuan deserts of North America.

We will now consider certain rhythmic environmental factors which influence both climate and ecological behavior of organisms, such as incoming solar radiation and length of day. The sharp contrast between incoming solar radiation at sea level (maximal theoretical values disregarding cloudiness) in low and middle latitudes is well known. At low latitudes daily insolation varies only slightly throughout the year, forming a bimodal curve in correspondence with the sun passing twice a year over that latitude. The total variation between the extremes of maximal and minimal solar radiation during the year is in the order of 50 percent. At middle latitudes the annual radiation curve is unimodal in shape and shows, at a latitude as low as 30° north where North American warm deserts are more widespread, a seasonal variation between extremes in the order of 300 percent.

Photoperiodicity is also inconspicuous in tropical latitudes. At 10° north or south the difference between the shortest and the longest day of the year is around one hour, while at 30° it is almost four hours. Summarizing the climatic data, we can see that the Caribbean dry region has semiarid and arid climates partly comparable to those found in middle-latitude American deserts, particularly insofar as permanent water deficiency is concerned; but these tropical climates differ from the subtropical dry climates by more uniform distribution of solar radiation, higher relative humidity, higher minimal temperatures, slight variation of monthly means, and shorter variation in the length of day throughout the year.

Physiognomy and Structure of Plant Formations

General

Beard (1955) has given the most valuable and widely used of the classifications of tropical American vegetation. Arid vegetation appears as the dry extreme of two series of plant formations: the Seasonal Evergreen Formation Series and the Dry Evergreen Formation Series. Seasonal formations were arranged in Beard's scheme along a gradient of increasing climatic seasonality (rainfall seasonality because all are isothermal climates), from an ever-wet regime without dry seasons to the most highly desert types. The successive terms of this series, beginning with the Tropical Rain Forest *sensu stricto* as the optimal plant formation in tropical America are Seasonal Evergreen Forest, Semideciduous Forest, Deciduous Forest, Thorn Woodland, Thorn or Cactus Scrub, and Desert. The first two units appear under slightly seasonal climates; Deciduous Forest together with savannas appear under tropical wet and dry climates; while the last three members of this series, Thorn Woodland, Thorn Scrub, and Desert, occur under dry climates with an extended rainless season and as such are common in the dry Caribbean area.

The Dry Evergreen Formation Series of Beard's classification, in contrast with the Seasonal Series, occurs under almost continuously dry climates but where monthly rainfall values are not as low as during the dry season of the seasonal climates. The driest formations on this series are the Thorn Scrub and the Desert formations, these two series being convergent in physiognomic and morphoecological features according to Beard and other authors, such as Loveless and Asprey (1957). The remaining less-dry type, next to the previous two, is the Dry Evergreen Bushland formation, which also occurs under the dry climates of the Caribbean area. In summary, dry vegetation in tropical America has been included in four plant formations: Dry Evergreen Bushland, Thorn Woodland, Thorn or Cactus Scrub, and Desert. Their structures according to the original definitions are represented in figure 4-4. All of them have open physiognomies, where the upper-layer canopy in the more structured and richer types does not surpass 10 m in height and a cover of 80 percent, decreasing then in height and cover as the environmental conditions become less favorable.

Plant formations occurring in the arid Caribbean region fit closely with Beard's classification and types, though it seems necessary to add a new formation: Deciduous Bushland, structurally equivalent to the Dry Evergreen Bushland, but with a predominance of deciduous woody species. Before going into some details about each dry plant formation in the Caribbean area, let me add a final remark about the evident difficulty met with when some of the vegetation is classified in one or another type, particularly in the case of some low and poor associations of the Tropical Deciduous Forest whose features overlap with those of the Dry Evergreen Bushland or the Thorn Woodland. Human interference, through wood cutting and heavy goat grazing, frequently makes subjective conclusions difficult, and in many instances only a thorough quantitative recording of vegetational features could allow an objective characterization and classification of the stand. At a preliminary survey level these doubts remain. A detailed study of dry plant formations, such as that of Loveless and Asprey (1957, 1958) in Jamaica, will emphasize the need for quantitative data on vegetation structure and species morphoecology in order to classify these difficult intermediate dry formations.

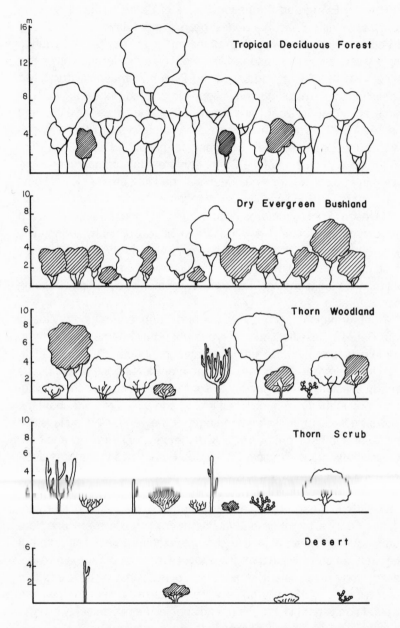

Fig. 4-4. Vegetation profiles of tropical American dry formations. Tropical Deciduous Forest has been included for comparison.

I will now very briefly consider each of the five dry plant formations as they occur nowadays in the Caribbean dry region.

Dry Evergreen Bushland

The Dry Evergreen Bushland formation has a closed canopy of low trees and shrubs at a height of about 2 to 4 m. Sparse taller trees and cacti, up to 10 m high, may emerge from this canopy. The two essential physiognomic features of this plant formation are, first, the closed nature of the plant cover, leaving no bare ground at all, and, second, the predominance of evergreen species, with a minor proportion of deciduous and succulent-aphyllous elements. The dominant woody species are evergreen low trees and shrubs, with sclerophyllous medium-sized leaves.

Floristically it is a rather rich plant formation, taking into account its dry nature, with an evident differentiation into various floristic associations. The most important families of this formation are the Euphorbiaceae, Boraginaceae, Capparidaceae, Leguminosae (Papilionoideae and Caesalpinoideae), Rhamnaceae, Polygonaceae, Rubiaceae, Myrtaceae, Flacourtiaceae, and Celastraceae. Cacti and agaves are also frequent, interspersed with a rich subshrubby and herbaceous flora.

This formation is widespread in the whole Caribbean dry region, being frequent in the islands, in the mainland coasts, and in the small Andean arid patches. Its physiognomy clearly differentiates this evergreen bushland from all other tropical dry formations; and from this physiognomic and structural viewpoint it looks more like the temperate scrubs in Mediterranean climates, such as the low chaparral of California and the garigue of southern France, than most other tropical types.

Deciduous Bushland

The Deciduous Bushland is quite similar in structure to the Dry Evergreen Bushland, but it differs mainly by the predominance of deciduous shrubs, while evergreen and aphyllous species only share a secondary role. This gives a highly seasonal appearance to the Deciduous Bushland, with two acutely contrasting aspects: one during the leafless period and the other when the dominant species are in full leaf. It also differs from other seasonal dry formations, such as the Thorn Woodland and the Thorn Scrub, because of its closed canopy of shrubs and low trees that leaves no bare ground. The most common families in this formation are the Leguminosae (mainly Mimosoideae), Verbenaceae, Euphorbiaceae, and Cactaceae. Floristically this type is not well known, but apparently it differs sharply from the Thorn Woodland and Thorn Scrub. Up to date the Deciduous Bushland has only been reported in the Lara-Falcón area (Smith, 1972).

Thorn Woodland

The distinctive physiognomic feature of the Thorn Woodland is a lack of a continuous canopy at any height, leaving large spaces of bare soil between the sparse trees and shrubs, particularly during dry periods when herbaceous annual cover is lacking. The upper layer of high shrubs and low trees and succulents is from 4 to 8 m high, with a variable cover, from less than 10 percent to a maximum of around 75 percent. A second woody layer 2 to 4 m high is generally the most important in cover, showing values ranging from 30 to 70 percent. The shrub layer of 0.5 to 2 m is also conspicuous, inversely related in importance to the two uppermost layers. The total cover of the herb and soil layers varies during the year because of the seasonal development of annual herbs, geophytes, and hemicryptophytes; the permanent biomass in these lowest layers is given by small cacti, like *Mammillaria*, *Melocactus*, and *Opuntia*.

As for the morphoecological features of its species, this formation is characterized by a high proportion of thorny elements, by many succulent shrubs, and by a total dominance of the smallest leaf sizes (lepto- and nanophyll), with a smaller proportion of aphyllous and microphyllous species together with rare mesophyllous elements; the last mentioned are generally highly scleromorphic. The relative proportion of evergreen and deciduous species is almost the same, with a good proportion of brevideciduous species.

From the floristic aspect this formation has a very characteristic flora, scarcely represented in wetter plant types. Among the most important families are the Leguminosae (particularly Mimosoideae and

Caesalpinoideae), Cactaceae, Capparidaceae, and Euphorbiaceae. Many floristic associations can be distinguished on the basis of the dominant species, but their distribution and ecology are scarcely known. The most important single species in this formation, distributed over its area, is undoubtedly *Prosopis juliflora*. When it is present, this low tree usually shares a dominant role in the community. This may probably be due, among other reasons, to its noteworthy ability for regrowth after cutting, as well as to its unpalatability to all domestic herbivores.

Thorn Scrub or Cactus Scrub

The Thorn Scrub, equivalent to the Semidesert formation of arid temperate areas, is still lower and more sparse than the Thorn Woodland, leaving a major part of bare ground, particularly during the driest period of the year. Low trees and columnar cacti from 4 to 8 m high appear widely dispersed or are completely lacking. Shrubs from 0.5 to 2 m high, though they form the closest plant layer, are also widely separated, as well as the subshrubs and herbs that form the scattered lower layer. Floristically the Thorn Scrub seems to be an impoverished Thorn Woodland, without significant additions to the flora of that formation. Cactaceae, Capparidaceae, Euphorbiaceae, and Mimosoideae continue to be the best-represented taxa. Even by its morphoecology and functionality this formation resembles the Thorn Woodland, showing a heterogeneous mixture of evergreen, deciduous, brevideciduous, and aphyllous species, with the smallest leaf sizes frequently being of sclerophyllous texture. Succulent species, particularly cacti, appear here at their optimum, frequently being the most noteworthy feature in the physiognomy of the plant formation.

This Thorn Scrub physiognomy is not so widely found in the Caribbean arid region as in the temperate deserts of North and South America. By structure and biomass it is comparable to the Semidesert formations of those arid regions, though the most extended associations of temperate American deserts, those formed by nonspiny low shrubs such as *Larrea*, are completely absent from the tropical American area. Thorn Scrub occurs in the Lara-Falcón region of Venezuela, in the northernmost part of the Guajira Peninsula, and in the driest islands like Coche and Cubagua.

Desert

Extremely desertic vegetational physiognomies are not uncommon in the Caribbean arid zone, but most of them seem determined by substratum-related factors and not primarily by climate. Thus, for example, one of the most widespread types of Desert formation occurs in the Lara-Falcón area, on sandstone hills of Tertiary age. Only four or five species of low shrubs grow there, such as species of *Cassia*, *Sida*, and *Heliotropium*, very widely interspersed with some woody *Capparis* and various Cactaceae and Mimosoideae. The total ground cover is less than 2 or 3 percent. To explain this extremely desertic vegetation in an area with enough rainfall to maintain thorn woodland in neighboring situations, Smith (1972) suggested the existence of heavy metals in the rock substrata; but there is not yet any further evidence to sustain this hypothesis, though undoubtedly the responsible factor is linked to a particular type of geological formation.

Another type of desert community that covers a wide extent of flat country in northern Venezuela appears on heavy soil developed on old Quaternary terraces. This desert community scarcely covers more than 5 percent of the ground and is composed mainly of species of *Jatropha*, *Opuntia*, *Lemaireocereus*, and *Ipomoea*, together with some annual herbs. Though this community is rather common in several parts of the Caribbean arid area, a satisfactory explanation for its occurrence has not been given for it, either.

Some more easily understood types of Desert formation are the salt deserts near the coast and the sand deserts of dunes and beaches. Salt deserts are almost everywhere characterized by low, shrubby Chenopodiaceae, such as *Salicornia* and *Heterostachys*; while sand deserts show a dominance of geophytes together with some shrubby species of *Lycium*, *Castela*, *Opuntia*, and *Acacia*.

Floristic Composition and Diversity

The floristic inventory of the Caribbean arid vegetation has not yet been made. My list of plant families and genera has been compiled from several sources (Boldingh, 1914; Tamayo, 1941 and 1967; Dugand, 1941 and 1970; Pittier et al., 1947; Croizat, 1954; Marcuzzi, 1956; Stoffers, 1956; Cuatrecasas, 1958; Trujillo, 1966) as well as from direct field knowledge of this vegetation and flora.

Table 4-1 presents a list of 94 families and 470 genera which have been reported from this area. Both figures must be taken as rough approximations of the regional total flora, because this arid flora is still not well known and in many areas plant collections are lacking; there are also some overrepresentations in the tabulated figures, because many of the listed taxa collected in the arid region surely belong to various riparian forests and therefore are not strictly part of the arid Caribbean flora. The total number of species is still more imprecisely known; a figure of 1,000 will give an idea of the magnitude of the species diversity in this vegetation.

If the floristic richness and diversity in more restricted areas is taken into consideration, the following figures are obtained: a thorough

---

Table 4-1. *Families and Genera of Flowering Plants Reported from the Caribbean Dry Region*

Family	Genera
Acanthaceae	*Anisacanthus*, *Anthacanthus*, *Dicliptera*, *Elytraria*, *Justicia*, *Odontonema*, *Ruellia*, *Stenandrium*
Achatocarpaceae	*Achatocarpus*
Aizoaceae	*Mollugo*, *Sesuvium*, *Trianthema*
Amaranthaceae	*Achyranthes*, *Alternanthera*, *Amaranthus*, *Celosia*, *Cyathula*, *Froelichia*, *Gomphrena*, *Iresine*, *Pfaffia*, *Philoxerus*
Amaryllidaceae	*Agave*, *Crinum*, *Fourcroya*, *Hippeastrum*, *Hymenocallis*, *Hypoxis*, *Zephyranthes*
Anacardiaceae	*Astronium*, *Mauria*, *Metopium*, *Spondias*
Apocynaceae	*Aspidosperma*, *Echites*, *Forsteronia*, *Plumeria*, *Prestonia*, *Rauvolfia*, *Stemmadenia*, *Thevetia*
Araceae	*Philodendron*
Aristolochiaceae	*Aristolochia*

Asclepiadaceae	*Asclepias, Calotropis, Cynanchum, Gomphocarpus, Gonolobus, Ibatia, Marsdenia, Metastelma, Omphalophthalmum, Sarcostemma*
Bignoniaceae	*Amphilophium, Anemopaegma, Arrabidaea, Bignonia, Clytostoma, Crescentia, Distictis, Lundia, Memora, Pithecoctenium, Tabebuia, Tecoma, Xylophragma*
Bombacaceae	*Bombacopsis, Bombax, Cavanillesia, Pseudobombax*
Boraginaceae	*Cordia, Heliotropium, Rochefortia, Tournefortia*
Bromeliaceae	*Aechmea, Bromelia, Pitcairnia, Tillandsia, Vriesia*
Burseraceae	*Bursera, Protium*
Cactaceae	*Acanthocereus, Cephalocereus, Cereus, Hylocereus, Lemaireocereus, Mammillaria, Melocactus, Opuntia, Pereskia, Phyllocactus, Rhipsalis*
Canellaceae	*Canella*
Capparidaceae	*Belencita, Capparis, Cleome, Crataeva, Morisonia, Steriphoma, Stuebelia*
Caryophyllaceae	*Drymaria*
Celastraceae	*Hippocratea, Maytenus, Pristimera, Rhacoma, Schaefferia*
Chenopodiaceae	*Atriplex, Chenopodium, Heterostachys, Salicornia*
Cochlospermaceae	*Amoreuxia, Cochlospermum*
Combretaceae	*Bucida, Combretum*
Commelinaceae	*Callisia, Commelina, Tripogandra*
Compositae	*Acanthospermum, Ambrosia, Aster, Baltimora, Bidens, Conyza, Egletes, Eleutheranthera, Elvira, Eupatorium, Flaveria,*

Gundlachia, Isocarpha, Lactuca, Lagascea, Lepidesmia, Lycoseris, Mikania, Oxycarpha, Parthenium, Pectis, Pollalesta, Porophyllum, Sclerocarpus, Simsia, Sonchus, Spilanthes, Synedrella, Tagetes, Trixis, Verbesina, Vernonia, Wedelia

Convolvulaceae	*Bonomia, Cuscuta, Evolvulus, Ipomoea, Jacquemontia, Merremia*
Cruciferae	*Greggia*
Cucurbitaceae	*Bryonia, Ceratosanthes, Corallocarpus, Doyerea, Luffa, Melothria, Momordica, Rytidostylis*
Cyperaceae	*Bulbostylis, Cyperus, Eleocharis, Fimbristylis, Hemicarpha, Scleria*
Elaeocarpaceae	*Muntingia*
Erythroxylaceae	*Erythroxylon*
Euphorbiaceae	*Acalypha, Actinostemon, Adelia, Argithamnia, Bernardia, Chamaesyce, Cnidoscolus, Croton, Dalechampsia, Ditaxis, Euphorbia, Hippomane, Jatropha, Julocroton, Mabea, Manihot, Pedilanthus, Phyllanthus, Sebastiania, Tragia*
Flacourtiaceae	*Casearia, Hecatostemon, Laetia, Mayna*
Gentianaceae	*Enicostemma*
Gesneriaceae	*Kohleria, Rechsteineria*
Goodeniaceae	*Scaevola*
Gramineae	*Andropogon, Anthephora, Aristida, Bouteloua, Cenchrus, Chloris, Cynodon, Dactyloctenium, Digitaria, Echinochloa, Eleusine, Eragrostis, Eriochloa, Leptochloa, Leptothrium, Panicum, Pappophorum, Paspalum, Setaria, Sporobolus, Tragus, Trichloris*
Guttiferae	*Clusia*

Hernandaceae	*Gyrocarpus*
Hydrophyllaceae	*Hydrolea*
Krameriaceae	*Krameria*
Labiatae	*Eriope, Hyptis, Leonotis, Marsypianthes, Ocimum, Perilomia, Salvia*
Lecythidaceae	*Chytroma, Lecythis*
Leguminosae (Caesalpinoideae)	*Bauhinia, Brasilettia, Brownea, Caesalpinia, Cassia, Cercidium, Haematoxylon, Schnella*
Leguminosae (Mimosoideae)	*Acacia, Calliandra, Cathormium, Desmanthus, Inga, Leucaena, Mimosa, Piptadenia, Pithecellobium, Prosopis*
Leguminosae (Papilionoideae)	*Abrus, Aeschynomene, Benthamantha, Callistylon, Canavalia, Centrosema, Crotalaria, Dalbergia, Dalea, Desmodium, Diphysa, Erythrina, Galactia, Geoffraea, Gliricidia, Humboldtiella, Indigofera, Lonchocarpus, Machaerium, Margaritolobium, Myrospermum, Peltophorum, Phaseolus, Piscidia, Platymiscium, Pterocarpus, Rhynchosia, Sesbania, Sophora, Stizolobium, Stylosanthes, Tephrosia*
Lennoaceae	*Lennoa*
Liliaceae	*Smilax, Yucca*
Loasaceae	*Mentzelia*
Loganiaceae	*Spigelia*
Loranthaceae	*Oryctanthus, Phoradendron, Phthirusa, Struthanthus*
Lythraceae	*Ammannia, Cuphea, Pleurophora, Rotala*
Malpighiaceae	*Banisteria, Banisteriopsis, Brachypteris, Bunchosia, Byrsonima, Heteropteris, Hiraea, Malpighia, Mascagnia, Stigmatophyllum, Tetrapteris*

Malvaceae	*Abutilon, Bastardia, Cienfuegosia, Hibiscus, Malachra, Malvastrum, Pavonia, Sida, Thespesia, Urena, Wissadula*
Melastomaceae	*Miconia, Tibouchina*
Meliaceae	*Trichilia*
Menispermaceae	*Cissampelos*
Moraceae	*Brosimum, Chlorophora, Ficus, Helicostylis*
Myrtaceae	*Anamomis, Pimenta, Psidium*
Nyctaginaceae	*Allionia, Boerhavia, Mirabilis, Naea, Pisonia, Torrubia*
Ochnaceae	*Sauvagesia*
Oenotheraceae	*Jussiaea*
Olacaceae	*Schoepfia, Ximenia*
Oleaceae	*Forestiera, Linociera*
Opiliaceae	*Agonandra*
Orchidaceae	*Bifrenaria, Bletia, Brassavola, Brassia, Catasetum, Dichaea, Elleanthus, Epidendrum, Gongora, Habenaria, Ionopsis, Maxillaria, Oncidium, Pleurothallis, Polystachya, Schombergkia, Spiranthes, Vanilla*
Oxalidaceae	*Oxalis*
Palmae	*Bactris, Copernicia*
Papaveraceae	*Argemone*
Passifloraceae	*Passiflora*
Phytolaccaceae	*Petiveria, Rivinia, Seguieria*
Piperaceae	*Peperomia, Piper*
Plumbaginaceae	*Plumbago*
Polygalaceae	*Bredemeyera, Monnina, Polygala, Securidaca*
Polygonaceae	*Coccoloba, Ruprechtia, Triplaris*

Portulacaceae	*Portulaca, Talinum*
Ranunculaceae	*Clematis*
Rhamnaceae	*Colubrina, Condalia, Gouania, Krugioden-dron, Zizyphus*
Rubiaceae	*Antirrhoea, Borreria, Cephalis, Chiococca, Coutarea, Diodia, Erithalis, Ernodea, Guet-tarda, Hamelia, Machaonia, Mitracarpus, Morinda, Psychotria, Randia, Rondeletia, Sickingia, Spermacoce, Strumpfia*
Rutaceae	*Amyris, Cusparia, Esenbeckia, Fagara, Helietta, Pilocarpus*
Sapindaceae	*Allophylus, Cardiospermum, Dodonaea, Paullinia, Serjania, Talisia, Thinouia, Urvillea*
Sapotaceae	*Bumelia, Dipholis*
Scrophulariaceae	*Capraria, Ilysanthes, Scoparia, Stemodia*
Simarubaceae	*Castela, Suriana*
Solanaceae	*Bassovia, Brachistus, Capsicum, Cestrum, Datura, Lycium, Nicotiana, Physalis, Solanum*
Sterculiaceae	*Ayenia, Buettneria, Guazuma, Helicteres, Melochia, Waltheria*
Theophrastaceae	*Jacquinia*
Tiliaceae	*Corchorus, Triumfetta*
Turneraceae	*Piriqueta, Turnera*
Ulmaceae	*Celtis, Phyllostylon*
Urticaceae	*Fleurya*
Verbenaceae	*Aegiphila, Bouchea, Citharexylon, Cleroden-drum, Lantana, Lippia, Phyla, Priva, Stachy-tarpheta, Vitex*
Violaceae	*Rinorea*
Vitaceae	*Cissus*

Zingiberaceae	*Costus*
Zygophyllaceae	*Bulnesia*, *Guaiacum*, *Kallstroemia*, *Tribulus*

---

floristic survey of a dry forest community in the lower Magdalena valley in Colombia, with an annual rainfall of 720 mm (Dugand, 1970), gives a total of 55 families, 154 genera, and 187 species of flowering plants in a stand of less than 300 ha. For the three small islands of Curaçao, Aruba, and Bonaire, with a total area of 860 km², Boldingh (1914) gives a list of 79 families, 239 genera, and 391 species of flowering plants, excluding the mangroves as the only local formation not belonging to the dry types.

As we can see in table 4-1, the best-represented families in total number of genera are the Leguminosae (50), Compositae (33), Euphorbiaceae (20), and Rubiaceae (19). Other well-represented families are the Amaranthaceae, Malvaceae, Malpighiaceae, Cactaceae, Verbenaceae, Orchidaceae, and Asclepiadaceae; almost all of them are typical of warm, arid floras everywhere.

If we compare now the floristic richness of this Caribbean dry region to the flora of North and South American middle-latitude deserts, we obtain roughly equivalent figures. In fact, Johnson (1968) gives a total of 278 genera and 1,084 species for the Mojave and Colorado deserts of California; Shreve (1951) reports 416 genera for the whole Sonoran desert, while Morello (1958) gives a list of 160 genera from the floristically less known Monte desert in Argentina. We can see then, that, in spite of a smaller total area, the Caribbean dry flora is as rich as other American desert floras.

Johnson (1968) gives a list of monotypic or ditypic genera of the Mojave-Colorado deserts, considered according to the ideas of Stebbins and Major (1965) to be old relict taxa. That list includes 60 species belonging to 56 genera. Applying this same criterion to the flora of the Caribbean desert I have recognized only 14 relict endemic species— a number that, even if it represents a gross underestimate, is significantly smaller than the preceding one (see table 4-2).

Concerning the geographic distribution and centers of diversification of the Caribbean arid taxa, it is not possible to proceed here to a detailed analysis because of the fragmentary knowledge of plant dis-

tribution in tropical America. However, I have tried to give a preliminary analysis based on only a few best-known families.

Taking the Compositae for instance, one of the most diversified
families within this vegetation, I took the data on its distribution from
Aristeguieta (1964) and Willis (1966). The species of 19 genera occurring in the Caribbean dry lands could be considered as widely distributed weeds, whose areas also extend to arid climates. Six genera
are very rich genera with a few species also occurring in arid vegetation: *Eupatorium*, *Vernonia*, *Mikania*, *Aster*, *Verbesina*, and *Simsia*;
2 genera (*Lepidesmia* and *Oxycarpha*) are monotypic taxa endemic
to the Caribbean coasts; while the remaining 6 genera (*Pollalesta*,
Egletes, *Baltimora*, *Gundlachia*, *Lycoseris*, and *Sclerocarpus*) are
small-to-medium-sized taxa restricted to tropical America, with some
species characteristic of arid plant formations. We see, then, that in
this family an important proportion of the species that occur in the
arid vegetation may be considered as weeds (19 out of 33 genera);
one part (6/33) has originated from widely distributed and very rich
genera, some of whose species have succeeded in colonizing arid
habitats also; while the remaining part, about a quarter of the genera
of Compositae occurring in arid vegetation, is formed of species belonging to genera of more restricted distribution and lesser adaptive
radiation, whose presence in this arid flora may be indicative of the
adaptation to arid conditions of an ancient Neotropical floristic stock—
in some cases, as in the two monotypic endemics, probably through
a rather long evolution in contact with similar environmental stress.

In all events, the Compositae, a very important family in the temperate and cold American deserts, neither shows a similar degree of
differentiation in the arid Caribbean flora nor occupies a prominent
role in these tropical plant communities.

Another family whose taxonomy and geographical distribution is
rather well known, the Bromeliaceae (Smith, 1971), has five genera
inhabiting the Caribbean arid lands; three of them, *Pitcairnia*, *Vriesia*,
and *Aechmea*, are very rich genera (150 to 240 species) mainly growing in humid vegetation types but with a few species also entering dry
plant formations. None of them is exclusive to the arid types. *Tillandsia*, a great and polymorphous genus of more than 350 species
adapted to nearly all habitat types from the epiphytic types in the rain
forests to the xeric terrestrial plants of extreme deserts, has 15

Table 4-2. *Relictual Endemic Species Occurring in the Caribbean Dry Region*

Family	Species
Asclepiadaceae	*Omphalophthalmum ruber* Karst.
Capparidaceae	*Belencita hagenii* Karst.
Capparidaceae	*Stuebelia nemorosa* (Jacq.) Dugand
Compositae	*Lepidesmia squarrosa* Klatt
Compositae	*Oxycarpha suaedaefolia* Blake
Cucurbitaceae	*Anguriopsis (Doyerea) margaritensis* Johnson
Leguminosae	*Callistylon arboreum* (Griseb.) Pittier
Leguminosae	*Humboldtiella arborea* (Griseb.) Hermann
Leguminosae	*Humboldtiella ferruginea* (H.B.K.) Harms.
Leguminosae	*Margaritolobium luteum* (Johnson) Harms.
Leguminosae	*Myrospermum frutescens* Jacq.
Lennoaceae	*Lennoa caerulea* (H.B.K.) Fourn.
Rhamnaceae	*Krugiodendron ferreum* (Vahl.) Urb.
Rubiaceae	*Strumpfia maritima* Jacq.

species recorded in the Caribbean arid lands; 14 of them are widely distributed species also occurring in dry formations. Only 1 species, *T. andreana*, growing on bare rock, seems strictly confined to dry plant formations. The fifth genus, *Bromelia*, a medium-sized genus of about 40 species, has 4 species growing in deciduous forests in the Caribbean that also extend their areas to the drier plant formations. As we can see by the distribution patterns of this old Neotropical family, the degree of speciation that has occurred in response to aridity in the Caribbean region seems to be minimal. This fact is in sharp contrast with the behavior of this family in other South American deserts, such as the Monte and the Chilean-Peruvian coastal deserts, where it has reached a good degree of diversification.

Let us take as a last example a typical family of arid lands, the Zygophyllaceae, recently studied by Lasser (1971) in Venezuela; it has four genera growing in the Caribbean arid region of which two,

Kallstroemia and *Tribulus*, are weedy genera of widely distributed species on bare soils and in dry habitats. The other two genera, *Bulnesia* and *Guaiacum*, are typical elements of arid and semiarid Neotropical plant formations. *Bulnesia* has its maximal diversification in semiarid and arid zones of temperate South America; while only one species, *B. arborea*, has reached the deciduous forests and thorn woodlands of northern South America, but without extending even to the nearby islands. But it is a dominant tree in many thorn woodland communities of northern South America. *Guaiacum* is a peri-Caribbean genus with several species from Florida to Venezuela, some of them exclusively restricted to arid coastal vegetation. In summary, this small family, whose species are frequently restricted to dry regions, does not show in the Caribbean arid flora the same degree of differentiation it has attained in southern South America, but it has nevertheless distinctly arid species, some originating from the south, such as *Bulnesia*, others from the north, such as *Guaiacum*.

Conclusions

As a conclusion, I wish to point out the most significant facts that follow from the preceding data. We have seen that in northern South America and in the nearby Caribbean islands a region of dry climates exists, which includes semiarid and arid climatic types, wherein five different plant formations occur. Considering the major environmental feature acting upon plant and animal life in this area, that is, the strong annual water deficit, these ecosystems seem subjected to water stress of comparable intensity and extension to that influencing living organisms in the extratropical South and North American deserts. If this water stress constitutes the directing selective force in the evolution of plant species and vegetation forms, the evolutionary framework would be comparable in tropical and extratropical American deserts. If, therefore, significant differences in speciation and vegetation features between these ecosystems could be detected, either they ought to be attributed to a different period of evolution under similar selective pressures, in which case the tropical and temperate American deserts would be of noncomparable geological age, or they could be attributed to the action on the evolution of these species of other environmental factors linked to the latitudinal difference between these deserts.

As many floristic and ecological features of these two types of ecosystems do not seem to be quite similar, even at a preliminary qualitative level of comparison, both previous hypotheses, that of differential age and that of divergent environmental selection, could probably be true. This supposes that the ancestral floristic stock feeding all dry American warm ecosystems was not so different as to explain the actual divergences on the basis of this sole historical factor.

The structure and physiognomy of plant formations occurring in the Caribbean area under a severe arid climate do not seem to correspond strictly to most semidesertic or desertic physiognomies of temperate North and South America. Several plant associations show undoubtedly a high degree of physiognomic convergence, also emphasized by a close floristic affinity, as is the case of the thorn scrub communities dominated in all these regions by species of *Prosopis*, *Cercidium*, *Cereus*, and *Opuntia*. But the most widespread plant associations in temperate American deserts, which are the scrub communities where a mixture of evergreen and deciduous shrubs prevail, like the *Larrea divaricata–Franseria dumosa* association of the Sonoran desert or the *Larrea cuneifolia* communities of the Monte desert; or the communities characterized by aphyllous or subaphyllous shrubs or low trees, such as the *Bulnesia retama–Cassia aphylla* communities of South America or the various *Fouquieria* associations in North America, do not have a similar physiognomic counterpart in tropical America.

As I have already noted in a previous paper (1972), even the degree of morphoecological adaptation in tropical American arid species is significantly smaller than that exhibited by the temperate American desert flora. Such plant features as succulence, spines, or aphylly are widely represented in the desert floras of North and South America, but they appear much more restricted quantitatively in the tropical American arid flora where, for instance, only one family of aphyllous plants occurs, the Cactaceae, in contrast to eleven families in the Monte region of Argentina.

Concerning floristic diversity, the dry Caribbean vegetation has a richness comparable to North American warm-desert floras and perhaps a richer flora than the warm deserts of temperate South America. The tropical arid flora is highly heterogeneous in origin and affinities, with the most significant contribution coming from neighboring less-dry formations, particularly the Tropical Deciduous Forest and the Dry Evergreen Woodland, with an important contribution from cosmopolitan or subcosmopolitan weeds, and a variety of floristic elements whose area of greater diversification occurs in northern or southern latitudes.

Among the elements of direct tropical descent reaching the dry formations from the contiguous less-arid types, the species of wide ecological spectrum predominate, whose ecological amplitude extends from subhumid or seasonally wet climates to semiarid and arid plant formations. On the other hand, few of them show a narrow ecological amplitude, appearing thus restricted only to arid plant communities; and in the majority of these cases the species thus restricted occur in particular types of habitats, like sand beaches, dunes, coral reefs, saline soils, and rock outcrops.

There exist in the Caribbean dry flora some species which are old relictual endemic taxa, in the sense considered by Stebbins and Major (1965), but they are neither as numerous as in North American deserts nor characteristic of "normal" habitats or typical communities; they are, rather, typical species of particular edaphic conditions or characteristics of the less-extreme types, such as the deciduous forests and dry evergreen woodlands.

In summary, then, the speciation of the autochthonous tropical taxa has been important in subhumid or semiarid plant formations as well as in restricted dry habitats, but the arid flora has received only a minor contribution from this source.

In spite of the actual occurrence of a chain of arid islands along the Andes connecting the dry areas of Venezuela and Peru, where neighboring patches occur no more than 200 to 300 km apart, southern floristic affinities are not conspicuous among the families analyzed. Further arguments are available to support this lack of connection between Caribbean and southern South American deserts on the basis of the distributions of all genera of Cactaceae (Sarmiento, 1973, unpublished). The representatives of this typical family that live in the Caribbean region show a closer phylogenetic affinity with the Mexican and West Indian cactus flora, a looser relationship with the Brazilian cactus flora, and a much more restricted affinity with the Peruvian and Argentinean cactus flora.

This slight affinity between tropical American and southern South American dry floras, in spite of more direct biogeographical and paleogeographical connections, is a rather difficult fact to explain, particularly if we consider that some species of disjunct area between North and South America, *Larrea divaricata*, for example, originated in South America and later expanded northward (Hunziker et al., 1972). These species have therefore crossed tropical America, but have not remained there.

In contrast to the loose affinity with southern South America, a stronger relationship with the North American arid flora is easily discernible. The most noteworthy cases are those of the genera *Agave*, *Fourcroya*, and *Yucca*, richly diversified in Mexico and southwestern United States, that reach their southern limits in the dry regions of northern South America. There are many other cases of North American genera, characteristic of dry regions, extending southward to Venezuela, Colombia, or less commonly to Ecuador and northern Brazil.

We can thus infer from the above information that the origin and age of the Caribbean arid vegetation certainly seems heterogeneous. Some elements evolved in tropical dry environments; many are almost cosmopolitan; others came from the north; and a few also came from the south. Several migratory waves along different routes probably occurred during a rather long evolutionary history under similar environmental conditions. Though the Central American connection does not actually offer a natural bridge for arid-adapted species, and there is no evidence of the former existence of this type of biogeographical bridge, the northern affinity of many Caribbean desert elements may be more easily explained by resorting to a dry bridge across the Caribbean islands, from Cuba and Hispaniola through the Lesser Antilles to Venezuela, instead of a more hypothetical Central American pass.

Axelrod's model (1950) of gradual evolution of the arid flora and vegetation in southwestern North America from a Madro-Tertiary geoflora, with the most arid forms and the maximal widespread of arid plant formations occurring only during the Quaternary, does not seem to fit well with the evidence provided by the analysis of the arid Caribbean flora and vegetation. On the contrary, the ideas of Stebbins and Major (1965) about the existence of small arid pockets along the western mountains from the late Mesozoic upward, together with a much more agitated evolutionary history from that time on to the Quaternary, are probably in better agreement with these data, which account for a heterogeneous and polychronic origin of these elements.

Acknowledgments

It is a great pleasure for me to acknowledge all the intellectual stimulus, material help, arduous criticism, and audacious ideas received through frequent and passionate discussions of these topics with my colleague, Maximina Monasterio.

Summary

Tropical American arid vegetation, particularly the formations occurring along the Caribbean coast of northern South America and the small nearby islands, is still not well known. However, within the framework of a comparative analysis of all American dry areas, this region provides not only the interest of knowing the features of plant cover in the driest region of tropical America, but also the knowledge that this possibly may clarify many obscure points of Neotropical biogeography, such as the evolutionary history of arid plant formations and the origin of their flora.

The major points of Caribbean dry ecosystems dealt with in this paper are (a) geographical distribution and climatic conditions, mainly the annual water deficiency and some differential features between low- and middle-latitude climates; (b) physiognomy, structure, and morphoecological traits of each of the five plant formations occurring in that area; and (c) floristic richness, origin, and affinities of floristic elements.

On this basis some relevant facts are discussed, such as the lack of correspondence between arid vegetation physiognomies in tropical

and temperate American dry regions; the comparable floristic diversification; and the varied origin of its taxa, where most elements evolved
on the spot from a tropical drought-adapted stock. Some others are
cosmopolitan taxa; many came from North America; and a few came
from the south. This brief analysis leads to the hypothesis that tropical
American desert flora is, at least in part, of considerable age and
shows a heterogeneous origin, probably brought about by several
migratory events. All these facts seem to support Stebbins and
Major's ideas about the complex evolution of American dry flora and
vegetation.

References

Aristeguieta, L. 1964. Compositae. In *Flora de Venezuela*, X, ed. T.
Lasser. Caracas: Instituto Botánico.
Axelrod, D. I. 1950. Evolution of desert vegetation in western North
America. *Publs Carnegie Instn* 590:1–323.
Beard, J. S. 1944. Climax vegetation in tropical America. *Ecology* 25:
127–158.
———. 1955. The classification of tropical American vegetation-
types. *Ecology* 36:89–100.
Boldingh, I. 1914. *The flora of Curaçao, Aruba and Bonaire*, vol. 2.
Leiden: E. J. Brill.
Croizat, L. 1954. La faja xerófila del Estado Mérida. *Universitas Emeri-
tensis* 1:100–106.
Cuatrecasas, J. 1958. Aspectos de la vegetación natural de Colombia. *Revta Acad. colomb. Cienc. exact. fís. nat.* 10:221–268.
Dugand, A. 1941. Estudios geobotánicos colombianos. *Revta Acad.
colomb. Cienc. exact. fís. nat.* 4:135–141.
———. 1970. Observaciones botánicas y geobotánicas en la costa
del Caribe. *Revta Acad. colomb. Cienc. exact. fís. nat.* 13:415–
465.
Hunziker, J. H.; Palacios, R. A.; de Valesi, A. G.; and Poggio, L. 1973.
Species disjunctions in *Larrea*: Evidence from morphology, cytogenetics, phenolic compounds and seed albumins. *Ann. Mo. bot.
Gdn.* 59:224–233.

Johnson, A. W. 1968. The evolution of desert vegetation in western North America. In *Desert Biology*, ed. G. W. Brown, vol.1, pp. 101–140. New York: Academic Press.

Koeppen, W. 1923. *Grundriss der Klimakunde*. Berlin and Leipzig: Walter de Gruyter & Co.

Lahey, J. F. 1958. *On the origin of the dry climate in northern South America and the southern Caribbean*. Ph.D. dissertation, University of Wisconsin.

Lasser, T. 1971. Zygophyllaceae. In *Flora de Venezuela, III*, ed. T. Lasser. Caracas: Instituto Botánico.

Loveless, A. R., and Asprey, C. F. 1957. The dry evergreen formations of Jamaica I. The limestone hills of the south coast. *J. Ecol.* 45:799–822.

Lowe, C.; Morello, J.; Goldstein, G.; Cross, J.; and Neuman, R. 1973. Análisis comparativo de la vegetación de los desiertos subtropicales de Norte y Sud América (Monte-Sonora). *Ecologia* 1:35–43.

McGinnies, W. G.; Goldman, B. J.; and Paylore, P. 1968. *Deserts of the world, an appraisal of research into their physical and biological environments*. Tucson: Univ. of Ariz. Press.

Marcuzzi, G. 1956. Contribución al estudio de la ecologia del medio xerófilo Venezolano. *Boln Fac. Cienc. for.* 3:8–42.

Meigs, P. 1953. World distribution of arid and semiarid homoclimates. In *Reviews of research on arid zone hydrology*, 1:203–209. Paris: Arid Zone Programme, Unesco.

Morello, J. 1958. La provincia fitogeográfica del Monte. *Op. lilloana* 2:1–155.

Pittier, H. 1926. *Manual de las plantas usuales de Venezuela*. 2d ed. Caracas: Fundación Eugenio Mendoza.

Pittier, H.; Lasser, T.; Schnee, L.; Luces de Febres, Z.; and Badillo, V. 1947. *Catálogo de la flora Venezolana*. Caracas: Litografía Vargas.

Porras, O.; Andressen, R.; and Pérez, L. E. 1966. *Estudio climatológico de las Islas de Margarita, Coche y Cubagua, Edo. Nueva Esparta*. Caracas: Ministerio de Agricultura y Cria.

Sarmiento, G. 1972. Ecological and floristic convergences between seasonal plant formations of tropical and subtropical South America. *J. Ecol.* 60:367–410.

————. 1973. The historical plant geography of South American dry vegetation. I. The distribution of the Cactaceae. Unpublished.

Shreve, F. 1951. Vegetation of the Sonoran desert. *Publs Carnegie Instn* 591:1–178.

Smith, L. B. 1971. Bromeliaceae. In *Flora de Venezuela, XII*, ed. T. Lasser. Caracas: Instituto Botánico.

Smith, R. F. 1972. La vegetación actual de la región Centro Occidental: Falcón, Lara, Portuguesa y Yaracuy de Venezuela. *Boln Inst. for lat.-am. Invest. Capacit*. 39–40:3–44.

Stebbins, G. L., and Major, J. 1965. Endemism and speciation in the California flora. *Ecol. Monogr.* 35:1–35.

Stehlé, H. 1945. Los tipos forestales de las islas del Caribe. *Caribb. Forester* 6:273–416.

Stoffers, A. L. 1956. The vegetation of the Netherlands Antilles. *Uitg. natuurw. Stud-Kring Suriname* 15:1–142.

Tamayo, F. 1941. Exploraciones botánicas en la Peninsula de Paraguaná, Estado Falcón. *Boln Soc. venez. Cienc. nat.* 47:1–90.

————. 1967. El espinar costanero. *Boln Soc. venez. Cienc. nat.* 111: 163–168.

Thornthwaite, C. W. 1948. An approach toward a rational classification of climate. *Geogr. Rev.* 38:155–194.

Trujillo, B. 1966. *Estudios botánicos en la región semiárida de la Cuenca del Turbio, Cejedes Superior*. Mimeographed.

Willis, J. C. 1966. *A dictionary of the flowering plants and ferns*. Cambridge: At the Univ. Press.

5. Adaptation of Australian Vertebrates to Desert Conditions   A. R. Main

Introduction

It is an axiom of modern biology that organisms survive in the places where they are found because they are adapted to the environmental conditions there. Current thinking has often associated the more subtle adaptations with physiological attributes, and the analysis of physiology has been widely applied to desert-dwelling animals in order to better understand their adaptation. Results of these inquiries frequently do not produce complete or satisfying explanations of why or how organisms survive where they do, and it is possible that explanations couched in terms of physiology alone are too simplistic. Clearly, while physiology cannot be ignored, other factors, including behavioral traits, need to be taken into account.

Accordingly, this paper sets out to interpret the adaptations of Australian vertebrates to desert conditions in the light of the physiological traits, the species ecology, and the geological and evolutionary history of the biota. To the extent that the components of the biota are integrated, its evolution can be conceived of as analogous to the evolution of a population; thus migrations and extinctions are analogous to genetic additions and deletions; and change in the ecological role of a component of the biota, the analogue of mutation.

The biota has changed and evolved mainly as a result of (a) changes in location and disposition of the land mass, (b) changes in the environment consequent on (a) above, and (c) extinctions and accessions. In the course of these changes strategies for survival will also change and evolve. It is the totality of these strategies which constitutes the adaptations of the biota.

Change in Location of Australia

The present continent of Australia appears to have broken away from East Antarctica in late Mesozoic times and to have moved to its present position adjacent to Asia in middle Tertiary (Miocene) times. In the course of these movements southern Australia changed its latitude from about 70°S in the Cretaceous to about 30°S at present (Brown, Campbell, and Crook, 1968; Heirtzler et al., 1968; Le Pichon, 1968; Vine, 1966, 1970; Veevers, 1967, 1971).

Changes in Environment

Prior to the fragmentation discussed above, the tectonic plate that is now Australia probably had a continental-type climate except when influenced by maritime air. As movement to the north proceeded, extensive areas were covered by epicontinental seas, and, later, extensive fresh-water lake systems developed in the central parts of the present continent. As Australia changed its latitude, the continental climate was influenced by the temperature of the surrounding oceans and particularly the temperature, strength, and origin of the ocean currents which bathed the shores. The ocean currents would in turn be driven by the global circulation, and the variations in the strength of the circulation and its cellular structure have affected not only the strength of the currents but also the climate of the continent. Frakes and Kemp (1972) suggested that for these reasons the Oligocene was colder and drier than the Eocene.

The present location of Australia across the global high-pressure belt, coupled with the fact that ocean currents driven by the west-wind storm systems pass south of the continent, has meant the inevitable drying of the central lake systems and the onset of desert conditions in the interior of the continent. In the absence of marine fossils or volcanicity the precise timing of the stages in the drying of the continent is not possible. Stirton, Tedford, and Miller (1961, p. 23; and see also Stirton, Woodburne, and Plane, 1967) used the morphological evolution shown by marsupial fossils to infer possible age in terms of Lyellian epochs of the sedimentary beds in which marsupial fossils have been found. Ride (1971) tabulated the fossil evidence as it relates to macropods.

Two other events associated with the changed position of the continent have occurred concurrently: (a) the development of weathering profiles, especially duricrust formation, on the land surface; and (b) changes in the composition of the flora.

Weathering profiles capped with duricrust are widespread throughout Australia, and Woolnough (1928) believed this duricrust to be synchronously developed over an enormous area. Since the Upper Cretaceous Winton formation was capped by duricrust, Woolnough believed the episode to be of Miocene age. The climate at this time of peneplanation and duricrust formation was thought to be marked by well-defined wet and dry seasons, so that the more soluble material was leached away in the wet season, and less soluble and particularly colloidal fraction of the weathering products was carried to the surface and precipitated during the dry season. Recent work in Queensland where basalts overlie deep weathering profiles indicates that deep weathering took place earlier than early Miocene (Exon, Langford-Smith, and McDougall, 1970). Other workers suggested that, as the climate becomes progressively drier, weathering processes follow a sequence from laterite formation through silcrete formation to aeolian processes and dune formation (Watkins, 1967).

Biologically the significant aspect of duricrust formation is, however, the removal from, or binding within, the weathering profile of soluble plant nutrients. Beadle (1962a, 1966) showed experimentally that the woodiness which is so characteristic of Australian plants is to some extent related to the low phosphorus status of the soil. Australian soils are well known for their low phosphorus status (Charley and Cowling, 1968; Wild, 1958). It has been argued that the low phosphorus is due to the low status of the parent rocks (Beadle, 1962b) or to the leaching which occurred during the process of laterization (Wild, 1958).

Changes in the floral composition are indicated by the fossil and pollen record. Early in the Tertiary, pollen of southern beech (Notho-fagus), in common with other pollen present in these deposits, suggests that a vegetation with a floral composition similar to that of present-day western Tasmania was widespread in southern Australia (see fig. 5-1), for example, at Kojonup (McWhae et al., 1958); Cool-gardie (Balme and Churchill, 1959); Nornalup, Denmark, Pidinga, and Cootabarlow, east of Lake Eyre (Cookson, 1953; Cookson and Pike, 1953, 1954); and near Griffith in New South Wales (Packham, 1969, p. 504).

Later the Lake Eyre deposits show a change, and the pollen record is dominated by myrtaceous and grass pollen (Balme, 1962). By Pliocene times the fossil record is restricted to eastern Australia and suggests a cool rain forest with *Dacrydium, Araucaria, Nothofagus*, and *Podocarpus*, which was later replaced by wet sclerophyll forest with *Eucalyptus resinifera* (Packham, 1969, p. 547). This record is consistent with a drying of the climate; however, in Tasmania comparable changes in the floral composition—that is, from *Nothofagus* forest to myrtaceous shrub or *Eucalyptus* woodland with a grass understory—result from fire (Gilbert, 1958; Jackson, 1965, 1968*a*, 1968*b*), and it seems highly likely that associated with the undeniable deterioration of the climate there occurred an increased incidence of fire.

Many authors have recognized that the present Australian flora not only is adapted to periodic fires but also includes many species which are dependent on fire for their persistence (Gardner, 1944, 1957; Mount, 1964, 1969; Cochrane, 1968). At present many wild or bush fires are intentionally lit or are the result of man's carelessness, but every year there are many fires which are caused by lightning strike (Wallace, 1966).

Fires are important in the Australian arid, semiarid, and seasonally arid environments because it is principally from the ash beds resulting from intense fires, and not from the slow decay of plant material, that nutrients are returned to the soil. It is thought that the oily nature of the common Australian shrubs and trees and their fire dependence reflect an evolutionary adaptation to fire. There is no doubt that in the past, in the absence of man, many intense fires were lit when lightning strike ignited ample and highly inflammable fuel.

Apart from returning nutrients to the soil, fire appears to be an important ecological factor in habitats ranging from the well-watered coastal woodlands dominated by *Eucalyptus* forests to the hummock grassland (dominated by *Triodia*) of the arid interior (Burbidge, 1960; Winkworth, 1967). Numerous postfire successions occur depending on the season of the burn, the quantity of fuel, and the frequency of burning. As an ecological factor, in arid Australia fire is as ubiquitous as drought.

Not all the changes in the biota have been due to fire and the increasing aridity. Numerous elements of the flora must have invaded Australia and then colonized the arid sandy interior by way of littoral sand dunes (Gardner, 1944; Burbidge, 1960). This invasion of Australia could only have occurred after the collision of the Australian plate with Asia in Miocene times. Simultaneously these migrant plant species would have been accompanied by rodents and other vertebrates of Asian affinities which also invaded through similar channels (Simpson, 1961).

To summarize the foregoing, Australia arrived in its present position from much higher southern latitudes, and the change in latitude was associated with a change in climate which passed from being mild and uniform in early Tertiary through marked seasonality to severe and arid by the end of the Pleistocene. Associated with climatic change two things occurred: first, a removal of plant nutrients and the probable development of a "woody" flora, and, second, the concurrent appearance of fire as a significant ecological factor.

Extinctions and Accessions

The climatic changes led to numerous extinctions in the old vertebrate fauna, for example, the Diprotodontidae; to a marked development in macropod marsupials (Stirton, Tedford, and Woodburne, 1968; Woodburne, 1967), which are adapted to the low-nutrient-status fibrous plants; and to the radiation of those Asian invaders which could exploit the progressive development of an arid climate in central Australia. As a result of the events outlined above, the fauna of arid Australia consists of two elements:

1. An older one originating in a cool, high-latitude climate now adapted to or at least persisting under arid conditions. Ride (1964), in his review of fossil marsupials, placed *Wynyardia bassiana*, the oldest diprotodont marsupial known, as of Oligocene age. This was at a time when Australia still occupied a southern location far distant from Asia (Brown et al., 1968, p. 308), suggesting that marsupials are part of the old fauna not derived from Asia.

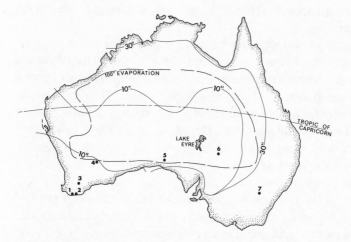

Fig. 5-1. Map of Australia showing approximate extent of arid zone as defined in Slatyer and Perry (1969). The northern boundary corresponds to the 30-inch (762 mm) isohyet, while the southern boundary corresponds with the 10-inch (242 mm) isohyet. The northern boundary of the 10-inch isohyet is also shown, as is the approximate boundary of the region experiencing evaporation of 100 inches (2,540 mm) or more per year. Localities from which fossil pollen recorded: 1. Nornalup; 2. Denmark; 3. Kojonup; 4. Coolgardie; 5. Pidinga; 6. Cootabarlow; 7. Lachlan River occurrence (near Griffith).

2. A younger element (not older than the time at which Australia collided with Asia) derived by evolution from migrants which established themselves on beaches. Rodents, agamid lizards, elapid snakes, and some bird groups fall in this category. These invasive episodes have continued up to the present.

The Australian Arid Environment

The extent of the Australian arid zone is shown in figure 5-1. This area is characterized by irregular rainfall, constant or seasonal shortage of water, high temperatures, and, for herbivores, recurrent seasonal inadequacy of diet. In many respects the arid area manifests in a more intense form the less intense seasonal or periodic droughts of the surrounding semiarid areas. Earlier it has been suggested that the present arid conditions are merely the terminal manifestation of climatic conditions which commenced to deteriorate in middle Tertiary times.

Desert Adaptation

A biota which has survived under increasingly arid conditions for such a long period of time might be expected to have evolved well-marked adaptations to high temperatures, shortage of water, and poor-quality diet. Yet one would not necessarily expect that all species would show similar or equal adaptive responses. The reason for this is that the incidence of intense drought is patchy; and, while some parts of arid Australia are always suffering drought, the areas suffering drought in the same season or between seasons are spatially discontinuous and often widely separated, so that it is conceivable that a mobile population could flee drought-stricken areas for more equable parts. Moreover, since fire as well as drought is ubiquitous in arid Australia, the benefits resulting from break of drought may be quite different according to whether an area has been recently burnt or not. Furthermore, these differences will be dissimilar depending on whether the biotic elements occur early or late in seral stages of the postfire succession. Indeed, many animal species which occupy the late seral or climax stages of the postfire succession could conceivably avoid much of the heat stress and resultant water shortage consequent upon evaporative cooling by behaviorally seeking out the cooler sites in the climax vegetation. All of the foregoing suggest that for the Australian arid biota we should not only look at the stressful factors (high temperature, water shortage, and quality of diet) but also determine the postfire seral stages occupied by the species.

Tolerance to the stressful factors is important because it affects the individual's ability to survive to reproduce. When individuals of a population reproduce successfully, the population persists. However, in its persistence a population will not maintain constant numerical abundance, because, for example, drought and fire will reduce numbers; and species favoring one seral stage of the postfire succession will, in any one locality, vary from being rare, then abundant, and finally again rare, as the preferred seral stage is passed through. For any species only detailed inquiry will show how population characteristics of reproductive capacity, age to maturity, longevity, and dispersal are related to the individual's ability to survive drought.

Individual Responses

Mammals

Among mammals, and marsupials especially, the macropods (kangaroos and wallabies) have received most study, but some work has been done on rodents, particularly *Notomys* (MacMillan and Lee, 1967, 1969, 1970). Both these groups are herbivores, and the macropods particularly show spectacular adaptations to the fibrous nature of their food plants and the attendant low-nutrition quality of the diet.

With the exception of the forest-dwelling *Hypsiprimnodon moschatus*, all kangaroos and wallabies so far investigated have a ruminantlike digestive system. This is an especially elaborated saccular development of the alimentary tract anterior to the true stomach. Within the sacculated "ruminal area" the fibrous ingesta are retained and fermented by a prolific bacterial flora and protozoan fauna. As a result of this activity, otherwise indigestible cellulose is broken down to material which can be metabolized by the kangaroo as an energy source (Moir, Somers, and Waring, 1956).

The bacteria of the gut need a nitrogen source in order to grow. In general, most natural diets in arid Australia are low in nitrogen; however, the bacteria of the kangaroo's gut are able to use urea as a nitrogen source (Brown, 1969) and so can supplement dietary nitrogen by recycling urea which would otherwise require water for its excretion in urine. The common arid-land kangaroo *Macropus robustus* (the euro) can remain in positive nitrogen balance on a diet which contains less than 1 g nitrogen per day for a euro of the average body weight of 12.8 kg (Brown and Main, 1967; Brown, 1968). However, it can supplement the dietary intake of nitrogen by recycling urea (Brown, 1969). In this connection the two common kangaroos (table 5-1) of arid Australia show contrasting solutions to the stresses imposed by heat, drought, and shortage of water, and in fact respond differently to seasonal stress (Main, 1971).

Body temperatures of marsupials are elevated at high ambient temperatures, but they are maintained below environmental tempera-

Table 5-1. *Comparison of Adaptations of Two Arid-Land Species of Kangaroos*

Characteristic	Red Kangaroo (*Megaleia rufa*)	Euro (*Macropus robustus*)
Coat	short, close; highly reflective	long, shaggy; not reflective
Preferred shelter	sparse shrubs	caves and rock piles
Diet	better fodder; higher nitrogen content	poorer fodder; lower nitrogen content
Temperature regulation	reflective coat; evaporative cooling	cool of caves or rock piles; evaporative cooling
Water shortage	acute shortage not demonstrated	unimportant except when shelter inadequate
Urea recycling	not pronounced; urea always high in urine	pronounced when fermentable energy as starch available, e.g., as seed heads of grasses
Electrolyte	higher in diet	lower in diet
Population characteristics	flock; locally nomadic	solitary; sedentary
Breeding	continuous	continuous

Source: Data from Dawson and Brown (1970), Storr (1968), and Main (1971).

tures. Temperature regulation under hot conditions appears to be costly in terms of water (Bartholomew, 1956; Dawson, Denny, and Hulbert, 1969; Dawson and Bennet, 1971).

MacMillan and Lee (1969) studied two species of desert-dwelling *Notomys* and interpreted their findings in terms of the contrasting habitats occupied, so that the salt-flat-dwelling *Notomys cervinus* has a kidney adapted to concentrating electrolytes while the sandhill-dwelling *N. alexis* was adapted to concentrating urea.

Birds

Because of their mobility, birds in an arid environment present a different set of problems to those presented by both mammals and lizards. They can and do fly long distances to watering places and, when water is available, may use it for evaporative cooling. Many of the adaptations are likely to be related to conservation of water so that the frequency of drinking is reduced. Fisher, Lindgren, and Dawson (1972) studied the drinking patterns of many species, including the zebra finch and budgerigar which have been shown under laboratory conditions to consume little or no water (Cade and Dybas, 1962; Cade, Tobin, and Gold, 1965; Calder, 1964; Greenwald, Stone, and Cade, 1967; Oksche et al., 1963).

It is likely that the ability of the budgerigar and the zebra finch to withstand water deprivation in the laboratory reflects their ability to survive in the field with minimum water intake and so exploit food resources which are distant from the available water. Johnson and Mugaas (1970) showed that both these species possess kidneys which are modified in a way which assists in water conservation.

Reptiles

Many responses of individuals to hot arid environments are readily measured: preferred temperatures, heat tolerance, rates of pulmonary and cutaneous water loss, and tolerance to dehydration.

Under field conditions nocturnal lizards, for example, geckos, have to tolerate the temperatures experienced in their daytime shelters. On the other hand, diurnal species, for example, agamids, such as *Amphibolurus*, have body temperatures higher than ambient temperatures during the cooler parts of the year and body temperatures cooler than ambient during the hotter season. The body temperatures recorded for field-caught animals indicate a specific constancy (Licht et al., 1966b) which comes about by a series of behavioral responses which range from body posture to avoidance reactions (Bradshaw and Main, 1968). A large series of data on body temperatures in the field has been presented by Pianka (1970, 1971a, and 1971b) and Pianka and Pianka (1970).

With the exception of *Diporophora bilineata*, with a mean body temperature in the field of 44.3°C (Bradshaw and Main, 1968), no species recorded a mean temperature above 39°C. However, arid-land species spend more of their time in avoidance reactions than do species from semiarid situations (Bradshaw and Main, 1968). Further information on preferred body temperature can be obtained by placing lizards in a temperature gradient and allowing them to choose a body temperature.

Data from neither the field-caught animals nor those selecting temperature in a gradient indicate any marked preference for exceptionally high temperatures on the part of most lizard species. However, in a situation where choice of temperature was not possible, it is conceivable that species from arid environments could tolerate higher temperatures for a longer period than species from less-arid situations. Bradshaw and Main (1968) compared *Amphibolurus ornatus*, a species from semiarid situations, with *A. inermis*, a species from arid areas, after acclimating them to 40°C and then exposing them to 46°C. Their mean survival times were 64 ± 5.6 and 62 ± 6.58 minutes respectively. There was no statistically significant difference in the survival time of each species. These results suggest that the major adaptation of *Amphibolurus* species to hot arid environments is likely to be the development of a pattern of behavioral avoidance of heat stress.

Not all lizards in hot arid situations show the pattern of *Amphibolurus* sp. and *Diporophora bilineata*, which when acclimated to 40°C can withstand an exposure of six hours to a body temperature of 46°C without apparent ill effect and survive for thirty minutes at 49°C (Bradshaw and Main, 1968). The nocturnal geckos, which must tolerate the temperature of their daytime refuge, show another pattern illustrated by *Heteronota binoei*, a species sheltering beneath litter; *Rhynchoedura ornata*, a species which frequently shelters in cracks and holes (deserted spider burrows) in bare open ground; and *Gehyra variegata*, a species sheltering beneath bark. Data for these three species are given in table 5-2. These data suggest that adaptation to high temperatures in Australian geckos is considerably modified by behavioral and habitat preferences and is not directly related to increased aridity in a geographical sense.

Bradshaw (1970) determined the respiratory and cutaneous components of the evaporative water loss in specimens of *Amphibolurus ornatus*, *A. inermis*, and *A. caudicinctus* (matched for body weight) held in the dry air at 35°C after being held under conditions which allowed them to attain their preferred body temperature by behavioral regulation. Bradshaw showed that evaporative water loss was greatest in *A. ornatus* and least in *A. inermis*. He also showed that losses by both pathways were reduced in the desert species. All differences were statistically significant. However, while the cutaneous component was greater than the respiratory in *A. ornatus*, it was less than the respiratory in *A. inermis*. Bradshaw also compared CO_2 production of uniformly acclimated *A. ornatus* with that of *A. inermis* and showed that CO_2 production and respiratory rate of *A. inermis* were significantly lower than *A. ornatus*. Bradshaw concluded that the greater water economy of desert-living *Amphibolurus* was achieved both by reduction in metabolic rate and change to a more impervious integument.

By means of a detailed field population study of *A. ornatus*, Bradshaw (1971) was able to show that individuals of the same cohort grew at different rates so that some animals matured in one, two, or three years. These have been referred to as fast- or slow-growing animals. Bradshaw, using marked animals of known growth history, showed that during summer drought there was a difference between fast- and slow-growing animals with respect to distribution of fluids and electrolytes. Slow-growing animals showed no difference when compared with fully hydrated animals except that electrolytes in plasma and skeletal muscle were elevated. Fast-growing animals, however, showed weight losses and changes in fluid volume. Weight losses greater than 20 percent of hydrated weight encroached upon the extracellular fluid volume; but the decrease in volume was restricted to the interstitial fluid, leaving the circulating fluid volume intact, that is, the blood volume and plasma volume remained constant. Earlier, Bradshaw and Shoemaker (1967) showed that the diet of *A. ornatus* consisted of sodium-rich ants and that during summer the lizards lacked sufficient water to excrete the electrolytes without using body water. Instead, the sodium ions were retained at an elevated level in the extracellular fluid which increased in volume by an isosmotic shift of fluid from the intracellular compartment. This sodium retention operates to protect fluid volumes when water is scarce and so enhances survival. Electrolytes were excreted following the occasional summer thunderstorm.

In his population study Bradshaw (1971) showed that only fast-growing animals died as a result of summer drought. Bradshaw (1970) extended his study of water and weight loss in field populations to other species including *A. inermis* and *A. caudicinctus*. As a result of this study he concluded that only males of *A. inermis* lost weight and that, in all species studied, fluid volumes were protected by the retention of sodium ions during periods when water was short. Bradshaw also showed that sodium retention occurred in *A. ornatus* in midsummer but only occurred in *A. inermis* and *A. caudicinctus* after long and intense drought.

Both *A. inermis* and *A. caudicinctus* complete their life cycle in a year (Bradshaw, 1973, personal communication; Storr, 1967). They are thus fast growing in the classification used to describe the life history of *A. ornatus*; but, either by a change in metabolic rate and integument or by some other means, they have avoided the deleterious effects associated with the rapid development of *A. ornatus*.

Population Response

The capacity and speed with which a species can occupy an empty but suitable habitat are related to its capacity to increase. Cole (1954) pointed out that time to maturity, litter size, and whether reproduction is a single episode or repeated throughout the female's life bear on the rate at which a population can increase; but he believed that reproduction early in life was most important for the population. No systematic recording of life-history data appears to have been undertaken in Australia, but such information is critical for understanding how populations persist in fluctuating environments. Whether the fluctuations are due to recurrent drought or to fire or seral stages of postfire succession is not too important, because following any of these events a population nucleus will have the opportunity to expand quickly into an empty but suitable habitat. Moreover, its chances of persisting are enhanced if it can very quickly occupy all favorable habitats at the maximum density because the random spatial distribution of the next drought or fire sequence will determine the sites of the next *refugium*.

The foregoing would suggest that modification of the life history, particularly early maturity, might be as important as physiological adaptation under Australian arid conditions. However, young or small animals are at a disadvantage because of the effects of metabolic body size compared with larger, older mature animals, and hence there is an advantage in late maturity and greater longevity, so that the risks of death which are related to metabolic body size are spread more favorably than they would be in a species in which each generation lived for only one year. Undoubtedly, natural selection will have produced adaptations of life history so that the foregoing apparent conflicts are resolved.

Several workers—MacArthur and Wilson (1967) and Pianka (1970) —have considered the response of populations to selection in terms of whether high fecundity and rapid development or individual fitness and competitive superiority have been favored. These two types of selection were referred to by MacArthur and Wilson (1967) as r-selection and K-selection. Pianka (1970) has tabulated the correlates of each type of selection.

King and Anderson (1971) pointed out that, if a cyclically changing environment varies over few generations, r-type selection factors will be dominant; on the other hand, in a changing environment which has a period of fluctuation many generations long, K-type selection will be dominant. In this connection we might consider quick maturity and large clutch size as manifesting the response to r-type selection; and slower maturity, smaller clutch size, well-marked display, and other devices for marking territory as representing responses to K-type selection.

Mertz (1971) showed that the response of a population to selection will be different depending on whether the population is increasing or declining. In the latter case selection favors the long-lived individual which continues to breed and is thus able to exploit any environmental amelioration even if it occurs late in life. This type of selection tends to produce long-lived populations.

Earlier it was suggested that Australia has been subjected to a prolonged climate and fire-induced deterioration of the environment
which might be expected to produce a response akin to that envisaged
by Mertz (1971) and unlike the advantageous rapid development and
early reproduction mentioned by Cole (1954). Selection for longevity
is a special case in which competitive superiority is principally expressed in terms of a long reproductive life. Murphy (1968) showed
this was as a consequence of uncertainty in survival of the prerepro-
ductive stages.

With the foregoing outline, it is possible to consider the little information known about the life histories of species from arid Australia in
terms of whether they reflect selection during the past for capacity to
increase, competitive efficiency related to carrying capacity, or longevity.

Mammals

Macropods. The fossil record suggests that in both Tertiary and post-
Pleistocene times the macropods have increased their dominance of
the fauna despite the general deterioration of the climate (Stirton et al.,
1968). It has already been suggested that the ruminantlike digestion
preadapted these species to the desert conditions. The highly developed ruminantlike digestion of macropods can be viewed as a device for delaying the death from starvation caused by a nutritionally inadequate diet. It is thus a device for maximizing physiological
longevity once adulthood is achieved.

Among the marsupials there is considerable diversity in their life
histories, but there appear to be tendencies toward longevity with
respect to populations in arid situations as indicated in the two cases
below:

1. In the typical mainland swampy situations the quokka (*Setonix
brachyurus*) matures early and breeds continuously and is apparently not long lived. On the other hand, a population of this
species on the relatively arid Rottnest Island is older than the mainland form when it first breeds. Breeding is seasonal, and so Rottnest
animals tend to produce fewer offspring per unit time than the mainland form (Shield, 1965). Moreover, individuals from the island population tend to live seven to eight years, with a few females present and

still reproducing in their tenth year. The pollen record on Rottnest indicates that the environment has declined from a woodland to a coastal heath and scrubland over the past 7,000 years (Storr, Green, and Churchill, 1959). Despite the difference in detail, the modification in the life history of the quokka on the semiarid Rottnest Island achieves the same end as the red kangaroo and euro discussed below.

2. Typical arid-land species, such as the red kangaroo, *Megaleia rufa*, and the euro, *Macropus robustus*, have no defined season of breeding, and females are always carrying young except under very severe drought. Both these species tend to be long lived (Kirkpatrick, 1965), and females may still be able to bear young when approaching twenty years of age.

The breeding of the red kangaroo and euro suggests that adaptation of life history has centered around the metabolic advantages of large body size in a long-lived animal which is virtually capable of continuous production of offspring, some of which must by chance be weaned into a seasonal environment which permits growth to maturity.

Numerous workers have shown that macropod marsupials have lowered metabolic rates with which are associated reduced requirements for water, energy, and protein and a slower rate of growth. The first three of these are of advantage during times of drought; and, should the last contribute to longevity, it will also be advantageous, insofar as offspring have the potential to be distributed into favorable environments whenever they occur.

Rodents. The Australian rodents appear to have a typical rodent-type reproductive pattern with a high capacity to increase. They appear to be able to survive through drought because of their small size and capacity to persist as small populations in minor, favorable habitats.

Birds

Most bird species which have been studied physiologically belong to taxonomic groups which also occur outside Australia, for example, finches, pigeons, caprimulgids, and parrots. The information on which a comparative study of modifications of the life histories of the Australian forms with their old-world relatives could be based has not been assembled. However, several observations—for example, Cade et al. (1965), that Australian and African estrildine finches are markedly different in physiology, and Dawson and Fisher (1969), that the spotted nightjar (*Eurostopodus guttatus*), like all caprimulgids, has a depressed metabolism—are suggestive that the life histories of some Australian species (finches) might be highly modified, while others show only slight modification from their old-world relatives (caprimulgids); and these may, in a sense, be thought of as being preadapted to survival in arid Australia.

Keast (1959), in a review of the life-history adaptations of Australian birds to aridity, showed that the principal adaptation is opportunistic breeding after the break of drought when the environment can provide the necessities for successful rearing of young. Longevity of individuals in unknown, but the breeding pattern is consistent with selection which has favored longevity.

Fisher et al. (1972) observed that honeyeaters (Meliphagidae), which are widespread and common throughout arid Australia, are surprisingly dependent on water. The growth of these birds to maturity and their metabolism are not known, but these authors speculated that the dependence may be due in large part to the water loss attendant on the activity associated with the high degree of aggressive behavior exhibited by all species of honeyeaters.

The following speculation would be consistent with the observations of Fisher et al. (1972): Most honeyeaters frequent late and climax stages when the vegetation is at its maximum diversity with numerous sources of nectar and insects. Such a habitat preference would suggest that K-type selection would have operated in the past, and the advantages of obtaining and maintaining an adequate territory by aggressive display may outweigh any disadvantages of individual high water needs which were consequent upon the aggressive display.

Reptiles

Table 5-2 has been compiled from the information available on lizard physiology and biology. The information is not equally complete for all species tested; however, it does suggest that Australian desert

Table 5-2. *Physiological, Ecological, and Life-History Information for Selected Species of Australian Lizards*

Species	Mean Preferred Temperature	Mean Survival		Water Loss (mg/g/hr)	Seral Stage
		Minutes	Temperature (°C)		
Amphibolurus inermis	36.4[a]	102.8 2.0	46.0 48.0	1.05 at 35°C	burrows in early seres
A. caudicinctus	37.7[a]	92.8 45.0	46.0 47.0	1.80 at 35°C	rock piles in climax hummock grassland
A. scutulatus	38.2[a]	40.8 28.0	46.0 47.0	?	shady climax woodland
Diporophora bilineata	44.3[b]	360.0 29.5	46.0 49.0	?	fire disclimax
Moloch horridus	36.7[a]	?	?	?	late seres and climax

Age to Maturity (yrs)	Reproduction		Longevity (yrs)	Reference
	No. of Clutches per Year	Eggs per Clutch (means)		
0.75	possibly 2	3.43	1	Licht et al., 1966a, 1966b; Pianka, 1971a; Bradshaw, personal communication
0.75	possibly 2	?	1	Licht et al., 1966a, 1966b; Storr, 1967; Bradshaw, 1970
?	possibly only 1	6.5	?	Licht et al., 1966a, 1966b; Pianka, 1971c
?	?	?	?	Bradshaw and Main, 1968
3–4	usually 1	6–7	6–20	Sporn, 1955, 1958, 1965; Licht et al., 1966b; Pianka and Pianka, 1970

Species	Mean Preferred Temperature	Mean Survival		Water Loss (mg/g/hr)	Seral Stag
		Minutes	Temperature (°C)		
Gehyra variegata	35.3[a]	72.8 2.0	43.5 46.0	2.07 at 25°C 3.37 at 30°C 3.80 at 35°C	climax and postclima> woodland
Heteronota binoei	30.0[ac]	162.0 0.0	40.5 43.5	0.27 at 30°C	climax witl litter
Rhynchoe- dura ornata	34.0[a]	55.3	46.0	?	holes in b soil in clin woodland

[a]In gradient.   [b]In field.   [c]May be too high—see Licht et al., 1966b.

species exhibit a wide range of tolerances to elevated temperatures. It is surprising, for example, that *Heteronota binoei* survives at all in the desert. Geckos, depending on the species, may have clutches of a single egg, but no species have clutches larger than two eggs; however, they may have one or two clutches each breeding season. *Heteronota binoei* and *Gehyra variegata* have respectively one and two clutches. *Heteronota binoei*, with an apparent preference for low temperatures and an inability to tolerate high temperatures, has adapted to the desert by its extremely low rate of water loss, behavioral attachment to sheltered climate situations, and, relative to *G. variegata*, early maturity and large clutch size (two eggs vs. one).

On the other hand, *G. variegata* is better adapted to high temperatures and, even though it is relatively poor at conserving water, is able to survive in the deteriorating and more exposed situations of the late climax and postclimax. Moreover, these physiological adapta-

Age to Maturity (yrs)	Reproduction		Longevity (yrs)	Reference
	No. of Clutches per Year	Eggs per Clutch (means)		
2; breed in 3rd	2	1	mean 4.4	Bustard and Hughes, 1966; Licht et al., 1966*a*, 1966*b*; Bustard, 1968*a*, 1969; Bradshaw, personal communication
1.6 or 2.5	usually 1	2	mean 1.9	Bustard and Hughes, 1966; Licht et al., 1966*a*; Bustard, 1968*b*
?	?	?	?	Licht et al., 1966*a*, 1966*b*

tions are associated with a long adult life and thus enhance the possibility of favorable recruitment in any season where conditions are ameliorated so that eggs and young have an enhanced survival.

Among the agamids the information is not nearly as complete. *Amphibolurus inermis* and *A. caudicinctus* are early maturing, short-lived species relying on a high rate of reproduction to maintain the population and are thus the analogue of *H. binoei*. *Moloch horridus* and *A. scutulatus*, on the other hand, appear to be the analogue of *G. variegata*; and it is unfortunate that information on age to maturity and longevity of *A. scutulatus* is not available. One can only speculate on age to maturity and longevity of *Diporophora*, but it seems likely that recruitment would only be successful in years when summer cyclonic rain ameliorated environmental conditions; and one might guess that it is a long-lived animal.

It is interesting that the fast-maturing species either have a cool refuge in which the small young can establish themselves (*H. binoei* in climax) or a cool season in which they can grow to almost adult size (*A. inermis*, *A. caudicinctus*). In addition, these species have another adaptation in producing twin broods in each breeding season. Should there be a drought, the young from the first brood will almost certainly be lost. However, should the young be born into a season in which thunderstorms are common, they would be able to thrive under almost ideal conditions. Since the offspring from the second clutch are born late in the summer or early autumn, they are almost certain to survive regardless of the preceding summer conditions.

Discussion

The foregoing suggests that early in Tertiary times Australia underwent a change in position from higher (southern) to lower (tropical) latitudes. Stemming from this there has been a prolonged and disastrous change in climate toward increasing aridity. This has been accompanied by the increased incidence of fire as an ecological factor.

Much of the original biota has become extinct as the result of these changes, but there have been some additions from Asia. Both the old and new elements of the biota that have survived to the present have done so because they have been able to accommodate their individual physiology and population biology to the stresses imposed by climatic deterioration (drought) and fire.

The foregoing has been achieved by a series of complementary strategies as follows:

1. Physiological strategies
   - *a.* Behavioral avoidance of stressful environmental factors
   - *b.* Heat tolerance
   - *c.* Ability to conserve water, including ability to handle electrolytes
   - *d.* Ability to survive on diets of low nutritional value (herbivores)
2. Reproductive strategies
   - *a.* High reproductive capacity, so enabling a population nucleus surviving after drought to rapidly repopulate the former range and to occupy all areas which could possibly form *refugia* in future droughts

   b. Increased competitive advantage by means of small well-
   tended broods of young and well-developed displays for holding territories
   c. Increased longevity, so that adults gain advantage from metabolic body size while young are produced over a span of years,
   so ensuring that at least some are born into a seasonal environment in which they can survive and become recruits to
   the adult population

It is thus apparent that vertebrates inhabiting arid parts of Australia
display a diversity of individual adaptations to single components of
the arid environment, and it is difficult to interpret the significance of
experimental laboratory findings achieved as the result of simple
single-factor experiments. For example, under experimental conditions, kangaroos and wallabies, if exposed to high ambient temperatures, use quantities of water in evaporative cooling (Bartholomew,
1956; Dawson et al., 1969; Dawson and Bennet, 1971).

Yet these arid-land species are capable of surviving intense
drought conditions when the environment provides the appropriate
shelter conditions. These may be postfire seral stages as needed by
the hare wallaby, *Lagorchestes conspicillatus* (Burbidge and Main,
1971), or rock piles needed by the euro, *Macropus robustus*. Given
that the euro and hare wallaby have shelter of the appropriate quality,
both species are apparently well adapted to grow and reproduce on
the low-quality forage which is available where they live. Moreover,
both species are capable of reproduction at all seasons so that, while
their reproductive potential is limited—because of having only one
young at a time—they do maximize their reproductive potential by
continuous breeding and by distributing the freshly weaned young at
all seasons, which is particularly important in a seasonally unpredictable environment.

In general, while it is true that some species show a highly developed degree of adaptation to arid conditions, it is difficult to find a
case which is unrelated to seral successional stages. A pronounced
example of this is afforded by the lizard *Diporophora bilineata* and the
gecko *Diplodactylus michaelseni*, which can withstand higher field
body temperatures than any other Australian species but which appear to be abundant only in excessively exposed fire disclimax situations.

Most of the vertebrates which survive in the desert appear to do so not solely because of well-developed individual adaptation (tolerance) to the hot dry conditions of arid Australia, but because of habitat preference and population attributes which permit the species to cope, first, with the ecological consequences of fire and, second, with drought. In a sense, adaptation to fire has preadapted the vertebrates to drought.

Desert species have had to choose whether the ability of a population to grow is equivalent to ability to persist. Two circumstances can be envisaged in which ability to grow is equivalent to persistence: when rapid repopulation of an area after drought will ensure that all potential future *refugia* are occupied and when rapid population growth excludes other species from a resource.

In the desert where drought conditions are the norm, however, persistence is achieved by females replacing themselves with other females in their lifetime. This requires that juveniles must withstand or avoid desert conditions until they reach reproductive age. Seasonal amelioration of conditions in desert environments is notoriously unpredictable, and it seems that many Australian desert animals persist as populations because of long reproductive lives during which some young will be produced and grow to maturity.

In considering the individual and the population aspects of survival we should envisage the space occupied by an animal as providing scope for minimizing the environmental stresses of heat, water shortage, and poor-quality diet. An animal will choose to live in places where the stresses are least; when these are not available, it will select sites or opt for physiological responses which allow it to prolong the time to death. Urea recycling by macropods should be viewed in this light. When environmental amelioration occurs, it is taken as an opportunity to replenish the population by recruiting young.

Acknowledgments

Financial assistance is acknowledged from the University of Western Australia Research Grants Committee, the Australian Research Grants Committee, and Commonwealth Scientific and Industrial Research Organization. Professor H. Waring, Dr. S. D. Bradshaw, and Dr. J. C. Taylor kindly read and criticized the manuscript.

Summary

It is suggested that the Australian deserts developed as a consequence of the movement in Tertiary times of the continental plate from higher latitudes to its present position. An increasing incidence of wild fire is associated with the development of dry conditions. The vertebrate fauna has adapted to the development of deserts and incidence of fire at two levels: (a) the individual or physiological, emphasizing such strategies as behavioral avoidance of stressful conditions, conservation of water, tolerance of high temperatures, and, with macropod herbivores, ability to survive on low-quality forage and through the supplementation of nitrogen by the recycling of urea; and (b) the population, emphasizing reproductive strategies and longevity, so that young are produced over a long period of time thus enhancing the possibility of successful recruitment.

It is further suggested that survival of individuals and persistence of the population are only possible when the environment, especially the postfire plant succession, provides the space and scope for the implementation of the strategies which have evolved.

References

Balme, B. E. 1962. Palynological report no. 98: Lake Eyre no. 20 Bore, South Australia. In *Investigation of Lake Eyre*, ed. R. K. Johns and N. H. Ludbrook. *Rep. Invest. Dep. Mines S. Aust.* No. 24, pts. 1 and 2, pp. 89–102.

Balme, B. E., and Churchill, D. M. 1959. Tertiary sediments at Coolgardie, Western Australia. *J. Proc. R. Soc. West. Aust.* 42:37–43.

Bartholomew, G. A. 1956. Temperature regulation in the macropod marsupial *Setonix brachyurus*. *Physiol. Zoöl.* 29:26–40.

Beadle, N. C. W. 1962a. Soil phosphate and the delimitation of plant communities in Eastern Australia, II. *Ecology* 43:281–288.

———. 1962b. An alternative hypothesis to account for the generally low phosphate content of Australian soils. *Aust. J. agric. Res.* 13: 434–442.

———. 1966. Soil phosphate and its role in molding segments of the Australian flora and vegetation, with special reference to xeromorphy and sclerophylly. *Ecology* 47:992–1007.

Bradshaw, S. D. 1970. Seasonal changes in the water and electrolyte metabolism of *Amphibolurus* lizards in the field. *Comp. Biochem. Physiol.* 36:689–718.

———. 1971. Growth and mortality in a field population of *Amphibolurus* lizards exposed to seasonal cold and aridity. *J. Zool., Lond.* 165:1–25.

Bradshaw, S. D., and Main, A. R. 1968. Behavioral attitudes and regulation of temperature in *Amphibolurus* lizards. *J. Zool., Lond.* 154: 193–221.

Bradshaw, S. D., and Shoemaker, V. H. 1967. Aspects of water and electrolyte changes in a field population of *Amphibolurus* lizards. *Comp. Biochem. Physiol.* 20:855–865.

Brown, D. A.; Campbell, K. S. W.; and Crook, K. A. W. 1968. *The geological evolution of Australia and New Zealand*. Oxford: Pergamon Press.

Brown, G. D. 1968. The nitrogen and energy requirements of the euro (*Macropus robustus*) and other species of macropod marsupials. *Proc. ecol. Soc. Aust.* 3:106–112.

———. 1969. Studies on marsupial nutrition. VI. The utilization of dietary urea by the euro or hill kangaroo, *Macropus robustus* (Gould). *Aust. J. Zool.* 17:187–194.

Brown, G. D., and Main, A. R. 1967. Studies on marsupial nutrition. V. The nitrogen requirements of the euro, *Macropus robustus*. *Aust. J. Zool.* 15:7–27.

Burbidge, A. A., and Main, A. R. 1971. Report on a visit of inspection to Barrow Island, November, 1969. *Rep. Fish. Fauna West. Aust.* 8:1–26.

Burbidge, N. T. 1960. The phytogeography of the Australian region. *Aust. J. Bot.* 8:75–211.

Bustard, H. R. 1968a. The ecology of the Australian gecko *Gehyra variegata* in northern New South Wales. *J. Zool., Lond.* 154:113–138.

———. 1968b. The ecology of the Australian gecko *Heteronota binoei* in northern New South Wales. *J. Zool., Lond.* 156:483–497.

———. 1969. The population ecology of the gekkonid lizard *Gehyra variegata* (Dumeril and Bibron) in exploited forests in northern New South Wales. *J. Anim. Ecol.* 38:35–51.

Bustard, H. R., and Hughes, R. D. 1966. Gekkonid lizards: Average ages derived from tail-loss data. *Science, N.Y.* 153:1670–1671.

Cade, T. J., and Dybas, J. A. 1962. Water economy of the budgerygah. *Auk* 79:345–364.

Cade, T. J.; Tobin, C. A.; and Gold, A. 1965. Water economy and metabolism of two estrildine finches. *Physiol. Zoöl.* 38:9–33.

Calder, W. A. 1964. Gaseous metabolism and water relations of the zebra finch *Taenopygia castanotis*. *Physiol. Zoöl.* 37:400–413.

Charley, J. L., and Cowling, S. W. 1968. Changes in soil nutrient status resulting from overgrazing in plant communities in semi-arid areas. *Proc. ecol. Soc. Aust.* 3:28–38.

Cochrane, G. R. 1968. Fire ecology in southeastern Australian sclerophyll forests. *Proc. Ann. Tall Timbers Fire Ecol. Conf.* 8:15–40.

Cole, La M. C. 1954. Population consequences of life history phenomena. *Q. Rev. Biol.* 29:103–137.

Cookson, I. C. 1953. The identification of the sporomorph *Phyllocladites* with *Dacrydium* and its distribution in southern Tertiary deposits. *Aust. J. Bot.* 1:64–70.

Cookson, I. C., and Pike, K. M. 1953. The Tertiary occurrence and distribution of *Podocarpus* (section *Dacrycarpus*) in Australia and Tasmania. *Aust. J. Bot.* 1:71–82.

———. 1954. The fossil occurrence of *Phyllocladus* and two other podocarpaceous types in Australia. *Aust. J. Bot.* 2:60–68.

Dawson, T. J., and Brown, G. D. 1970. A comparison of the insulative and reflective properties of the fur of desert kangaroos. *Comp. Biochem. Physiol.* 37:23–38.

Dawson, T. J.; Denny, M. J. S.; and Hulbert, A. J. 1969. Thermal balance of the macropod marsupial *Macropus eugenii* Desmarest. *Comp. Biochem. Physiol.* 31:645–653.

Dawson, W. R., and Bennet, A. F. 1971. Thermoregulation in the marsupial *Lagorchestes conspicillatus*. *J. Physiol., Paris* 63:239–241.

Dawson, W. R., and Fisher, C. D. 1969. Responses to temperature by the spotted nightjar (*Eurostopodus guttatus*). *Condor* 71:49–53.

Exon, N. R.; Langford-Smith, T.; and McDougall, I. 1970. The age and geomorphic correlations of deep-weathering profiles, silcrete, and basalt in the Roma-Amby Region Queensland. *J. geol. Soc. Aust.* 17:21–31.

Fisher, C. D.; Lindgren, E.; and Dawson, W. R. 1972. Drinking patterns and behaviour of Australian desert birds in relation to their ecology and abundance. *Condor* 74:111–136.

Frakes, L. A., and Kemp, E. M. 1972. Influence of continental positions on early Tertiary climates. *Nature, Lond.* 240:97–100.

Gardner, C. A. 1944. Presidential address: The vegetation of Western Australia. *J. Proc. R. Soc. West. Aust.* 28:xi–lxxxvii.

———. 1957. The fire factor in relation to the vegetation of Western Australia. *West. Aust. Nat.* 5:166–173.

Gilbert, J. M. 1958. Forest succession in the Florentine Valley, Tasmania. *Pap. Proc. R. Soc. Tasm.* 93:129–151.

Greenwald, L.; Stone, W. B.; and Cade, T. J. 1967. Physiological adjustments of the budgerygah (*Melopsettacus undulatus*) to dehydrating conditions. *Comp. Biochem. Physiol.* 22:91–100.

Heirtzler, J. R.; Dickson, G. O.; Herron, E. M.; Pitman, W. C.; and Le Pichon, X. 1968. Marine magnetic anomalies, geomagnetic field reversals, and motions of the ocean floor and continents. *J. geophys. Res.* 73:2119–2136.

Jackson, W. D. 1965. Vegetation. In *Atlas of Tasmania*, ed. J. L. Davis, pp. 50–55. Hobart, Tasm.: Mercury Press.

———. 1968a. Fire and the Tasmanian flora. In *Tasmanian year book no. 2*, ed. R. Lakin and W. E. Kellend. Hobart, Tasm.: Commonwealth Bureau of Census and Statistics, Hobart Branch.

———. 1968b. Fire, air, water and earth: An elemental ecology of Tasmania. *Proc. ecol. Soc. Aust.* 3:9–16.

Johnson, O. W., and Mugaas, J. N. 1970. Quantitative and organizational features of the avian renal medulla. *Condor* 72:288–292.

Keast, A. 1959. Australian birds: Their zoogeography and adaptation to an arid continent. In *Biogeography and ecology in Australia*, ed. A. Keast, R. L. Crocker, and C. S. Christian, pp. 89–114. The Hague: Dr. W. Junk.

King, C. E., and Anderson, W. W. 1971. Age specific selection, II. The interaction between r & K during population growth. *Am. Nat.* 105:137–156.

Kirkpatrick, T. H. 1965. Studies of Macropodidae in Queensland. 2. Age estimation in the grey kangaroo, the eastern wallaroo and the red-necked wallaby, with notes on dental abnormalities. *Qd J. agric. Anim. Sci.* 22:301–317.

Le Pichon, X. 1968. Sea-floor spreading and continental drift. *J. geophys. Res.* 73:3661–3697.

Licht, P.; Dawson, W. R.; and Shoemaker, V. H. 1966*a*. Heat resistance of some Australian lizards. *Copeia* 1966:162–169.

Licht, P.; Dawson, W. R.; Shoemaker, V. H.; and Main, A. R. 1966*b*. Observations on the thermal relations of Western Australian lizards. *Copeia* 1966:97–110.

MacArthur, R. H., and Wilson, E. O. 1967. *The theory of island biogeography.* Monographs in Population Biology, 1. Princeton: Princeton Univ. Press.

MacMillan, R. E., and Lee, A. K. 1967. Australian desert mice: Independence of exogenous water. *Science, N.Y.* 158:383–385.

————. 1969. Water metabolism of Australian hopping mice. *Comp. Biochem. Physiol.* 28:493–514.

————. 1970. Energy metabolism and pulmocutaneous water loss of Australian hopping mice. *Comp. Biochem. Physiol.* 35:355–369.

McWhae, J. R. H.; Playford, P. E.; Lindner, A. W.; Glenister, B. F.; and Balme, B. E. 1958. The stratigraphy of Western Australia. *J. geol. Soc. Aust.* 4:1–161.

Main, A. R. 1971. Measures of well-being in populations of herbivorous macropod marsupials. In *Dynamics of populations*, ed. P. J. den Boer and G. R. Gradwell, pp. 159–173. Wageningen: PUDOC.

Mertz, D. B. 1971. Life history phenomena in increasing and decreasing population. In *Statistical ecology, volume II: Sampling and modeling biological populations and population dynamics*, ed. G. P. Patil, E. C. Pielou, and W. E. Waters, pp. 361–399. University Park: Pa. St. Univ. Press.

Moir, R. J.; Somers, M.; and Waring, H. 1956. Studies in marsupial nutrition: Ruminant-like digestion of the herbivorous marsupial *Setonix brachyurus* (Quoy and Gaimard). *Aust. J. biol. Sci.* 9:293–304.

Mount, A. B. 1964. The interdependence of eucalypts and forest fires in southern Australia. *Aust. For.* 28:166–172.

————. 1969. Eucalypt ecology as related to fire. *Proc. Ann. Tall Timbers Fire Ecol. Conf.* 9:75–108.

Murphy, G. I. 1968. Pattern in life history and the environment. *Am. Nat.* 102:391–404.

Oksche, A.; Farner, D. C.; Serventy, D. L.; Wolff, F.; and Nicholls,

C. A. 1963. The hypothalamo-hypophysial neurosecretory system of the zebra finch, *Taeniopygia castanotis. Z. Zellforsch. mikrosk. Anat.* 58:846–914.

Packham, G. H., ed. 1969. The geology of New South Wales. *J. geol. Soc. Aust.* 16:1–654.

Pianka, E. R. 1969. Sympatry of desert lizards (*Ctenotus*) in Western Australia. *Ecology* 50:1012–1030.

———. 1970. On r and K selection. *Am. Nat.* 104:592–597.

———. 1971a. Comparative ecology of two lizards. *Copeia* 1971:129–138.

———. 1971b. Ecology of the agamid lizard *Amphibolurus isolepis* in Western Australia. *Copeia* 1971:527–536.

———. 1971c. Notes on the biology of *Amphibolurus cristatus* and *Amphibolurus scutulatus*. *West. Aust. Nat.* 12:36–41.

Pianka, E. R., and Pianka, H. D. 1970. The ecology of *Moloch horridus* (Lacertilia: Agamidae) in Western Australia. *Copeia* 1970:90–103.

Ride, W. D. L. 1964. A review of Australian fossil marsupials. *J. Proc. R. Soc. West. Aust.* 47:97–131.

———. 1971. On the fossil evidence of the evolution of the Macropodidae. *Aust. Zool.* 16:6–16.

Shield, J. W. 1965. A breeding season difference in two populations of the Australian macropod marsupial *Setonix brachyurus*. *J. Mammal.* 45:616–625.

Simpson, G. G. 1961. Historical zoogeography of Australian mammals. *Evolution* 15:431–446.

Slatyer, R. O., and Perry, R. A., eds. 1969. *Arid lands of Australia*. Canberra: Aust. Nat. Univ. Press.

Sporn, C. C. 1955. The breeding of the mountain devil in captivity. *West. Aust. Nat.* 5:1–5.

———. 1958. Further observations on the mountain devil in captivity. *West. Aust. Nat.* 6:136–137.

———. 1965. Additional observations on the life history of the mountain devil (*Moloch horridus*) in captivity. *West. Aust. Nat.* 9:157–159.

Stirton, R. A.; Tedford, R. D.; and Miller, A. H. 1961. Cenozoic stratigraphy and vertebrate palaeontology of the Tirari Desert, South Australia. *Rep. S. Aust. Mus.* 14:19–61.

Stirton, R. A.; Tedford, R. H.; and Woodburne, M. O. 1968. Australian Tertiary deposits containing terrestrial mammals. *Univ. Calif. Publs geol. Sci.* 77:1–30.

Stirton, R. A.; Woodburne, M. O.; and Plane, M. D. 1967. A phylogeny of Diprotodontidae and its significance in correlation. *Bull. Bur. Miner. Resour. Geol. Geophys. Aust.* 85:149–160.

Storr, G. M. 1967. Geographic races of the agamid lizard *Amphibolurus caudicinctus. J. Proc. R. Soc. West. Aust.* 50:49–56.

———. 1968. Diet of kangaroos (*Megaleia rufa* and *Macropus robustus*) and merino sheep near Port Hedland, Western Australia. *J. Proc. R. Soc. West. Aust.* 51:25–32.

Storr, G. M.; Green, J. W.; and Churchill, D. M. 1959. The vegetation of Rottnest Island. *J. Proc. R. Soc. West. Aust.* 42:70–71.

Veevers, J. J. 1967. The Phanerozoic geological history of northwest Australia. *J. geol. Soc. Aust.* 14:253–271.

———. 1971. Phanerozoic history of Western Australia related to continental drift. *J. geol. Soc. Aust.* 18:87–96.

Vine, F. J. 1966. Spreading of the ocean floor: New evidence. *Science, N.Y.* 154:1405–1415.

———. 1970. Ocean floor spreading. *Rep. Aust. Acad. Sci.* 12:7–24.

Wallace, W. R. 1966. Fire in the Jarrah forest environment. *J. Proc. R. Soc. West. Aust.* 49:33–44.

Watkins, J. R. 1967. The relationship between climate and the development of landforms in the Cainozoic rocks of Queensland. *J. geol. Soc. Aust.* 14:153–168.

Wild, A. 1958. The phosphate content of Australian soils. *Aust. J. agric. Res.* 9:193–204.

Winkworth, R. E. 1967. The composition of several arid spinifex grasslands of central Australia in relation to rainfall, soil water relations, and nutrients. *Aust. J. Bot.* 15:107–130.

Woodburne, M. O. 1967. Three new diprotodontids from the Tertiary of the Northern Territory. *Bull. Bur. Miner. Resour. Geol. Geophys. Aust.* 85:53–104.

Woolnough, W. G. 1928. The chemical criteria of peneplanation. *J. Proc. R. Soc. N.S.W.* 61:17–53.

6. Species and Guild Similarity of North American Desert Mammal Faunas: A Functional Analysis of Communities   James A. MacMahon

Introduction

A major thrust of current ecological and evolutionary research is the analysis of patterns of species diversity or density in similar or vastly dissimilar community types. Such studies are believed to bear on questions concerned with the nature of communities and their stability (e.g., MacArthur, 1972), the concept of ecological equivalents or ecospecies (Odum, 1969 and 1971; Emlen, 1973), and, of course, the nature of the "niche" (Whittaker, Levin, and Root, 1973).

An approach emerging from this plethora is that of functional analysis of community components: attempts to compare the functionally similar community members, regardless of their taxonomic affinities. Root (1967, p. 335) coined the term *guild* to define "a group of species that exploit the same class of environmental resources in a similar way. This term groups together species without regard to taxonomic position, that overlap significantly in their niche requirements."

Guild is clearly differentiated from *niche* and *ecotope*, recently redefined and defined respectively (Whittaker et al., 1973, p. 335) as "applying 'niche' to the role of the species within the community, 'habitat' to its distributional response to intercommunity environmental factors, and 'ecotope' to its full range of adaptations to external factors of both niche and habitat." *Guild* groups parts of species' niches permitting intercommunity comparisons.

Without referring to the semantic problems, Baker (1971) used such a "functional" approach when he compared nutritional strategies of North American grassland myomorph rodents. Wiens (1973) developed a similar theme in his recent analysis of grassland bird communities, as did Wilson (1973) with an analysis of bat faunas, and Brown (1973) with rodents of sand-dune habitats.

This paper is an attempt to compare species and functional analyses of the small mammal component of North American deserts and to use these analyses to discuss some aspects of the broader ecological and evolutionary questions of "similarity" and function of communities.

Sites and Techniques

Sites

The data base for this study is simply the species lists for a number of desert or semidesert grassland localities in the western United States. The list for a locality represents those species that occur on a piece of landscape of 100 ha in extent. This size unit allows the inclusion of spatial heterogeneity.

The localities used, the data source, and the abbreviations to be used subsequently are Jornada Bajada (*j*), a Chihuahuan desert shrub community near Las Cruces, New Mexico, operated by New Mexico State University as part of the US/IBP Desert Biome studies; Jornada Playa (*jp*), a desert grassland and mesquite area a few meters from *j* operated under the same program; Portal, Arizona (*cc*), a semidesert scrub area studied extensively by Chew and Chew (1965, 1970); Santa Rita Experimental Range (*sr*), south of Tucson, Arizona, an altered desert grassland studied by University of Arizona personnel for the US/IBP Desert Biome; Tucson Silverbell Bajada (*t*), a typical Sonoran desert (*Larrea-Cereus-Cercidium*) locality north west of Tucson, Arizona, operated as *sr*; Big Bend National Park (*bb*), a Chihuahuan desert shrub community near the park headquarters typified by Denyes (1956) and K. L. Dixon (1974, personal communication); Deep Canyon, California (*dc*), studied by Ryan (1968) and Joshua Tree National Monument (*jt*) studied by Miller and Stebbins (1964)—both *Larrea*-dominated areas in a transition from a Sonoran desert subdivision (Coloradan) but including many Mojave desert elements; Rock Valley (*rv*), northwest of Las Vegas, Nevada, on the

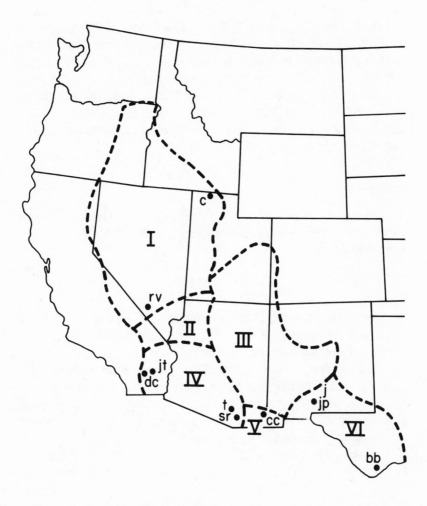

Fig. 6-1. The location of sites discussed (abbreviations explained in text) and outline of mammal provinces adapted from Hagmeier and Stults (1964): I. Artemisian; II. Mojavian; III. Navajonian; IV. Sonoran; V. Yaquinian; VI. Mapimi.

Atomic Energy Commission's Nevada Test Site, a Mojave desert shrub site operated as part of the US/IBP Desert Biome by personnel of the Environmental Biology Division of the Laboratory of Nuclear Medicine and Radiation Biology of the University of California, Los Angeles; and Curlew Valley (*c*), a Great Basin desert, sagebrush site of the US/IBP Desert Biome operated by Utah State University. The positions of the sites are summarized in figure 6-1. All sites have been visited and observed by me.

Analyses

Similarity was calculated using a modified form of Jaccard analysis (community coefficients) (Oosting, 1956; see also MacMahon and Trigg, 1972):

$$\frac{2w}{a+b} \times 100$$

where *w* is the number of species common to both faunas being compared, *a* is the number of species in the smaller fauna, and *b* is the number of species in the larger fauna.

Species similarity merely uses different taxa as units for calculations. Functional similarity uses functional units (guilds) based mainly on food habits and adult size of nonflying mammals, jack rabbit in size or smaller. The twelve desert guilds recognized, with examples of species from a single locality, include five granivores (two possible dormant-season divisions) (*Dipodomys spectabilis*, *D. merriami*, *Perognathus penicillatus*, *P. baileyi*, *P. amplus*); a "carnivorous" mouse (*Onychomys torridus*); a large and small browser (*Lepus californicus*, *Sylvilagus auduboni*); two micro-omnivores (*Peromyscus eremicus*, *P. maniculatus*); a "pack rat" (*Neotoma albigula*); and a diurnal medium-sized omnivore (*Citellus tereticaudus*). When grassland guilds are mentioned, two grazers are added to the above. Data for all pair-wise comparisons of sites are summarized in figure 6-2.

The list of mammals for all sites includes forty-seven species in fifteen genera. An additional fifteen or so species occur near the sites but were not collected on the prescribed areas.

SPECIES

FUNCTIONS	j	jp	cc	sr	t	bb	dc	jt	rv	c
j		67	67	41	41	56	33	30	20	14
jp	67		63	52	39	46	32	23	19	19
cc	79	63		60	56	54	32	29	19	14
sr	63	60	52		63	38	29	27	15	25
t	85	79	56	55		48	33	30	20	04
bb	80	62	85	69	85		40	44	48	08
dc	85	67	67	63	71	80		63	60	09
jt	86	69	69	65	73	89	73		73	08
rv	85	67	67	55	85	80	85	86		14
c	60	67	56	55	50	72	60	53	71	

Fig. 6-2. Similarity analysis (%) matrix derived from Jaccard analysis (see text): species comparisons above the diagonal, guild comparisons below the diagonal.

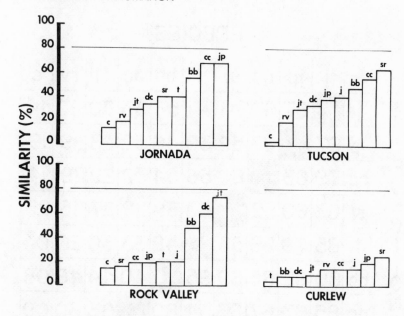

Fig. 6-3. Comparison of the similarity (%) of lists of nonflying mammal species on all sites to those of four "typical" North American desert sites: Jornada (j), Chihuahuan desert; Tucson (t), Sonoran desert; Rock Valley (rv), Mojave Desert; Curlew (c), Great Basin (cold desert).

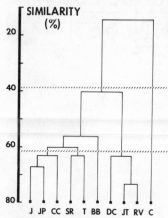

Fig. 6-4. Dendrogram showing relationships between species composition (maximum percent similarity, using Jaccard analysis) at North American sites (see text for abbreviations). The levels of significance for provinces (about 62%) and for superprovinces (about 39%), as defined by Hagmeier and Stults (1964), are marked.

Results and Discussion

Species Density

Figure 6-3 depicts the comparison of the species composition of all sites with that of each of the four "typical" desert sites of the US/IBP Desert Biome. These sites represent each of the four North American deserts: three "hot" deserts—Chihuahuan (*j*), Sonoran (*t*), Mojave (*rv*); one "cold" desert—Great Basin (*c*). Sites *jp*, *sr*, and *cc* are considered to have strong desert grassland affinities; all were "good" grasslands in historical times (Gardner, 1951; Lowe, 1964; Lowe et al., 1970). It is clear that none of the comparisons indicates high similarity (operationally defined as 80%). Comparison of figure 6-3 and figure 6-1 indicates that what similarity exists seems to be due to geographic proximity.

Maximum Species Similarity

The maximum species similarity of all sites is used to develop a dendrogram of relationships of sites similar to those of Hagmeier and Stults (1964) (fig. 6-4). The five groupings derived (using a 60% similarity level as was used by Hagmeier and Stults)—that is, *j*, *jp*, *cc*; *sr*, *t*; *bb*; *dc*, *jt*, *rv*; and *c*—do not follow closely the mammal provinces erected by Hagmeier and Stults (1964) and are redrawn here (fig. 6-1).

There is agreement between my data and those of Hagmeier and Stults at the superprovince level (about 38% similarity) which sets *c* (Artemisian) apart from all hot desert sites. The Artemisian province is equivalent to the Great Basin desert or cold desert in extent. The failure of this study and that of Hagmeier and Stults to agree may be due to the animal size limit used herein (smaller than jack rabbits) or the confined areal sample (100 ha vs. larger areas).

The groups defined here (at the province level) do seem to have some faunal meaning: there is a distinct Big Bend (*bb*) assemblage, a Sonoran desert group (*sr*, *t*), a Chihuahuan desert group (*j*, *jp*, *cc*), a Mojave desert group (*dc*, *jt*, *rv*), and a Great Basin desert group (*c*).

Some groups can be explained utilizing the evolution-biogeography discussion of Findley (1969) which differentiated eastern and western

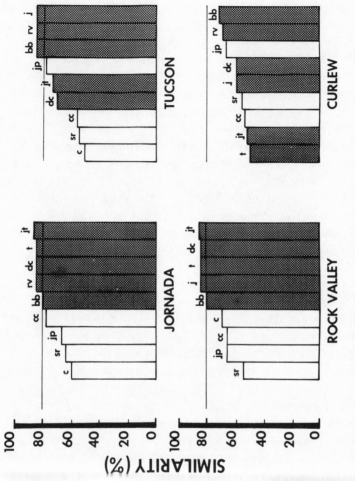

Fig. 6-5. Comparison of similarity in guilds of nonflying mammals at all sites. Presentation as in figure 6-3, except that the bars of all "hot" desert localities are shaded.

desert components meeting in southeastern Arizona—this coincides with the Sonoran desert versus the Chihuahuan desert above, with the cc site being intermediate. Figure 6-2 supports the intermediacy of cc. Findley postulated that the Deming Plain was a barrier to desert mammal movement in pluvial times, that it limited gene flow, and that it permitted speciation to the west and east. There is sharp differentiation of these two components from the Mojave and Great Basin, which is expected on the basis of their different geological histories, and also from the Big Bend, which might be more representative of the major portion of the Chihuahuan desert mammal fauna in Mexico.

Functional Diversity

When the forty-seven species of mammals are placed in guilds, rather than being treated as taxa, there is a high degree of similarity among all hot desert sites, but significant differences persist between the cold desert (c) and hot deserts (fig. 6-5).

A further indication of the biological soundness of guilds follows from a comparison of each desert with its geographically closest desert or destroyed grassland (jp, cc, or sr). This comparison generally indicates no increase in similarity whether using taxa or guilds (table 6-1). If some distinctly grassland guilds (grazers) are added

---

Table 6-1. *Similarity between Desert Grasslands and Closest Deserts*

Sites Compared	Similarity Index By Species	By Guilds
j-jp	67	67
j-cc	67	79
t-cc	56	56
t-sr	63	55

Note: Figures represent percent similarity coefficient (Jaccard analysis) of each desert (altered) grassland with its geographically closest desert on the basis of both species composition of fauna and functional groups (guilds).

(table 6-2), similarity of grasslands to each other rises from low levels to those considered significant. Nonsimilarity was then a problem of not including enough specifically grassland guilds.

Comparison of levels of similarity is significant and explains the operational definition of adequate similarity. Using mean similarity values (mean ± standard error) calculated from data in figure 6-2, functional (guild) similarity among hot deserts is 81.87 ± 1.43 percent; grassland to other grasslands, 87.33 ± 0.33 percent; hot deserts to grasslands, 66.32 ± 1.88 percent; cold desert to hot desert, 61.0 ± 3.69 percent; and cold desert to grassland, 59.33 ± 3.85 percent. Three functional categories are clear: hot desert, cold desert, and grassland. Cross comparison of the means of functional categories versus nonfunctional ones demonstrates significant (.001 level) differences. The Behrens-Fisher modification of the t test and Cochran's approximation of t' were used because of the heterogeneity of sample variances (Snedecor and Cochran, 1967).

Table 6-2. *Similarity among Altered Grassland Sites*

Sites Compared	Similarity Indexes		
	By Species	By Guild	By Guilds (including two grassland guilds)
jp-cc	63	63	87
cc-sr	60	52	88
sr-jp	52	60	87

Note: Figures represent percent similarity coefficient (Jaccard analysis) between destroyed or altered grassland sites, using species composition and functional units (guilds) and adding two specifically grassland guilds.

General Discussion

Causes of Species Mixes

An implication of the data presented here is that any one of a number of species of small mammals may be functionally similar in a particular example of desert scrub community. It is well known that various species of mice in the genera *Microdipidops*, *Peromyscus*, and *Perognathus* may have overlapping ranges and be desert adapted, but seem to replace each other in specific habitats of various soil-texture characteristics (sandy to rocky) (Hardy, 1945; Ryan, 1968; Ghiselin, 1970). Soil surface strength, and vegetation height and density, explain relative densities of some desert small mammals (Rosenzweig and Winakur, 1969; Rosenzweig, 1973). Interspecific behavior is part of the partitioning of habitats by a number of desert rodents: *Neotoma* (Cameron, 1971); *Dipodomys* (Christopher, 1973); and a seven-species "community" (MacMillan, 1964). Brown (1973) and Brown and Lieberman (1973) attribute species diversity of sand-dune rodents to a mix of ecological, biogeographic, and evolutionary factors, a position similar to that I take for a broader range of desert conditions.

Many other factors are also involved in defining various axes of specific niches (*sensu* Whittaker et al., 1973) of desert mammals; these specific niche differences do not exclude the functional overlap of other niche components. These examples and others merely show how finely genetically plastic organisms can subdivide the environment. These differences need not change the basic role of the species.

If the important shrub community function is seed removal, it does not matter to the community what species does it. The removal could be by any one of several rodents differing in soil-texture preferences or perhaps by a bird or even ants. Ecological equivalents may or may not be genetically close. All niche axes of a species or population are not equally important to the functioning of the community.

Importance of Functional Approach

Since all three hot deserts have similar functional diversity but vary in basic ways (e.g., more rain, more vegetational synusia, biseasonal rainfall pattern in the Sonoran desert as compared to the Mojave), functional diversity does not correlate well with those factors thought to relate animal species diversity to community structure—that is, vertical and horizontal foliage complexity (Pianka, 1967; MacArthur, 1972).

The crux of the problem is that most measures of species diversity include some measure of abundance (Auclair and Goff, 1971). The analysis herein counts only presence or absence—a level of abstraction more general than species diversity measures.

This greater generality is justified, I believe, because it strikes closer to the problem of comparing community types which are basically similar (e.g., hot deserts) but include units each having undergone specific development in time and space (e.g., Mojave vs. Sonoran vs. Chihuahuan hot deserts). The analysis seeks to elucidate various levels of the least-common denominators of community function.

The guild level of abstraction has potential applied value. The generalization of forty-seven mammal species into only twelve functional groups may permit the development of suitable predictive computer models of "North American Hot Deserts." The process of accounting for the vagaries of every species makes the task of modeling cumbersome and may prohibit rapid expansion of this ecological tool. The abstraction of a large number of species into biologically determined "black boxes" is an acceptable compromise.

There is clear evidence that temporal variations in the density of mammal species may so affect calculations of species diversity that their use and interpretation in a community context are difficult (M'Closkey, 1972). While such variations are intrinsically interesting biologically, they should not prevent us from seeking more generally applicable, albeit less detailed, predictors of community organization.

Guilds, Niches, Species, Stability

The guilds chosen here were selected on the basis of subjective familiarity with classical mammals. These guilds may not be requisite for the community as a whole to operate. If one were able to perceive the *requisite* guilds of a community, several things seem reasonable. First, to be stable (able to withstand perturbations without changing basic structure and function) a community cannot lose a requisite guild. Species performing functions provide a stable milieu if, first, they are themselves resilient to a wide range of perturbations (i.e., no matter what happens they survive) or, second, each requisite function can be performed by any one of a number of species, despite niche differences among these species—that is, the community contains a high degree of functional redundancy, preventing or reducing changes in community characteristics. The importance of this was alluded to by Whittaker and Woodwell (1972), who cited the case of oaks replacing chestnuts after they were wiped out by the blight in the North American eastern deciduous forests, and on theoretical grounds by Conrad (1972).

Any stable community is some characteristic mix of resilient and redundant species. Species diversity per se may not then correlate well with stability. Stability might come with a number of species diversities as long as the requisite guilds are represented. Tropics and deserts might represent extremes of a large number of redundant species as opposed to fewer more resilient species, both mixes conferring some level of stability as witnessed by the historical persistence of these community types.

None of this implies that all species are requisite to a community, or that coevolution is the only path for community evolution. Many species may be "tolerated" by communities just because the community is well enough buffered that minor species have no noticeable effect. As long as they pass their genetic make-up on to a new generation, species are successful; they need not do anything for the community.

Acknowledgments

These studies were made possible by a National Science Foundation grant (GB 32139) and are part of the contributions of the US/IBP Desert Biome Program. I am indebted to the following people for data collected by them or under their supervision: W. Whitford, E. L. Cockrum, K. L. Dixon, F. B. Turner, B. Maza, R. Anderson, and D. Balph. F. B. Turner and N. R. French kindly commented on a version of the manuscript.

Summary

The nonflying, small mammal faunas of western United States deserts were compared (coefficient of community) on the basis of their species and guild (functional) composition. Guilds were derived from information on animal size and food habits.

It is concluded that the similarity among sites with respect to guilds, though the species may differ, is a result of a complex of evolutionary events and particular contemporary community characteristics of the specific sites. Functional similarity, based on functional groups (guilds), is rather constant among hot deserts and different between hot deserts and either cold desert or desert grassland.

The functional analysis describes only a part of the niche of an organism, but perhaps an important part. Such abstractions and generalizations of the details of the community's complexities permit mathematical modeling to progress more rapidly and allow address to the general question of community "principles."

Guilds required by communities to maintain community integrity against perturbations may be better correlates to community stability than the various measures of species diversity currently popular.

References

Auclair, A. N., and Goff, F. G. 1971. Diversity relations of upland forests in the western Great Lakes area. *Am. Nat.* 105:499–528.

Baker, R. H. 1971. Nutritional strategies of myomorph rodents in North American grasslands. *J. Mammal.* 52:800–805.

Brown, J. H. 1973. Species diversity of seed-eating desert rodents in sand dune habitats. *Ecology* 54:775–787.

Brown, J. H., and Lieberman, G. A. 1973. Resource utilization and co-existence of seed-eating desert rodents in sand dune habitats. *Ecology* 54:788–797.

Cameron, G. N. 1971. Niche overlap and competition in woodrats. *J. Mammal.* 52:288–296.

Chew, R. M., and Chew, A. E. 1965. The primary productivity of a desert shrub (*Larrea tridentata*) community. *Ecol. Monogr.* 35:355–375.

———. 1970. Energy relationships of the mammals of a desert shrub *Larrea tridentata* community. *Ecol. Monogr.* 40:1–21.

Christopher, E. A. 1973. Sympatric relationships of the kangaroo rats, *Dipodomys merriami* and *Dipodomys agilis*. *J. Mammal.* 54:317–326.

Conrad, M. 1972. Stability of foodwebs and its relation to species diversity. *J. theoret. Biol.* 32:325–335.

Denyes, H. A. 1956. Natural terrestrial communities of Brewster County, Texas, with special reference to the distribution of mammals. *Am. Midl. Nat.* 55:289–320.

Emlen, J. M. 1973. *Ecology: An evolutionary approach.* Reading, Mass: Addison-Wesley.

Findley, J. S. 1969. Biogeography of southwestern boreal and desert mammals. *Univ. Kans. Publs Mus. nat. Hist.* 51:113–128.

Gardner, J. L. 1951. Vegetation of the creosotebush area of the Rio Grande Valley in New Mexico. *Ecol. Monogr.* 21:379–403.

Ghiselin, J. 1970. Edaphic control of habitat selection by kangaroo mice (*Microdipodops*) in three Nevadan populations. *Oecologia* 4:248–261.

Hagmeier, E. M., and Stults, C. D. 1964. A numerical analysis of the distributional patterns of North American mammals. *Syst. Zool.* 13:125–155.

Hardy, R. 1945. The influence of types of soil upon the local distribution of some small mammals in southwestern Utah. *Ecol. Monogr.* 15:71–108.

Lowe, C. H. 1964. Arizona landscapes and habitats. In *The vertebrates of Arizona*, ed. C. H. Lowe, pp. 1–132. Tucson: Univ. of Ariz. Press.

Lowe, C. H.; Wright, J. W.; Cole, C. J.; and Bezy, R. L. 1970. Natural hybridization between the teiid lizards *Cnemidophorus sonorae* (parthenogenetic) and *Cnemidophorus tigris* (bisexual). *Syst. Zool.* 19:114–127.

MacArthur, R. H. 1972. *Geographical ecology.* New York: Harper & Row.

M'Closkey, R. T. 1972. Temporal changes in populations and species diversity in a California rodent community. *J. Mammal.* 53:657–676.

MacMahon, J. A., and Trigg, J. R. 1972. Seasonal changes in an old-field spider community with comments on techniques for evaluating zoosociological importance. *Am. Midl. Nat.* 87:122–132.

MacMillan, R. E. 1964. Population ecology, water relations and social behavior of a southern California semidesert rodent fauna. *Univ. Calif. Publs Zool.* 71:1–66.

Miller, A. H., and Stebbins, R. C. 1964. *The lives of desert animals in Joshua Tree National Monument.* Berkeley and Los Angeles: Univ. of Calif. Press.

Odum, E. P. 1969. The strategy of ecosystem development. *Science, N.Y.* 164:262–270.

————. 1971. *Fundamentals of ecology*. 3d ed. Philadelphia: W. B. Saunders Co.

Oosting, H. J. 1956. *The study of plant communities*. 2d ed. San Francisco: Freeman Co.

Pianka, E. R. 1967. On lizard species diversity: North American flatland deserts. *Ecology* 48:333–351.

Root, R. B. 1967. The niche exploitation pattern of the blue-gray gnatcatcher. *Ecol. Monogr.* 37:317–350.

Rosenzweig, M. L. 1973. Habitat selection experiments with a pair of co-existing heteromyid rodent species. *Ecology* 54:111–117.

Rosenzweig, M. L., and Winakur, J. 1969. Population ecology of desert rodent communities: Habitats and environmental complexity. *Ecology* 50:558–572.

Ryan, R. M. 1968. *Mammals of Deep Canyon*. Palm Springs, Calif.: Desert Museum.

Snedecor, G. W., and Cochran, W. G. 1967. *Statistical methods*. 6th ed. Ames: Iowa State Univ. Press.

Whittaker, R. H., and Woodwell, G. M. 1972. Evolution of natural communities. In *Ecosystem structure and function*, ed. J. Wiens, pp. 137–156. Corvallis: Oregon State Univ. Press.

Whittaker, R. H.; Levin, S. A.; and Root, R. B. 1973. Niche, habitat, and ecotope. *Am. Nat.* 107:321–338.

Wiens, J. A. 1973. Pattern and process in grassland bird communities. *Ecol. Monogr.* 43:237–270.

Wilson, D. E. 1973. Bat faunas: A trophic comparison. *Syst. Zool.* 22:14–29

7. The Evolution of Amphibian Life Histories in the Desert Bobbi S. Low

Introduction

Among desert animals amphibians are especially intriguing because at first glance they seem so obviously unsuited to arid environments. Most amphibians require the presence of free water at some stage in the life cycle. Their skin is moist and water permeable, and their eggs are not protected from water loss by any sort of tough shell. Perhaps the low number of amphibian species that live in arid regions reflects this.

Some idea of the variation in life-history patterns which succeed in arid regions is necessary before examining the environmental parameters which shape those life histories. Consider three different strategies. Members of the genus *Scaphiopus* found in the southwestern United States frequent short-grass plains and alkali flats in arid and semiarid regions and are absent from high mountain elevations and extreme deserts (Stebbins, 1951). Species of the genus breed in temporary ponds and roadside ditches, often on the first night after heavy rains. *Scaphiopus bombifrons* lays 10 to 250 eggs in a number of small clusters; *Scaphiopus hammondi* lays 300 to 500 eggs with a mean of 24 eggs per cluster; and *S. couchi* lays 350 to 500 eggs in a number of small clusters. All three species burrow during dry periods of the year.

The genus *Bufo* is widespread, and a large number of species live in arid and semiarid regions. *Bufo alvarius* lives in arid regions but, unlike *Scaphiopus*, appears to be dependent on permanent water. Stebbins (1951) notes that, while summer rains seem to start seasonal activity, such rains are not always responsible for this activity.

The mating call has been lost. Also, unlike *Scaphiopus*, this toad lays between 7,500 and 8,000 eggs in one place at one time.

Eleutherodactylus latrans occurs in arid and semiarid regions and does not require permanent water. It frequents rocky areas and canyons and may be found in crevices, caves, and even chinks in stone walls. In Texas, Stebbins (1951) reported that *E. latrans* becomes active during rainy periods from February to May. About fifty eggs are laid on land in seeps, damp places, or caves; and, unlike either *Scaphiopus* or *Bufo*, the male may guard the eggs.

These three life histories diverge in degree of dependence on water, speed of breeding response, degree of iteroparity, and amount and kind of reproductive effort per offspring or parental investment (Trivers, 1972). All three strategies may have evolved, and at least are successful, in arid regions.

Two forces will shape desert life histories. The first is the relatively high likelihood of mortality as a result of physical extremes. Considerable work has been done on the mechanics of survival in amphibians which live in arid situations (reviewed by Mayhew, 1968). Most of the research concentrated on physiological parameters like dehydration tolerance, ability to rehydrate from damp soil, and speed of rehydration (Bentley, Lee, and Main, 1958; Main and Bentley, 1964; Warburg, 1965; Dole, 1967); water retention and cocoon formation (Ruibal, 1962a, 1962b; Lee and Mercer, 1967); and the temperature tolerances of adults and tadpoles (Volpe, 1953; Brattstrom, 1962, 1963; Heatwole, Blasina de Austin, and Herrero, 1968). Bentley (1966), reviewing adaptations of desert amphibians, gave the following list of characteristics important to desert species:

1. No definite breeding season
2. Use of temporary water for reproduction
3. Initiation of breeding behavior by rainfall
4. Loud voices in males, with marked attraction of both males and females, and the quick building of large choruses
5. Rapid egg and larval development
6. Ability of tadpoles to consume both animal and vegetable matter
7. Tadpole cannibalism
8. Production of growth inhibitors by tadpoles

9. High heat tolerance by tadpoles
10. Metatarsal spade for burrowing
11. Dehydration tolerance
12. Nocturnal activity

Bentley's list consists mostly of physiological or anatomical characteristics associated with survival in the narrow sense. Only two or three items involve special aspects of life cycles, and some characteristics as stated are not exclusive to desert forms. Most investigators have emphasized the problems of survival for desert amphibians for the obvious reason that the animal and its environment seem so ill-matched, and most investigators have emphasized morphological and physiological attributes because they are easier to measure. A notable exception to this principally anatomical or physiological approach is the work of Main and his colleagues (Main, 1957, 1962, 1965, 1968; Main, Lee, and Littlejohn, 1958; Main, Littlejohn, and Lee, 1959; Lee, 1967; Martin, 1967; Littlejohn, 1967, 1971) who have discussed life-history adaptations of Australian desert anurans. Main (1968) has summarized some general life-history phenomena that he considered important to arid-land amphibians including high fecundity, short larval life, and burrowing. However, as he implied, the picture is not simple. A surprising variety of successful life-history strategies exists in arid and semiarid amphibian species, far greater than one would predict from attempts (Bentley, 1966; Mayhew, 1968) to summarize desert adaptations in amphibians. If survival were the critical focus of selection, one might predict fewer successful strategies and more uniformity in the kinds of life histories successful in arid-land amphibians.

But succeeding in the desert, as elsewhere, is a matter of balancing risk of mortality against optimization of reproductive effort so that realized reproduction is optimized. As soon as survival from generation to generation occurs, selection is then working on differences in reproduction among the survivors, an important point emphasized by Williams (1966a) in arguing that adaptations should most often be viewed as the outcomes of better-versus-worse alternatives rather than as necessities in any given circumstance. The focus I wish to develop here is on the critical parameters shaping the evolution of life-history strategies and the better-versus-worse alternatives in each of a number of situations. Adaptations of desert amphibians have scarcely been examined in this light.

Life-History Components and Environmental Parameters

Wilbur, Tinkle, and Collins (1974), in an excellent paper on the evolution of life-history strategies, list eight components of life histories: juvenile and adult mortality schedules, age at first reproduction, reproductive life span, fecundity, fertility, fecundity-age regression, degree of parental care, and reproductive effort. The last two are included in Trivers's (1972) concept of parental investment. For very few, if any, anurans are all these parameters documented.

I will concentrate here on problems of parental investment, facultative versus nonfacultative responses, cryptic versus clumping responses, and shifts in life-history stages. How does natural selection act on these traits in different environments? What environmental parameters are actually significant?

Classifying ranges of environmental variation may seem at first like a job for geographers; but, even when one acknowledges that deserts may be hot or cold, seasonal or nonseasonal, that they may possess temporary or permanent waters, different vegetation, and different soils, I think a few parameters can be shown to have overriding importance. These are (a) the *range* of the variation in the environmental attributes I have just described—temperature, humidity, day length, and so on; (b) the *predictability* of these attributes; and (c) their *distribution*—the patchiness or grain of the environment (Levins, 1968). Wilbur et al. (1974) consider trophic level and successional position also as life-history determinants in addition to environmental uncertainty. At the intraordinal level, these effects may be more difficult to sort out; and, at present, data are really lacking for anurans.

Range

Obviously the overall range of variation in environmental parameters is important in shaping patterns of behavior or life history. The same life-history strategy will not be equally successful in an environment where, for instance, temperature fluctuates only 5° daily, as in an environment in which fluctuations may be as much as 20° to 30°C. The range of fluctuations may strongly affect selection on physiological adaptations and differences in survival. Ranges of variation, particularly in temperature and water availability, are extreme in the environments of desert amphibians; but such effects have been dealt with more fully than the others I wish to discuss, and so I will concentrate on other factors.

Predictability

It is probably true that deserts are less predictable than either tropical or temperate mesic situations; Bentley's (1966) list of adaptations reflected this characteristic. The terms *uncertainty* and *predictability* are generally used for physical effects—seasonality and catastrophic events, for instance—but may include both spatial and biotic components. In fact, both patchiness and the distribution of predation mortality modify uncertainty.

Two aspects of predictability must be distinguished, for they affect the relative success of different life-history strategies quite differently. Areas may vary in reliability with regard to when or where certain events occur, such as adequate rainfall for successful breeding. Further, the suitability of such events may vary—a rain or a warm spell, whenever and wherever it may occur, may or may not be suitable for breeding. In a northern temperate environment the succession of the seasons is predictable. For a summer-breeding animal some summers will be better than others for breeding; this is reflected, for example, in Lack's (1947, 1948) results on clutch-size variation in English songbirds from year to year (see also Klomp, 1970; Hussell, 1972). Most summers, however, will be at least minimally suitable, and relatively few temperate—mesic-area organisms appear to have evolved to skip breeding in poor years. On the other hand, in most deserts rain is less predictable not only in regard to when and where it occurs, but also in regard to its effectiveness. Perhaps this latter aspect of environmental predictability has not been sufficiently emphasized in terms of its role in shaping life histories.

It is probably sufficient to distinguish four classes of environments with regard to predictability.

1. Predictable and relatively unchanging environments, such as caves and to a lesser extent tropical rain forests.
2. Predictably fluctuating or cyclic environments, areas with diurnal and seasonal periodicities, like temperate mesic areas.
3. Acyclic environments, unpredictable with reference to the timing and frequency of important events like rain, but predictable in terms of their effectiveness. If an event occurs, either it always is effective or the organism can judge the effectiveness.
4. Noncyclic environments that give few clues as to effectiveness of events: for example, rainfall erratic in spacing, timing, and amount. Areas like the central Australian desert present this situation for most frogs.

Optimal life-history strategies will differ in these environments, and desert amphibians must deal not only with extremes of temperature and aridity that seem contrary to their best interests but also with high degrees of unpredictability in those same environmental features and with localized and infrequent periods suitable for breeding.

Environmental uncertainty may have significant effects on shifts in life histories and on phenotypic similarities between life-history stages. If the duration of habitat suitable for adults is uncertain, or frequently less than one generation, the evolution of very different larval stages, not dependent on duration of the adult habitat, will be favored. The very fact that anurans show complex metamorphosis, with very different larval and adult stages, suggests this has been a factor in anuran evolution. Wilbur and Collins (1973) have discussed ecological aspects of amphibian metamorphosis and the role of uncertainty in the evolution of metamorphosis. An effect of complex metamorphosis is to increase independence of variation in the likelihoods of success in different life stages. Selective forces in the various habitats occupied by the different life stages are more likely to change independently of one another. As I will show later, this situation has profound effects on life-cycle patterns.

Predictable seasonality will favor individuals which breed seasonally during the most favorable period. Those who breed early in the good season will produce offspring with some advantage in size and feeding ability, and perhaps food availability, over the offspring of later breeders. Females which give birth or lay eggs early may, furthermore, increase their fitness and reduce their risk of feeding and improving their condition during the good season (Tinkle, 1969). Fisher (1958) has shown that theoretical equilibrium will be reached when the numbers of individuals breeding per day are normally distributed, if congenital earliness of breeding and nutritional level are also normally distributed. Predation (see below) on either eggs or breeding adults may cause amphibian breeding choruses to become clumped in space (Hamilton, 1971) and time. The timing, then, of the breeding peaks will depend on the balance between the time required after conditions become favorable for animals to attain breeding condition and the pressure to breed early. Both seasonal temperature and seasonal rainfall differences may limit breeding, and most amphibians in North American mesic areas and seasonally dry tropics (Inger and Greenberg, 1956; Schmidt and Inger, 1959) appear to breed seasonally.

In predictable unchanging environments, two strategies may be effective, depending on the presence or absence of predation. If no predation existed, individuals in "uncrowded" habitats would be selected to mature early and breed whenever they mature, maximizing egg numbers and minimizing parental investment per offspring, while individuals in habitats of high interspecific competition would be selected for the production of highly competitive offspring. That is, neither climatic change nor predation would influence selection, and MacArthur and Wilson's (1967) suggestion of r- and K- trends may hold. The result would be that adults would be found in breeding condition throughout the year. In a study by Inger and Greenberg (1963), reproductive data were taken monthly from male and female *Rana erythraea* in Sarawak. Rain and temperature were favorable for breeding throughout the year. From sperm and egg counts and assessment of secondary sex characters, they determined that varying proportions of both sexes were in breeding condition throughout the year. The proportion of breeding bore no obvious relation to climatic factors. Inger and Greenberg suggested that this situation represented the "characteristic behavior of most stock from which modern species of frogs arose." If predation exists in nonseasonal environments, year-round breeding with cryptic behavior may be successful; but if predation is erratic or predictably fluctuating (rather than constant), a "selfish herd" strategy may be favored.

Situations in which important events are unpredictable lead to other strategies. Life where the environment is unpredictable not only as to when or where events will occur but also as to whether or not they will be effective is comparable to playing roulette on a wheel weighted in an unknown fashion. Two strategies will be at a selective advantage:

1. Placing a large number of small bets will be favored, rather than placing a small number of large bets, or placing the entire bet on one spin of the wheel. In other words, in such an unpredictable situation, one expects iteroparous individuals who will lay a few eggs each time there is a rain. A corollary to this prediction is that, when juvenile mortality is unpredictable, longer adult life as well as iteroparity will be favored (cf. Murphy, 1968).

2a. Any strategy will be favored which will help an individual to judge the effectiveness of an event (i.e., to discover the weighting of the wheel). The central Australian species of *Cyclorana*—in fact, most of the Australian deep-burrowing frogs—may represent such a case. During dry periods, *Cyclorana platycephalus*, for instance, burrows three to four feet deep in clay soils. Light rains have no effect on dormant frogs even when rain occurs right in the area, since much of it runs off and does not percolate through to the level where the frogs are burrowed. Any rain reaching the frogs, we may suppose, is likely to be sufficient for tadpoles to mature and metamorphose. Thus, whatever functions (*sensu* Williams, 1966a) burrowing may serve in *Cyclorana*, one effect is that selective advantage accrues to those burrowing deeply because reproductive effort is not expended on unsuitable events.

2b. Any behavior which makes events less random, enhancing positive effects or reducing the effects of catastrophic events, will be favored. For example, parents may be favored who lay their eggs in some manner that tends to reduce the impact of flooding on their offspring, such as by laying their eggs out of the water and in rocky crevices or up on leaves or in burrows. A number of leptodactylid frogs do this (table 7-1). Obviously, such a strategy would only be favored when it had the effect of making mortality nonrandom. In deserts, where humidity is low and evapotranspiration high, it would not appear to be a particularly effective strategy; in fact only one *Eleutherodactylus* (Stebbins, 1951) and one species of *Pseudophryne* (Main, 1965) living in arid regions appear to follow strategies of hiding their eggs (table 7-1).

When the timing of events is unpredictable, but their effectiveness is not, individuals who only respond to suitable events will obviously be favored. This situation probably never exists a priori but only because organisms living in environments unpredictable both as to timing and effectiveness will evolve to respond only to suitable events, as in the burrowing *Cyclorana*. Thus, environments in the No. 4 category above will slowly be transformed into No. 3 environments by changes in the organisms inhabiting them. This emphasizes the importance of describing environments in terms of the organisms.

In the evolution of life cycles in uncertain environments, one kind of evidence of "learning the weighting of the wheel" is the capability of quickly exploiting unpredictable breeding periods—for example, ability to start a reproductive investment quickly after a desert rainfall. Another is the ability to terminate inexpensively an investment that has become futile, such as the care of offspring begun during a rainfall that turns out to be inadequate. These are adaptations over and above iteroparity as such, which is a simpler strategy.

Uncertainty and Parental Care. The effect of uncertainty on degree and distribution of parental investment varies with the type of unpredictability. Some kinds of uncertainty, such as prey availability, apparently can be ameliorated by increased parental investment. Types of uncertainty arising from biotic factors, rather than physical factors, comprise most of this category. Thus, vertebrate predators as a rule should show lengthened juvenile life and high degree of parental care because the biggest and best-taught offspring are at an advantage.

Uncertainties which are catastrophic or otherwise not density dependent appear to favor minimization or delay of parental investment such that the cost of loss at any point before the termination of parental care is minimized. The limited distribution of parental investment in desert amphibians supports this suggestion (fig. 7-3), and it appears to be true not only for anurans, in which parental care varies but is

Table 7-1. *Habitat, Clutch, and Egg Sizes of Various Anurans*

Species	Habitat[a]	Adult Size (mm)	Site of Deposition[c]	Number of Eggs[d]
Ascaphidae				
Ascaphus truei	1D	30–40	2b	28–50/
Leiopelma hochstetteri	8D		4[f]	6–18/
Pelobatidae				
Scaphiopus bombifrons	1B	35	2a	10–250/ 10–50
S. couchi	1B	80	2a	350–500/ 6–24
S. hammondi	1B	38	2a	300–500/24
Bufonidae				
Bufo alvarius	1B	180	1,2a	7,500– 8,000/
B. boreas	1F	95	2a	16,500/
B. cognatus	1G	85	1,2a,2b	20,000/
B. punctatus	1G	55	2a	30–5,000/ 1–few

[a]Habitat: 1 = North America A = Temporary ponds
2 = Central America B = Permanent water, xeric areas
3 = South America C = Permanent water, mesic areas
4 = Europe D = Permanent streams
5 = Asia E = Caves
6 = Africa F = Mesic   F+ = Cloud or tropical rain forest
7 = Australia G = Grasslands, savannahs, or subhumid corridor
8 = New Zealand
[b]Size of adult female.

Egg Size (mm)	Time to Hatch (hours)	Time to Metamorphose (days)	Time to Mature (years)	Reference
4.0–5.0	720	365+		Noble and Putnam, 1931; Slater, 1934; Stebbins, 1951
	30 days[e]			
	<48	36–40		Stebbins, 1951
1.4–1.6	9–72	18–28		Ortenburger and Ortenburger, 1926; Stebbins, 1951; Gates, 1957
1.0–1.62	38–120	51		Little and Keller, 1937; Stebbins, 1951; Sloan, 1964
1.4		30		Stebbins, 1951; Mayhew, 1968
1.5–1.7	48			Stebbins, 1951
1.2	53	30–45		Stebbins, 1951
1.0–1.3	72	40–60		Stebbins, 1951

[b]Deposition site:	1	= Temporary ponds	3b	= Burrows, not requiring rain to hatch	
	2a	= Permanent ponds	4a	= Terrestrial (seeps, etc.)	
	2b	= Permanent streams	4b	= On leaves above water	
	3a	= Burrows, requiring rain to hatch	4c	= On submerged leaves	
			5	= With parent: brood pouch, on back, etc.	

[d]When eggs are laid in several clusters, figures represent total number laid/number per cluster.
[e]Larval development completed in egg.
Tending behavior.
[f](W): winter (S): summer.

[h]Tadpoles burrow to water.
[i]Female digs tunnel to water.

Species	Habitat[a]	Adult Size (mm)	Site of Deposition[c]	Number of Eggs[d]
B. woodhousei	1G	130	1	25,600/
B. compactilis	1G	70	1	
B. microscaphus	1B	65	1	several thousand
B. regularis	6G	65[b]	1	23,000
B. rangeri	6G	105[b]	1,2a	
B. carens	6G	74–92[b]		10,000
B. angusticeps	6	65		650–850
B. gariepensis	6	55		100+
B. vertebralis	6	30		
Ansonia muellari	5D	31[b]		150
Phrynomeridae				
Phrynomerus bifasciatus bifasciatus	6G	65	1,2a	400–1,500
Microhylidae				
Gastrophryne carolinensis	1,2	20	2a	850
G. mazatlanensis	3	20	4	175–200
Breviceps adspersus adspersus	6	38	4a,3b	20–46

[a]Habitat: 1 = North America   A = Temporary ponds
2 = Central America   B = Permanent water, xeric areas
3 = South America   C = Permanent water, mesic areas
4 = Europe   D = Permanent streams
5 = Asia   E = Caves
6 = Africa   F = Mesic   F+ = Cloud or tropical rain forest
7 = Australia   G = Grasslands, savannahs, or subhumid corridor
8 = New Zealand
[b]Size of adult female.

Egg Size (mm)	Time to Hatch (hours)	Time to Metamorphose (days)	Time to Mature (years)	Reference
1.0–1.5	48–96	34–60		Mayhew, 1968; Blair, 1972
1.4	48			Stebbins, 1951
1.75–1.9				Stebbins, 1951
1.0	24–48	72–143		Power, 1927; Wager, 1965; Stewart, 1967
1.3	96	35–42		Stewart, 1967
1.6	72–96			Stewart, 1967
2.0				Wager, 1965
2.2	48			Wager, 1965
<1.0				Wager, 1965
2.15				Inger, 1954
.3–1.5	96	30		Stewart, 1967
	48	20–70	2	Stebbins, 1951
.2–1.4				Stebbins, 1951
1.5		28–42 days[e]		Wager, 1965

[c]Deposition site:

1	= Temporary ponds	3b	= Burrows, not requiring rain to hatch
2a	= Permanent ponds	4a	= Terrestrial (seeps, etc.)
2b	= Permanent streams	4b	= On leaves above water
3a	= Burrows, requiring rain to hatch	4c	= On submerged leaves
		5	= With parent: brood pouch, on back, etc.

[d]When eggs are laid in several clusters, figures represent total number laid/number per cluster.
[e]Larval development completed in egg.
[f]Tending behavior.
[g](W): winter (S): summer.
[h]Tadpoles burrow to water.
[i]Female digs tunnel to water.

Species	Habitat[a]	Adult Size (mm)	Site of Deposition[c]	Number of Eggs[d]
B. a. pentheri	6	38	4a,3b	20
Hypopachus variolosus	2G	29–53[b]	1	30–50

Ranidae
Pyxicephalus adspersus	6G	115	1	3,000–4,000
P. delandii	6G	65	1	2,000–3,000
P. natalensis	6G	51	1	hundreds / 1–6
Ptychadena anchietae	6G	48–58[b]	1	200–300
P. oxyrhynchus	6G	57[b]	1	300–400
P. porosissima	6G	44[b]	1	?/1
Hildebrandtia ornata	6G	63.5	2	?/1
Rana fasciata fuellborni	6C	44.5[b]	4a	64/1–12
R. f. fasciata	6G	51	1,2	?/1
R. angolensis	6	76		thousands
R. fuscigula	6	127	2	1,000–15,000
R. wageri	6	51[b]	4a	100 1,000/ 12–100

[a]Habitat: 1 = North America   A = Temporary ponds
             2 = Central America B = Permanent water, xeric areas
             3 = South America   C = Permanent water, mesic areas
             4 = Europe   D = Permanent streams
             5 = Asia   E = Caves
             6 = Africa   F = Mesic   F+ = Cloud or tropical rain forest
             7 = Australia   G = Grasslands, savannahs, or subhumid corridor
             8 = New Zealand
[b]Size of adult female.

Egg Size (mm)	Time to Hatch (hours)	Time to Metamorphose (days)	Time to Mature (years)	Reference
5.0	28–42 days[e]			Wager, 1965
	24			Wager, 1965
2.0	48	49		Stewart, 1967
1.5	72	35		Wager, 1965
1.2	96			Wager, 1965; Stewart, 1967
1.0	30			Wager, 1965; Stewart, 1967
1.3	48	42–56		Wager, 1965
1.0	48			Wager, 1965
1.4				Wager, 1965
2.0–3.0		730		Stewart, 1967
1.65		28–35		Wager, 1965
1.5	168			Wager, 1965
1.5	168–240	1,095		Wager, 1965
2.8	192–216			Wager, 1965

[c]Deposition site:

1 = Temporary ponds	3b = Burrows, not requiring rain to hatch
2a = Permanent ponds	4a = Terrestrial (seeps, etc.)
2b = Permanent streams	4b = On leaves above water
3a = Burrows, requiring rain to hatch	4c = On submerged leaves
	5 = With parent: brood pouch, on back, etc.

[d]When eggs are laid in several clusters, figures represent total number laid/number per cluster.
[e]Larval development completed in egg.
[f]Tending behavior.
[g](W): winter (S): summer.
[h]Tadpoles burrow to water.
[i]Female digs tunnel to water.

Species	Habitat[a]	Adult Size (mm)	Site of Deposition[c]	Number of Eggs[d]
R. grayi	6B	45	3a	few hundred/1–few
R. catesbiana	1	205	2	10,000–25,000
R. pipiens	1	90	2	1,200–6,500
R. temporaria	1	90	1	1,500–4,000
R. tarahumarae	1,2	115	1,2	2,200
R. aurora aurora	1C	102	2a	750–1,300
R. a. cascadae	1C	95	2a	425
R. a. dratoni	1C	95	2a	2,000–4,000
R. boylei	1B,C	70	2a, b	900–1,000
R. clamitans	1B,C	102	2a, b	1,000–5,000
R. pretiosa pretiosa	1F	90	2	1,100–1,500
R. p. lutiventris	1F	90	2	2,400
R. sylvatica	1F	60	2a,1	2,000–3,000
Phrynobatrachus natalensis	6G	28–30[b]	1	200–400/ 25–50
P. ukingensis	6G	16[b]	1	
Anhydrophryne rattrayi	6	20[b]	3b	11–19
Natalobatrachus bonegergi	6F	38	4b	75–100
Arthroleptis stenodactylus	6	29–44[b]	3b	100/33

[a]Habitat: 1 = North America   A = Temporary ponds
2 = Central America   B = Permanent water, xeric areas
3 = South America   C = Permanent water, mesic areas
4 = Europe   D = Permanent streams
5 = Asia   E = Caves
6 = Africa   F = Mesic   F+ = Cloud or tropical rain forest
7 = Australia   G = Grasslands, savannahs, or subhumid corridor
8 = New Zealand
[b]Size of adult female.

Egg Size (mm)	Time to Hatch (hours)	Time to Metamorphose (days)	Time to Mature (years)	Reference
1.5	5–10	90–120		Wager, 1965
1.3	4–5	120–365	2–3	Stebbins, 1951
1.7	312–480	60–90	1–3	Stebbins, 1951
	336–504	90–180	3–5	Stebbins, 1951
2–2.2				Stebbins, 1951
3.04	192–480		3–4	Stebbins, 1951
2.25	192–480			Stebbins, 1951
2.1	192–480			Stebbins, 1951
2.2		90–120		Stebbins, 1951
1.5	72–144	90–360		Stebbins, 1951
2–2.8	96		2+	Stebbins, 1951
1.97				Stebbins, 1951
1.7–1.9	336–504	90		Stebbins, 1951
1.0(W)[g] 0.7(S)	48	28		Wager, 1965; Stewart, 1967
0.9		35		Stewart, 1967
2.6	28 days[e]			Wager, 1965
2.0	144–240	270		Wager, 1965; Stewart, 1967
2.0	e			Stewart, 1967

[c]Deposition site:

1 = Temporary ponds	3b = Burrows, not requiring rain to hatch
2a = Permanent ponds	4a = Terrestrial (seeps, etc.)
2b = Permanent streams	4b = On leaves above water
3a = Burrows, requiring rain to hatch	4c = On submerged leaves
	5 = With parent: brood pouch, on back, etc.

[d]When eggs are laid in several clusters, figures represent total number laid/number per cluster.
[e]Larval development completed in egg.
[f]Tending behavior.   [h]Tadpoles burrow to water.
[g](W): winter (S): summer.   [i]Female digs tunnel to water.

Species	Habitat[a]	Adult Size (mm)	Site of Deposition[c]	Number of Eggs[d]
A. wageri	6	25	3b	11–30
Arthroleptella lightfooti	6F	20	3b	40/5–8
A. wahlbergi	6F	28	3b	11–30
Cacosternum n. nanum	6G	20	4c	8–25/5–8
Chiromantis xerampelina	6F	60–87b	4b	150
Hylambates maculatus	6B	54–70b	4c	few hundred/1
Kassina wealii	6	40	1,4c	500/1
K. senegalensis	6	35–43b	1	400/1–few
Hemisus marmoratum	6F	38b	3fh	200
H. guttatum	6F	64b	3fi	2,000
Leptopelis natalensis	6F	64	4a	200
Afrixalus spinifrons	6F	22	4c	?/10–50
A. fornasinii	6F	30–40b	4b	40
Hyperolius punticulatus	6	32–43b	1	?/19
H. pusillus	6	22.0–30b	4b	?/60–90
H. tuberilinguis	6	36–39b	4b	350–400

[a]Habitat:
  1 = North America   A = Temporary ponds
  2 = Central America B = Permanent water, xeric areas
  3 = South America   C = Permanent water, mesic areas
  4 = Europe   D = Permanent streams
  5 = Asia   E = Caves
  6 = Africa   F = Mesic   F+ = Cloud or tropical rain forest
  7 = Australia   G = Grasslands, savannahs, or subhumid corridor
  8 = New Zealand
[b]Size of adult female.

Egg Size (mm)	Time to Hatch (hours)	Time to Metamorphose (days)	Time to Mature (years)	Reference
2.5	28 days[e]			Wager, 1965
4.5	10 days[e]			Stewart, 1967
2.5	e			Wager, 1965
0.9	48	5		Wager, 1965
1.8	120–144			Wager, 1965
1.5	144	300		Wager, 1965; Stewart, 1967
2.4	144	60		Wager, 1965; Stewart, 1967
1.5	144	90		Stewart, 1967
2.0	240			Wager, 1965
2.5				Wager, 1965
3.0				Wager, 1965
1.2	168	42		Wager, 1965
1.6–2.0				Wager, 1965; Stewart, 1967
2.5				Stewart, 1967
2.0	432	56		Stewart, 1967
1.3–1.5	96–120	60		Stewart, 1967

[c]Deposition site:

1 = Temporary ponds	3b = Burrows, not requiring rain to hatch
2a = Permanent ponds	4a = Terrestrial (seeps, etc.)
2b = Permanent streams	4b = On leaves above water
3a = Burrows, requiring rain to hatch	4c = On submerged leaves
	5 = With parent: brood pouch, on back, etc.

[d]When eggs are laid in several clusters, figures represent total number laid/number per cluster.
[e]Larval development completed in egg.
[f]Tending behavior.
[g](W): winter (S): summer.
[h]Tadpoles burrow to water.
[i]Female digs tunnel to water.

Species	Habitat[a]	Adult Size (mm)	Site of Deposition[c]	Number of Eggs[d]
H. pusillus	6	17–21[b]	1,2a	500/1–76
H. nasutus nasutus	6	20.6–23.8[b]	2	200/2–20
H. marmoratus nyassae	6	29–31[b]	2a	370
H. horstocki	6		2a	?/10–30
H. semidiscus	6	35	2a, 4c	200/30
H. verrucosus	6	29	2a	400/4–20
Leptodactylidae				
Eleutherodactylus rugosus	2G		1	several thousand
Limnodynastes tasmaniensis	7F	39.4[b]		1,100
L. dorsalis dumerili	7F	61.5[b]		3,900
Leichriodus fletcheri	7	46.5[b]		300
Adelotus brevus	7	33.5[b]		270
Philoria frosti	7	49.2[b]		95
Helioporus albopunctatus	7	73.3[b]		480
H. eyrei	7F	54.0[b]		265–270
H. psammophilis	7	42–52		160

[a]Habitat: 1 = North America   A = Temporary ponds
2 = Central America   B = Permanent water, xeric areas
3 = South America   C = Permanent water, mesic areas
4 = Europe   D = Permanent streams
5 = Asia   E = Caves
6 = Africa   F = Mesic   F+ = Cloud or tropical rain forest
7 = Australia   G = Grasslands, savannahs, or subhumid corridor
8 = New Zealand
[b]Size of adult female.

Egg Size (mm)	Time to Hatch (hours)	Time to Metamorphose (days)	Time to Mature (years)	Reference
1.4–1.5	120	42		Stewart, 1967
0.8–2.2	120			Wager, 1965; Stewart, 1967
2.0	192			Wager, 1965; Stewart, 1967
1.0				Wager, 1965
1.0	108	60		Wager, 1965
1.3				Wager, 1965
4.0	24			Wager, 1965
1.47				Martin, 1967
1.7				Martin, 1967
1.7				Martin, 1967
1.5				Martin, 1967
3.9				Martin, 1967
2.75				Main, 1965; Lee, 1967
2.50–3.28				Main, 1965; Lee, 1967; Martin, 1967
3.75				Lee, 1967

cDeposition site:
1 = Temporary ponds
2a = Permanent ponds
2b = Permanent streams
3a = Burrows, requiring rain to hatch
3b = Burrows, not requiring rain to hatch
4a = Terrestrial (seeps, etc.)
4b = On leaves above water
4c = On submerged leaves
5 = With parent: brood pouch, on back, etc.

dWhen eggs are laid in several clusters, figures represent total number laid/number per cluster.
eLarval development completed in egg.
fTending behavior.
g(W): winter (S): summer.
hTadpoles burrow to water.
iFemale digs tunnel to water.

Species	Habitat[a]	Adult Size (mm)	Site of Deposition[c]	Number of Eggs[d]
H. barycragus	7	68–80		430
H. inornatus	7	55–65		180
Crinea rosea	7F	24.8[b]		26–32
C. leai	7F	21.1[b]		52–96
C. georgiana	7F	21.1[b]		70
C. insignifera	7F	19–21[b]		

Hylidae

Hyla arenicolor	1	37	1	several hundred/1
H. regilla	1	55	1	500–1,250/ 20–25
H. versicolor	1		2	1,000–2,000
H. verrucigera	2		1	200
H. lancasteri	2F+	41.1[b]	2b,4b	20–23
H. myotympanum	2F+	51.6[b]		120
H. thorectes	2F+	70[b]	2	10
H. ebracata	2F+	36.5[b]	4b	24–76
H. rufelita	2F	60[b]	2	75–80
H. loquax	2F	45[b]	2	250
H. crepitans	2G	52.6[b]	1	
H. pseudopuma	2F	41.3b	4b	???10
H. tica	2F+	38.9		

[a]Habitat: 1 = North America   A = Temporary ponds
2 = Central America   B = Permanent water, xeric areas
3 = South America   C = Permanent water, mesic areas
4 = Europe   D = Permanent streams
5 = Asia   E = Caves
6 = Africa   F = Mesic   F+ = Cloud or tropical rain forest
7 = Australia   G = Grasslands, savannahs, or subhumid corridor
8 = New Zealand
[b]Size of adult female.

Egg Size (mm)	Time to Hatch (hours)	Time to Metamorphose (days)	Time to Mature (years)	Reference
2.60				Lee, 1967
3.75				Lee, 1967
2.35		60+ days[e]		Main, 1957
1.66–2.03		149–174 days[e]	2	Main, 1957
0.97–1.3		130+ days[e]	1	Main, 1957
				Main, 1957
2.1		40–70		Stebbins, 1951
1.3	168–336		2	Stebbins, 1951
	96–120	45–65	1–3	Stebbins, 1951
2.0		89		Trueb and Duellman, 1970
5.0				Duellman, 1970
2.25				Duellman, 1970
1.22				Duellman, 1970
1.2–1.4				Duellman, 1970; Villa, 1972
1.8				Villa, 1972
				Villa, 1972
1.8				Villa, 1972
1.71	24	65–69		Villa, 1972
2.0				Villa, 1972

[c]Deposition site: 1 = Temporary ponds   3b = Burrows, not requiring rain to hatch
                    2a = Permanent ponds   4a = Terrestrial (seeps, etc.)
                    2b = Permanent streams   4b = On leaves above water
                    3a = Burrows, requiring   4c = On submerged leaves
                         rain to hatch   5   = With parent: brood pouch, on back, etc.

[d]When eggs are laid in several clusters, figures represent total number laid/number per cluster.
[e]Larval development completed in egg.
[f]Tending behavior.   [h]Tadpoles burrow to water.
[g](W): winter (S): summer.   [i]Female digs tunnel to water.

Species	Habitat[a]	Adult Size (mm)	Site of Deposition[c]	Number of Eggs[d]
Agalychnis colli-dryas	2	71	4b	40–110/ 11–78
A. annae	2	82.9b	4b	47–162
A. calcarifer	2	65.0b	4b	16
Smilisca cyanosticta	2F+	70b	2	1,147
S. baudinii	2G	76–90	1	2,620–3,32●
S. phaeola	2G	80		1,870–2,01●
Pachymedusa dacnicolor	2G	103.6b	4b	100–350
Hemiphractus panimensis	2F	58.7b	5f	12–14
Gastrotheca ceratophryne	2F	74.2	5f	9

Centrolenellidae				
Centrolenella fleischmanni	2F	19.2	4b	17–28

[a]Habitat: 1 = North America   A = Temporary ponds
            2 = Central America   B = Permanent water, xeric areas
            3 = South America   C = Permanent water, mesic areas
            4 = Europe   D = Permanent streams
            5 = Asia   E = Caves
            6 = Africa   F = Mesic   F+ = Cloud or tropical rain forest
            7 = Australia   G = Grasslands, savannahs, or subhumid corridor
            8 = New Zealand
[b]Size of adult female.

generally low, but also for groups with high parental care, such as mammals. For example, marsupials have flourished in uncertain desert environments in central Australia where indigenous and introduced eutherians have not, even though the eutherian species prevail in areas of more predictable climate. In uncertain areas a premium

Egg Size (mm)	Time to Hatch (hours)	Time to Metamorphose (days)	Time to Mature (years)	Reference
2.3–5.0	96–240	50–80		Duellman, 1970; Villa, 1972
3.41				Villa, 1972
3.5				Villa, 1972
.22				Duellman, 1970
.3				Trueb and Duellman, 1970
				Duellman, 1970
				Duellman, 1970
5.0				Duellman, 1970
2.0				Duellman, 1970
.5	24	9		Villa, 1972

cDeposition site: 1   = Temporary ponds   3b   = Burrows, not requiring rain to hatch
                    2a = Permanent ponds   4a   = Terrestrial (seeps, etc.)
                    2b = Permanent streams   4b   = On leaves above water
                    3a = Burrows, requiring   4c   = On submerged leaves
                          rain to hatch   5   = With parent: brood pouch, on back, etc.
dWhen eggs are laid in several clusters, figures represent total number laid/number per cluster.
eLarval development completed in egg.
fTending behavior.   hTadpoles burrow to water.
g(W): winter (S): summer.   iFemale digs tunnel to water.

is set on strategies which will make breeding response facultative and reduce the cost of loss of offspring at any point. Facultative, rather than seasonal, delayed implantation (Sharman, Calaby, and Poole, 1966) and anoestrus condition during drought (Newsome, 1964, 1965, 1966) are examples. Also, I think, is the shape of the parental investment curve for marsupials, which is depressed to a remarkable degree in the initial stages (my unpublished data). This whole constellation of attributes provides facultativeness of response, capabilities for quick initiation of new investments, and less expense of termination at any point. While the classical arguments about marsupial proliferation in Australia have claimed that introduced eutherians "outcompete" marsupials (Frith and Calaby, 1969), they are probably able to do so only because they evolved their reproductive behavior in other kinds of environments. Most Australian environments may have consistently favored marsupialism over any step-by-step transitions toward placentalism. It may be worthwhile to reexamine the question in the light of a new framework.

Distribution

A third important environmental aspect is patchiness or graininess. Wet tropical areas, seemingly ideal from an amphibian's point of view, are basically rather fine grained environments. For instance, ponds, fields, and forest areas may interdigitate so that a single frog spends some time in each and may spend time in more than one pond. From an amphibian's point of view, most deserts are comparatively coarse grained. This does not mean that all the environmental patches are physically large (as may be implied in Levins's [1968] discussion) but that the suitable patches, of whatever size, are likely to be separated by large unsuitable or uninhabitable areas. Thus an individual is likely to spend its entire life in the same patch. For amphibians, widely separated permanent water holes in desert environments are islands and subject to the same selective pressures (MacArthur and Wilson, 1967).

Degrees of patchiness will have two major sorts of effects, on divergence rates and life-history strategies. In a coarse-grained or island model, as in the desert I have described, rates of speciation and extinction will both be higher than in a fine-grained environment. Thus, in some uncertain environments, if they are continually minimally inhabitable and also coarse grained, speciation and extinction rates, contrary to Slobodkin and Sanders's (1969) prediction, may be higher than in predictable environments, if those predictable areas are fine grained. This point, not considered by Slobodkin and Sanders, was raised by Lewontin (1969). Environmental uncertainty will affect populations in the coarse-grained situation much more than those in the fine-grained areas to the extent that there are differences in population sizes and isolation of populations. Slobodkin and Sanders considered only predictability, but predictability and patchiness, and their interaction, will influence the rate of speciation.

In very coarse grained models, because isolation is much more complete than in the fine-grained situation, immigration and emigration may be virtually nonexistent. The number of species in any suitable grain at any time will depend on infrequent past immigrations and will be lower than in the fine-grained model. Selection will be strong on several parameters, to be discussed below, but may be relaxed on characters, such as premating isolating mechanisms. Selection on these characters will be strongest in the fine-grained model where the number of sympatric species is higher. The desert coarse-grained situation is a model for the occurrence of character release (MacArthur and Wilson, 1967; Grant, 1972): populations founded by few individuals and on which selection on interspecific discrimination is relaxed. Thus, in the isolated desert populations described, one might predict that the variations in call characters (in males) and in call discrimination (in females) would be greater.

The distribution of suitable resources and the duration of this distribution will affect strategies of dispersal and competition. While density-dependent effects will operate here, the "r" and "K" parameters of Pianka (1970) and others are not sufficient indicators—a point made by Wilbur et al. (1974) for other groups of organisms.

Consider a pond suitable for breeding: it may be effectively isolated from other suitable areas, or other good ponds may be close or easy to reach. Dispersal ability will evolve to the degree that the cost-benefit ratio is favorable between the relative goodness of another pond and the risk incurred in getting there. Goodness relative to the home pond may be measured by a number of criteria: physical parameters, amount of competition from other species, and other conspecifics (Wilbur et al., 1974), amount of predation, and so on. The cost of reaching another pond and the probability of success in doing so may be correlated with distance, but other classic "barriers" (mountains, very dry areas) are also relevant. Both distance and barriers of low humidity and little free water are likely to be greater in arid regions than in tropical and temperate mesic areas.

If ponds are not totally isolated from each other and are relatively unchanging in "value," migration strategies will be more favored in finer-grained areas because the cost of migration is lower. If ponds are not isolated from each other, and their relative values fluctuate, the evolution of emigration strategies will depend in part on the persistence of ponds relative to the generation length of the frog. If ponds are temporary, and others are likely to be available, migration will be advantageous. The longer ponds last, the closer the situation approaches the "permanent pond" situation, where migration will be favored only in periods of high local population density. Some invertebrate groups, such as migratory locusts and crickets (Alexander, 1968), show phenotypic flexibility supporting this generalization; they increase the proportion of long-winged migratory offspring as the habitat deteriorates and in periods of high population density. Frog morphology does not alter in a comparable way, but dispersal behavior may show flexibility. I know of no pertinent data or studies, however.

In good patches like permanent waters, isolated from others, emigration will be disfavored. Increased parental investment will be favored only when it increases predictability in ways relevant to offspring success. Examination of table 7-1 shows that species with parental care and species laying large-yolked eggs occur in tropical and temperate areas but not generally in unpredictable areas. Since some of these species lay foamy masses not permeable to water, the aridity of desert areas alone is not sufficient to explain this distribution of strategies.

Two arid-region species do show parental investment in the form of larger or protected eggs. As previously described, *Eleutherodactylus latrans* females lay about fifty large eggs of 6–7.5 mm diameter on land or in caves (table 7-1; Stebbins, 1951); the males may guard the eggs. Since this frog lives largely in caves and rocky crevices, the microenvironment is far more stable and predictable than the zoogeography would suggest. The Australian *Pseudophryne occidentalis* lives by permanent waters with muddy rather than sandy soils. Eggs are laid in mud burrows near the edge of the water (Main, 1965). In both cases it appears that the nature of mortality is such that increased

parental investment is successful. This may be related to the relatively higher physical stability of the microhabitat when compared to
desert environments in general. The proportion of mortality due to
catastrophes which parental care is ineffective to combat is relatively
lower.

Mortality

Mortality may arise from a number of factors: foot shortages, predators (including parasites and diseases), and climate. An important
consideration in what life-history strategy will prevail is whether the
mortality is random (unpredictable) or nonrandom (predictable). Any
cause of mortality could be either random or nonrandom in its effects,
but mortality from biotic causes is probably less often random than
mortality from physical factors and may be more effectively countered
by strategies of parental investment.

Catastrophic mortality, which is essentially random rather than selective (even though it may be density dependent), will be more frequent in the coarse-grained desert environments I have described
than in the tropics. An example would be heavy sudden floods which
frequently occur after heavy rains in areas like central Australia and
the southwestern United States. This kind of flood may wash eggs,
tadpoles, and adults to flood-out areas which then dry up. The result
may be devastating sporadic mortality for populations living in the path
of such floods. Further, in terms of the animals themselves, environments may be predictable for certain stages in the life history and
unpredictable for others. In animals like amphibians with complex
metamorphosis, this difference can be particularly significant.

If any stage encounters significant uncertainty, one of two strategies should evolve: physical avoidance, such as hiding or development of protection in that stage, or a shift in life history to spend
minimal time in the vulnerable stage (table 7-2). If survivorship is high
for adults but uncertain and sometimes very low for tadpoles, one
predicts strategies of: (a) long adult life, iteroparity, and reduced investment per clutch; (b) long egg periods and short tadpole periods;
or (c) increased parental investment through hiding or tending behavior. Evolution of behavior like that of *Rinoderma darwini* may re-

sult from such pressure. The males appear to guard the eggs; when development reaches early tadpole stage, the males snap up the larvae, carrying them in the vocal sac until metamorphosis. Perhaps the extreme case is represented by the African *Nectophrynoides*, in which birth is viviparous.

In temporary waters in desert environments much uncertainty will be concentrated on aquatic stages, and two principal strategies should be evident in desert amphibians: increased iteroparity, longer adult life, and lower reproductive effort per clutch; and shifts in time spent in different stages, reducing time spent in the vulnerable stages. Short, variable lengths in egg and juvenile stages (table 7-1) will result.

Even in climatically more predictable areas, uncertainty of mortality may be concentrated on one stage. In some temperate urodele forms, Salthe (1969) suggested that success at metamorphosis correlated with size—that larger offspring were more successful. This in turn selected for lengthened time spent in aquatic stages.

Some generalizations are apparent from table 7-2. The important differences appear to be between uncertainty in juvenile stages and adult stages. All conditions of uncertain adult survival will lead to concentration of reproductive effort in one or a few clutches (semelparity or reduction of iteroparity). Uncertainty of survivorship in adult stages when combined with high predictability in juvenile stages may lead to the extreme conditions of neoteny and paedogenesis. Uncertainty in either or both juvenile stages leads to increased iteroparity and reduced reproductive effort per clutch.

Predation

Because predation is usually nonrandom, its effects on prey life histories will frequently differ from the effects of climate and other sources of mortality. An important point frequently overlooked is that, because predation and competition arise from biotic components of the system, they are not simply subsets of uncertainty. Their effects are more thoroughly related to density-dependent parameters. Some strategies will be effective which would not be advantageous in situations rendered uncertain solely by physical factors. Consider predation: strategies frequently effective in reducing predation-caused un-

Table 7-2. *Relative Uncertainty in Different Life-History Stages and Strategies of Selective Advantage*

Likelihood of Survival			Strategy
Egg	Tadpole	Adult	
high	high	low	semelparity or reduced iteroparity; large numbers of small eggs; no parental care
low	high	low	semelparity or reduced iteroparity; neoteny; quick hatching
high	low	low	semelparity or reduced iteroparity; large numbers of small eggs; no parental care; quick metamorphosis
high	low	high	iteroparity; large eggs, fewer eggs; avoidance of aquatic tadpole stage; parental care of tadpoles
low	high	high	iteroparity; tending, hiding of eggs; fewer eggs; viviparity
low	low	high	iteroparity; parental care, tending strategies; viviparity

certainty are those of spatial (Hamilton, 1971) and temporal clumping, increased parental investment (Trivers, 1972), and allelochemical effects. These strategies would be far less effective in increasing predictability of an environment rendered uncertain by physical factors.

Predation pressure may lead to hiding or tending eggs and consequent lowering of clutch size. Whether this is true or whether responses of increased fecundity (Porter, 1972; Szarski, 1972) prevail will depend on the nature of the predation. In the unusual case of a predator whose effect is limited, such as one which could eat no more than x eggs per nest, parents would gain by increased fecundity, making $(x+2)$ rather than $(x+1)$ eggs. However, m, the genotypic rate of increase, will be higher for these more fecund genotypes even in the absence of predation. Further, an increase in numbers of eggs laid implies either smaller eggs (in which case the predator may be able to eat $[x+2]$ eggs) or an increase in the size of the parent. In most cases, high fecundity carries a greater risk under increased predation—for example, by laying more eggs which are then lost or, in species like altricial birds with parental care, by incurring greater risk attempting to feed more offspring if they are not protected. In these cases, lowered fecundity and increased parental investment in caring for fewer eggs will be favored.

The strategies of hiding or protection and life-history shifts, which may follow from increased uncertainty in any stage, are also favored in the special case of uncertainty induced by predation. Predation concentrated on certain stages in the life cycle—on eggs, tadpoles, newly metamorphosed animals, or breeding adults—may lead to (a) quick hatching, tending, or hiding of eggs, as in *Scaphiopus* or *Helioporus* (table 7-1); (b) quick metamorphosis or tending of tadpoles, as in *Rhinoderma*; (c) cryptic behavior by newly metamorphosed animals (many species) or lengthened egg or tadpole stages with consequent greater size (and possibly reduced predation vulnerability) on metamorphosis, as in *Rana catesbiana* (table 7-1); or (d) cryptic behavior by adults or very clumped patterns of breeding behavior.

Length of the breeding season may also be strongly affected by the presence of predation. In fact, I think that the general shape of breeding-curve activities of many vertebrates may be related to predation. Fisher (1958) has shown that, if there is an optimal brooding time, a symmetrical curve will result. While restriction of resource availability, such as food or breeding resources, limits the seasonality of breeding and produces some clumping, such seasonal differences seem not to be sharp enough to explain the extreme temporal clumping of breeding and birth in many species. Temporary ponds of very short duration in arid regions are commonly assumed to show clumping for climatic reasons, but this is not certain; at any rate, the addition of predation to such a system should follow the same pattern as in any seasonal situation. In seasonal conditions a breeding-activity or birth curve may approach a normal curve, perhaps with a slight right-hand skew because earlier birth will give a size and food advantage to offspring and a risk advantage to parents. When predation on breeding adults or new young exists, however, two other pressures may cause both an increased right-hand skew and a sharper peak:

1. The advantage to those individuals which have offspring early before a generalized predator develops a specific search image.
2. The advantage to those individuals which breed and give birth or lay eggs when everyone else does—when, in other words, the predator food market is flooded. This constitutes a temporal "selfish herd" effect (Hamilton, 1971). Thus, if seasonality of resource availability exists so that thoroughly cryptic breeding is not of advantage, the curve of breeding or birth activity will tend under predation pressure to shift from a fairly normal distribution to a kurtotic curve with an abrupt beginning shoulder and a gentler trailing edge.

Despite their importance, predation effects on life histories have largely been ignored. This may be, in part, because the physical factors are so extreme that it seems sufficient to examine their effects on amphibian physiology and survival. Another reason predation effects may be slighted is that one ordinarily sees the end product of organisms which evolved with predation pressure, and the present-day descendents represent the most successful of the antipredation strategies. As a simple example, consider the large variety of substances found in the skin of most amphibians (Michl and Kaiser, 1963). A great variety exists, including such disparate compounds as urea, the bufadienolides, indoles, histamine derivatives, and polypeptides like caerulein (Michl and Kaiser, 1963; Erspamer, Vitali, and Roseghini, 1964; Anastasi, Erspamer, and Endean, 1968; Cei, Erspamer, and Roseghini, 1972; Low, 1972). The production of some of these compounds is energetically expensive; others are costly in terms of water economy (Cragg, Balinsky, and Baldwin, 1961; Balinsky, Cragg, and Baldwin, 1961). Why, then, do so many amphibians produce a wide variety of such costly compounds? Despite wide chemical variety most of these compounds share one striking attribute: they are either distasteful or have unpleasant physiological effects. Most irritate the mucous membranes. Bufadienolides and other cardiac glycosides have digitalislike effects on such predators as snakes as well as on mammals (Licht and Low, 1968). Caerulein differs in only two amino acids from gastrin and has similar effects (Anastasi et al., 1968), including the induction of vomiting.

Although I know of no good study of predation mortality in any desert amphibian, and demography data on amphibians are generally sparse (Turner, 1962), predation has been reported in every life-history stage (Surface, 1913; Barbour, 1934; Brockelman, 1969; Littlejohn, 1971; Szarski, 1972). It is obvious that there is selective advantage to tasting vile or being poisonous, and scattered studies show that successful predators on amphibians show adaptations of increased tolerance (Licht and Low, 1968) or avoidance of the poisonous parts (Miller, 1909; Wright, 1966; Schaaf and Garton, 1970).

Predation concentrated on adults will lead to the success of individuals which show cryptic behavior and color patterns as well as those which concentrate unpleasant compounds in their skins. Particularly poisonous or distasteful individuals with bright or striking color patterns may also be favored (Fisher, 1958). Two apparently opposite breeding strategies may succeed, depending on other factors discussed below. These are cryptic breeding behavior and temporally and spatially clumped breeding behavior.

Several strategies may evolve as a response to predation on eggs: eggs with foam coating, as in a number of *Limnodynastes* species (Martin, 1967, Littlejohn, 1971); eggs containing poisonous substances, as in *Bufo* (Licht, 1967, 1968); eggs hatching quickly, as in *Scaphiopus* (Stebbins, 1951; Bragg, 1965, summarizing earlier papers); and a clumping of egg laying or hiding or tending of eggs, as is done by a number of New World tropical species (table 7-1). If adults become poisonous and effectively invulnerable, they concomitantly become good protectors of the eggs.

The strategies of hiding or tending eggs involve a greater parental investment per offspring and result in a decrease in the total number of eggs laid (figs. 7-1 and 7-2). That a general correlation exists between strategies of parental care and numbers of eggs has been recognized for some time; but no pattern has been recognized, and explanations by herpetologists have verged on the teleological, such as those of Porter (1972).

Figures 7-1, 7-2, and 7-3 show the relationships of egg sige, female size, litter size, and predictability of habitat. Indeed, as the size of egg relative to the female increases, the clutch size decreases (table 7-1, fig. 7-1). This is as expected and correlates with results from other groups (Williams, 1966a, 1966b; Salthe, 1969; Tinkle, 1969). When habitat or egg-laying locality is shown on a graph plotting the ratio of egg size to female size against litter size (fig. 7-3), it is apparent that most of those species showing some increase in parental care, such as laying eggs in burrows or leaves or tending the eggs or tadpoles, lay fewer, larger eggs; these species without exception live in habitats of relatively high environmental predictability—tropical rain forests, caves, and so on (table 7-1). No species laying eggs in temporary ponds show such behavior. The species in areas of high predictability possess a variety of strategies of high parental investment per offspring. As mentioned above, *Rhinoderma darwini* males carry the eggs in the vocal sac (Porter, 1972, and others). *Leiopelma hochstetteri* eggs are laid terrestrially and tended by one of the parents.

Females of several species of *Helioporus* lay eggs in a burrow excavated by the male, and the eggs await flooding to hatch (Main, 1965; Martin, 1967). Eggs of *Pipa pipa* are essentially tended by the female, on whose back they develop. Barbour (1934) and Porter (1972) reviewed a number of cases of parental tending and hiding strategies.

In situations (such as physical uncertainty or unpredictable predation) where increased parental investment per offspring is ineffective in decreasing the mortality of an individual's offspring, the minimum investment per offspring will be favored. In these cases, individuals which win are those which lay eggs in the peak laying period and in the middle of a good area being used by others. Any approaching predator should encounter someone else's eggs first. This strategy should be common in deserts and indeed appears to be (table 7-1). The costs of playing this temporal and spatial variety of "selfish herd" game (Hamilton, 1971) are that some aspects of intraspecific competition are maximized and predators may evolve to exploit the conspicuous "herd."

Three strategies would appear to be of selective advantage if predation is concentrated on the tadpole stage. One is the laying of larger or larger-yolked eggs producing larger and less-vulnerable tad-

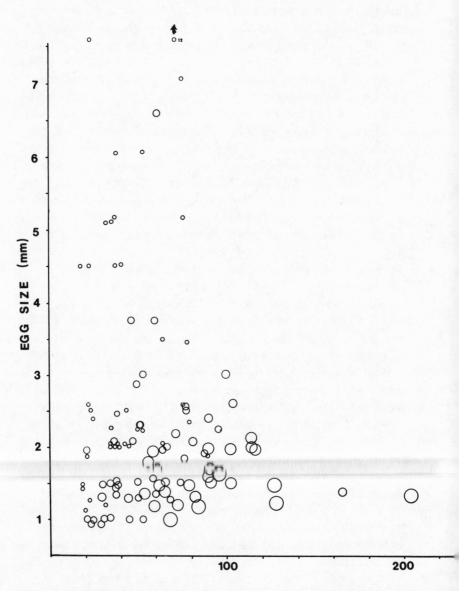

Fig. 7-1. Relationship of egg size to size of adult female for species from table 7-1. Size of circle indicates size of clutch:

o = ⟨ 500   ◯ = 1,000–10,000   ◯ = ⟩ 10,000
◯ = 500–1,000

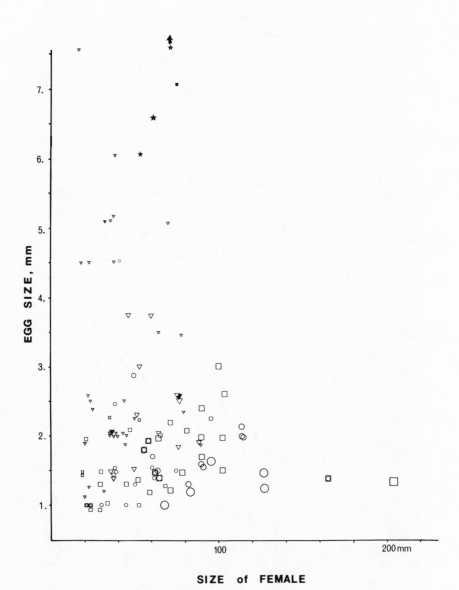

SIZE of FEMALE

Fig. 7-2. Relationship of egg size to size of adult female. As in figure 7-1, clutch size is shown by size of symbol. Solid symbols indicate tending behavior by a parent. Habitat of eggs:

 ○ = temporary water   Δ = laid in burrows, wrapped in leaves, etc.
 □ = permanent water   ★ = carried in brood pouch, on back, in
              vocal sac

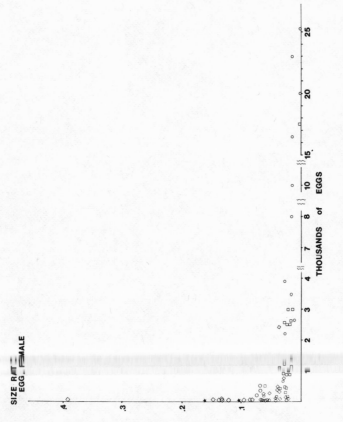

Fig. 7-3 Relationship of egg habitat to clutch size and relative size of eggs.
Habitat of eggs:

○ = temporary water   ◇ = laid on leaves, in burrows

□ = permanent water   ★ = carried by parent

poles at hatching. If seasonal or environmental conditions permit, producing offspring which spend a longer time as eggs may be successful. This would frequently involve strategies of hiding or tending eggs. In some species the entire development is completed while secreted so that on emergence offspring, in fact, are adults (*Leiopelma*, *Rhinoderma*). A third strategy is that of facultatively quick metamorphosis (*Bufo*, *Scaphiopus*), a strategy one might also expect to be favored in temporary ponds in desert situations. However, this advantage is balanced by intraspecific competition with its contingent selective advantage on size. So, in fact, what one would predict whenever genetic cost is not too great (Williams, 1966a) are facultative lengths of egg and larval periods and facultative hatching and metamorphosis. Thus, under strong predation and in more uncertain environments, one predicts an increase in facultativeness in these parameters. While this is predictable for both factors, studies of predation have not been able to separate out effects (De Benedictis, 1970). In amphibians of desert temporary ponds, either length of egg and larval periods are short or there is a large variation in reported lengths. Lengths of time to hatch in *S. couchi*, for example, range from 9 hours (Ortenburger and Ortenburger, 1926) to 48–72 hours (Gates, 1957); and in *S. hammondi*, from 38 (Little and Keller, 1937) to 120 hours (Sloan, 1964). The sizes at which these species metamorphose are highly variable (Bragg, 1965), suggesting that the strongest pressures of uncertainty center on the tadpole stages.

If predation is concentrated on juveniles, there will be an advantage to cryptic behavior by newly metamorphosed individuals. If predation is nonrandom and size related (as appears likely at metamorphosis when major predators are fish and other frogs), larger-sized individuals will be favored. If laying larger-yolked eggs results in larger offspring and increased offspring survivorship, this strategy will win. Certainly spending a longer time in the egg and tadpole stages, if predation is not heavily concentrated on these stages, will be favored. In species like *Rana catesbiana*, length of larval life is facultative and, for late eggs, is greater than a year. This appears to be involved with time required to reach a large enough size to be relatively invulnerable as a juvenile. While lengthened juvenile life cannot be selected for directly, conditions like predation on newly metamorphosed individuals are precisely those rendering lengthened periods in the tadpole stage advantageous.

Suggestions for Future Research

We can see that the interplay of these conditions is complex, and it is not necessarily a simple undertaking to predict strategies favored in each situation. Presently observed situations reflect the summation of a number of possibly conflicting selective advantages. Further, even though some biotic factors may be partially predictable from physical factors (e.g., in seasonal situations it is predictable that not only will food and breeding suitability be greater at some periods than others, but also at those same times predation will increase), others are not, and there is no single simple pattern.

The questions raised here are difficult to answer without further data, which are skimpy for anurans. Studies like Tinkle's (1969) on lizards or Inger and Greenberg's (1963) would afford comparative data for examination in the theoretical approach put forward here. For the most part, work on life histories in anurans has been zoogeographic and anecdotal. We haven't asked the right questions. Needed now are comparative studies similar to Tinkle's, between similar species in different habitats, and, in wide-ranging species, conspecific comparisons between habitats. We need to have:

1. Demographic data including length of life, time to maturity, age-specific fecundity, and degree of iteroparity, including number of eggs per clutch and number of clutches per year.
2. Ratio of egg size to female size.
3. Behavior: territoriality, tending behavior. (For example, Porter [1972] reported that *Rhinoderma darwini* males tend eggs that may not be their own. Such genetic altruism seems unlikely and needs further examination.)
4. Within wide-ranging species, comparative studies including, in addition to the above, work on mating-call parameters of males and discrimination of females.

Only when we begin to ask the above kinds of questions will we be able to develop an overall theoretical framework within which to view amphibian life histories. Many of the predictions and speculations discussed here seem obvious or trivial, but perhaps such attempts are necessary first steps toward a conceptual treatment of amphibian life histories.

Summary

Despite their normal requirement for an aqueous environment during the larval stage, a considerable number of amphibian species have adapted successfully to the desert environment. Possible methods of adaptation are considered, and their occurrence is reviewed. A number depend on modification of life histories, and attention is concentrated on these. Success depends on balancing the risk of mortality against the cost of reproductive effort.

Since desert environments are often less predictable than others, life-history strategies must take this uncertainty into account. This implies repeated small but prompt reproductive efforts and long adult life; behavior which enhances positive effects of random events, or reduces their negative effects, will be favored. Reduction in parental investment is generally advantageous in conditions of uncertainty in the physical environment.

From the amphibian point of view, the desert environment is patchy —coarse-grained—with high rates of speciation and extinction. Migration is favored where ponds are temporary and disfavored where they are permanent.

Mortality in the deserts is much more random than in mesic environments where it is dominated by predation. Reduction in the duration of vulnerable stages will then be advantageous. Responses to predation, however, have helped to shape amphibian life histories in the desert, as well as leading to production of noxious substances in many species. Differences in egg size and number per clutch may depend on likelihood of predation as against other hazards.

The importance of increased information about amphibian demography, and aspects of behavior related to it, is emphasized.

References

Alexander, R. D. 1968. Life cycle origins, speciations, and related phenomena in crickets. *Q. Rev. Biol.* 43:1–41.

Anastasi, A.; Erspamer, V.; and Endean, R. 1968. Isolation and amino acid sequence of caerulein, the active decapeptide of the skin of *Hyla caerulea*. *Archs Biochem. Biophys.* 125:57–68.

Balinsky, J. B.; Cragg, M. M.; and Baldwin, E. 1961. The adaptation of amphibian waste nitrogen excretion to dehydration. *Comp. Biochem. Physiol.* 3:236–244.

Barbour, T. 1934. *Reptiles and amphibians: Their habits and adaptations*. Boston and New York: Houghton Mifflin.

Bentley, P. J. 1966. Adaptations of Amphibia to desert environments. *Science, N.Y.* 152:619–623.

Bentley, P. J.; Lee, A. K.; and Main, A. R. 1958. Comparison of dehydration and hydration in two genera of frogs (*Helioporus* and *Neobatrachus*) that live in areas of varying aridity. *J. exp. Biol.* 35: 677–684.

Blair, W. F., ed. 1972. *Evolution in the genus "Bufo."* Austin: Univ. of Texas Press.

Bragg, A. N. 1965. *Gnomes of the night: The spadefoot toads*. Philadelphia: Univ. of Pa. Press.

Brattstrom, B. H. 1962. Thermal control of aggregation behaviour in tadpoles. *Herpetologica* 18:38–46.

———. 1963. A preliminary review of the thermal requirements of amphibians. *Ecology* 24:238–255.

Brockelman, W. Y. 1969. An analysis of density effects and predation in *Bufo americanus* tadpoles. *Ecology* 50:632–644.

Cei, J. M.; Erspamer, V.; and Roseghini, M. 1972. Biogenic amines. In *Evolution in the genus "Bufo,"* ed. W. F. Blair. Austin: Univ. of Texas Press.

Cragg, M. M.; Balinsky, J. B.; and Baldwin, E. 1961. A comparative study of the nitrogen secretion in some Amphibia and Reptilia. *Comp. Biochem. Physiol.* 3:227–236.

De Benedictis, P. A. 1970. "Interspecific competition between tadpoles of *Rana pipiens* and *Rana sylvatica*: An experimental field study." Ph.D. dissertation, University of Michigan.

Dole, J. W. 1967. The role of substrate moisture and dew in the water economy of leopard frogs, *Rana pipiens*. *Copeia* 1967:141–150.

Duellman, W. E. 1970. The hylid frogs of Middle America. *Monogr. Univ. Kans. Mus. nat. Hist.* 1:1–753.

Erspamer, V.; Vitali, T.; and Roseghini, M. 1964. The identification of new histamine derivatives in the skin of *Leptodactylus*. *Archs Biochem. Biophys.* 105:620–629.

Fisher, R. A. 1958. *The genetical theory of natural selection*. 2d rev. ed. New York: Dover.

Frith, H. J., and Calaby, J. H. 1969. *Kangaroos*. Melbourne: F. W. Cheshire.

Gates, G. O. 1957. A study of the herpetofauna in the vicinity of Wickenburg, Maricopa County, Arizona. *Trans. Kans. Acad. Sci.* 60:403–418.

Grant, P. R. 1972. Convergent and divergent character displacement. *J. Linn. Soc. (Biol.)* 4:39–68.

Hamilton, W. D. 1971. Geometry for the selfish herd. *J. theoret. Biol.* 31:295–311.

Heatwole, H.; Blasina de Austin, S.; and Herrero, R. 1968. Heat tolerances of tadpoles of two species of tropical anurans. *Comp. Biochem. Physiol.* 27:807–815.

Hussell, D. J. T. 1972. Factors affecting clutch-size in Arctic passerines. *Ecol. Monogr.* 42:317–364.

Inger, R. F. 1954. Systematics and zoogeography of Philippine Amphibia. *Fieldiana, Zool.* 33:185–531.

Inger, R. F., and Greenberg, B. 1956. Morphology and seasonal development of sex characters in two sympatric African toads. *J. Morph.* 99:549–574.

―――. 1963. The annual reproductive pattern of the frog *Rana erythraea* in Sarawak. *Physiol. Zoöl.* 36:21–33.

Klomp, H. 1970. The determination of clutch-size in birds. *Ardea* 58: 1–124.

Lack, D. 1947. The significance of clutch-size. Pts. I and II. *Ibis* 89: 302–352.

―――. 1948. The significance of clutch-size. Pt. III. *Ibis* 90:24–45.

Lee, A. K. 1967. Studies in Australian Amphibia. II. Taxonomy, ecology, and evolution of the genus *Helioporus* Gray (Anura: Leptodactylidae). *Aust. J. Zool.* 15:367–439.

Lee, A. K., and Mercer, E. H. 1967. Cocoon surrounding desert-dwelling frogs. *Science, N.Y.* 157:87–88.

Levins, R. 1968. *Evolution in changing environments*. Monographs in Population Biology, 2. Princeton: Princeton Univ. Press.

Lewontin, R. C. 1969. Comments on Slobodkin and Sanders "Contribution of environmental predictability to species diversity." *Brookhaven Symp. Biol.* 22:93.

Licht, L. E. 1967. Death following possible ingestion of toad eggs. *Toxicon* 5:141–142.

———. 1968. Unpalatability and toxicity of toad eggs. *Herpetologica* 24:93–98.

Licht, L. E., and Low, B. S. 1968. Cardiac response of snakes after ingestion of toad parotoid venom. *Copeia* 1968:547–551.

Little, E. L., and Keller, J. G. 1937. Amphibians and reptiles of the Jornada Experimental Range, New Mexico. *Copeia* 1937:216–222.

Littlejohn, M. J. 1967. Patterns of zoogeography and speciation by southeastern Australian Amphibia. In *Australian inland waters and their fauna*, ed. A. H. Weatherley, pp. 150–174. Canberra: Aust. Nat. Univ. Press.

———. 1971. Amphibians of Victoria. *Victorian Year Book* 85:1–11.

Low, B.S. 1972. Evidence from parotoid gland secretions. In *Evolution in the genus "Bufo,"* ed. W. F. Blair. Austin: Univ. of Texas Press.

MacArthur, R. H., and Wilson, E. O. 1967. *The theory of island biogeography*. Monographs in Population Biology, 1. Princeton: Princeton Univ. Press.

Main, A. R. 1957. Studies in Australian Amphibia. I. The genus *Crinia tschudi* in south-western Australia and some species from southeastern Australia. *Aust. J. Zool.* 5:30–55.

———. 1962. Comparisons of breeding biology and isolating mechanisms in Western Australian frogs. In *The evolution of living organisms*, ed. G. W. Leeper. Melbourne: Melbourne Univ. Press.

———. 1965. *Frogs of southern Western Australia.* Perth: West Australian Nat. Club.

———. 1968. Ecology, systematics, and evolution of Australian frogs. *Adv. ecol. Res.* 5:37–87.

Main, A. R., and Bentley, P. J. 1964. Water relations of Australian burrowing frogs and tree frogs. *Ecology* 45:379–382.

Main, A. R.; Lee, A. K.; and Littlejohn, M. J. 1958. Evolution in three genera of Australian frogs. *Evolution* 12:224–233.

Main, A. R.; Littlejohn, M. J.; and Lee, A. K. 1959. Ecology of Australian frogs. In *Biogeography and ecology in Australia*, ed. A. Keast, R. L. Crocker, and C. S. Christian. The Hague: Dr. W. Junk.

Martin, A. A. 1967. Australian anuran life histories: Some evolutionary and ecological aspects. In *Australian inland waters and their fauna*, ed. A. H. Weatherley, pp. 175–191. Canberra: Aust. Nat. Univ. Press.

Mayhew, W. W. 1968. Biology of desert amphibians and reptiles. In *Desert biology*, ed. G. W. Brown, vol. 1, pp. 195–356. New York and London: Academic Press.

Michl, H., and Kaiser, E. 1963. Chemie and Biochemie de Amphibiengifte. *Toxicon* 1963:175–228.

Miller, N. 1909. The American toad. *Am. Nat.* 43:641–688.

Murphy, G. I. 1968. Pattern in life history and the environment. *Am. Nat.* 102:391–404.

Newsome, A. E. 1964. Anoestrus in the red kangaroo, *Megaleia rufa*. *Aust. J. Zool.* 12:9–17.

———. 1965. The influence of food on breeding in the red kangaroo in central Australia. *CSIRO Wildl. Res.* 11:187–196.

———. 1966. Reproduction in natural populations of the red kangaroo *Megaleia rufa* in central Australia. *Aust. J. Zool.* 13:735–759.

Noble, C. K., and Putnam, P. G. 1931. Observations on the life history of *Ascaphus truei* Stejneger. *Copeia* 1931:97–101.

Ortenburger, A. I., and Ortenburger, R. D. 1926. Field observations on some amphibians and reptiles of Pima County, Ariz. *Proc. Okla. Acad. Sci.* 6:101–121.

Pianka, E. R. 1970. On r and K selection. *Am. Nat.* 104:592–597.

Porter, K. R. 1972. *Herpetology*. Philadelphia: W. B. Saunders Co.

Power, J. A. 1927. Notes on the habits and life histories of South African Anura with descriptions of the tadpoles. *Trans. R. Soc. S. Afr.* 14:237–247.

Ruibal, R. 1962a. The adaptive value of bladder water in the toad, *Bufo cognatus*. *Physiol. Zoöl.* 35:218–223.

———. 1962b. Osmoregulation in amphibians from heterosaline habitats. *Physiol. Zoöl.* 35:133–147.

Salthe, S. N. 1969. Reproductive modes and the number and size of ova in the urodeles. *Am. Midl. Nat.* 81:467–490.

Schaaf, R. T., and Garton, J. S. 1970. Racoon predation on the American toad, *Bufo americanus*. *Herpetologica* 26:334–335.

Schmidt, K. P., and Inger, R. F. 1959. Amphibia. *Explor. Parc natn. Upemba Miss. G. F. de Witt* 56.

Sharman, G. B.; Calaby, J. H.; and Poole, W. E. 1966. Patterns of reproduction in female diprotodont marsupials. *Symp. zool. Soc. Lond.* 15:205–232.

Slater, J. R. 1934. Notes on northwestern amphibians. *Copeia* 1934: 140–141.

Sloan, A. J. 1964. Amphibians of San Diego County. *Occ. Pap. S Diego Soc. nat. Hist.* 13:1–42.

Slobodkin, L. D., and Sanders, H. L. 1969. On the contribution of environmental predictability to species diversity. *Brookhaven Symp. Biol.* 22:82–96.

Stebbins, R. C. 1951. *Amphibians of western North America*. Berkeley and Los Angeles: Univ. of Calif. Press.

Stewart, M. M. 1967. *Amphibians of Malawi*. Albany: State Univ. of N.Y. Press.

Surface, H. A. 1913. The Amphibia of Pennsylvania. *Bi-m. zool. Bull. Pa Dep. Agric.* May–July 1913:67–151.

Szarski, H. 1972. Integument and soft parts. In *Evolution in the genus "Bufo,"* ed. W. F. Blair. Austin: Univ. of Texas Press.

Tinkle, D. W. 1969. The concept of reproductive effort and its relation to the evolution of life histories of lizards. *Am. Nat.* 103:501–514.

Trivers, R. L. 1972. Parental investment and sexual selection. In *Sexual selection and the descent of man*, ed. B. Campbell, pp. 136–179. Chicago: Aldine.

Trueb, L., and Duellman, W. E. 1970. The systematic status and life history of *Hyla verrucigera* Werner. *Copeia* 1970:601–610.

Turner, F. B. 1962. The demography of frogs and toads. *Q. Rev. Biol.* 37:303–314.

Villa, J. 1972. *Anfibios de Nicaragua*. Managua: Instituto Geográfico Nacional, Banco Central de Nicaragua.

Volpe, E. P. 1953. Embryonic temperature adaptations and relationships in toads. *Physiol. Zoöl.* 26:344–354.

Wager, V. A. 1965. *The frogs of South Africa*. Capetown: Purnell & Sons.

Warburg, M. R. 1965. Studies on the water economy of some Australian frogs. *Aust. J. Zool.* 13:317–330.

Wilbur, H. M., and Collins, J. P. 1973. Ecological aspects of amphibian metamorphosis. *Science, N.Y.* 182:1305.

Wilbur, H. M.; Tinkle, D. W.; and Collins, J. P. 1974. Environmental certainty, trophic level, and successional position in life history evolution. *Am. Nat.* 108:805–818.

Williams, G. C. 1966a. *Adaptation and natural selection: A critique of some current evolutionary thought.* Princeton: Princeton Univ. Press.

———. 1966b. Natural selection, the costs of reproduction, and a refinement of Lack's principle. *Am. Nat.* 100:687–692.

Wright, J. W. 1966. Predation on the Colorado River toad, *Bufo alvarius. Herpetologica* 22:127–128.

8. Adaptation of Anurans to Equivalent Desert Scrub of North and South America

W. Frank Blair

Introduction

The occurrence of desertic environments at approximately the same latitudes in western North America and in South America provides an excellent opportunity to investigate comparatively the structure and function of ecosystems that have evolved under relatively similar environments. A multidisciplinary investigation of these ecosystems to determine just how similar they are in structure and function is presently in progress under the Origin and Structure of Ecosystems Program of the U.S. participation in the International Biological Program. The specific systems under study are the Argentine desert scrub, or Monte, as defined by Morello (1958) and the Sonoran desert of southwestern North America.

In this paper I will discuss the origins and nature of one component of the vertebrate fauna of these two xeric areas, the anuran amphibians. Pertinent questions are (a) How do the two areas compare in the degree of desert adaptedness of the fauna? (b) How do the two areas compare with respect to the size of the desert fauna? (c) What are the geographical origins of the various components of the fauna? and (d) What are the mechanisms of desert adaptation?

The comparison of the two desert faunas must take into account a number of major factors that have influenced their evolution. The most important among these would seem to be:

1. The nature of the physical environment of physiography and climate
2. The degree of similarity of the vegetation in general ecological aspect and in plant species composition

3. The size of each desert area
4. Possible sources of desert-invading species and the nature of adjacent biogeographic areas
5. The past history of the area through Tertiary and Pleistocene times
6. The evolutionary-genetic capabilities of available stocks for desert colonization

The Physical Environment

As defined by Morello (1958), the Monte extends through approximately 20° of latitude from 24°35′S in the state of Salta to 44°20′S in the state of Chubut and through approximately 7° of longitude from 69°50′W in Neuquen to 62°54′W on the Atlantic coast. The Sonoran desert occupies an area lying approximately between lat. 27° and 34°N and between long. 110° and 116°W (Shelford, 1963, fig. 15-1). Both areas are characterized by lowlands and mountains. The present discussion will deal principally with the lowland fauna.

Rainfall in both of the areas is usually less than 200 mm annually (Morello, 1958; Barbour and Díaz, 1972). Thus, availability of water is the most important factor determining the nature of the vegetation and the most important control limiting the invasion of these areas by terrestrial vertebrates.

The Vegetation

A more precise discussion of the vegetation of the Monte will be found elsewhere in this volume (Solbrig, 1975), so I will point out only that the general aspect is very similar in the two areas. The genera *Larrea*, *Prosopis*, and *Acacia* are among the most important components of the lowland vegetation and are principally responsible for this similarity of aspect. Various other genera are shared by the two areas. Some notably desert-adapted genera are found in one area but not in the other (Morello, 1958; Raven, 1963; Axelrod, 1970).

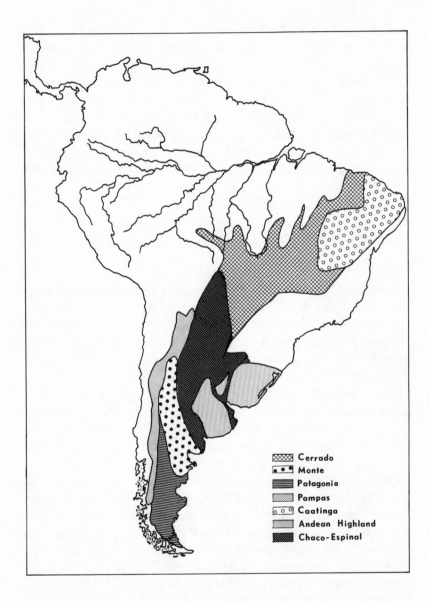

Fig. 8-1. Approximate distribution of xeric and subxeric areas in eastern and southern South America (adapted from Cabrera, 1953; Veloso, 1966; Sick, 1969).

Size of Area

The present areas of the Sonoran desert and the Monte are roughly similar in size. However, in considering the evolution of the desert-adapted fauna of the two continents, it is important to consider all contiguous desert areas. In this context the desertic areas of North America far exceed those that exist east of the Andes in South America. In South America there is only the Patagonian area with a cold desertic climate and the cold Andean Puna. In North America the addition of the Great Basin desert, the Mojave, and the Chihuahuan desert provides a much greater geographical expanse in which desert adaptations are favored.

Potential Sources of Stocks

The probability of any particular taxon of animal contributing to the fauna of either desert area obviously can be expected to decrease with the distance of that taxon's range from the desert area in question. This should be true not only because of the mere matter of distance but also because the more distant taxa would be expected to be adapted to the more distant and, hence, usually more different environments.

The nature of the adjacent ecological areas is, therefore, important to the process of evolution of the desert faunas. The Monte lies east of the Andean cordillera, which is a highly effective barrier to the interchange of lowland biota. To the south is the cold, desertic Patagonia, smaller in area than the Monte itself. To the east the Monte grades into the semixeric thorn forest of the Chaco, which extends into Paraguay and Uruguay and merges into the Cerrado and Caatinga of Brazil. East of the Chaco are the pampa grasslands between roughly lat. 31° and 38°S (fig. 8-1). With the huge area of Chaco, Cerrado, and Caatinga to the east and northeast, and with the Chaco showing a strong gradient of decreasing moisture from east to west, we might expect this eastern area to be a likely source for the evolution of Monte species of terrestrial vertebrates.

The geographical relationship of the Sonoran desert to possible source areas for invading species is very different from that of the Monte. Mountains are to the west, but beyond that little similarity exists. For one thing, the Sonoran desert is part of a huge expanse of desertic areas that stretches over 3,000 km from the southern part of the Chihuahuan desert in Mexico to the northern tip of the Great Basin desert in Oregon. To the east of these deserts in the United States, beyond the Rocky Mountain chain, are the huge central grasslands extending from the Gulf of Mexico into southern Canada. A similarity to the South American situation is seen, however, in the presence of a thorny vegetation type (the Mesquital), comparable to the Chaco, on the Gulf of Mexico lowlands of Tamaulipas and southern Texas. As in Argentina, a gradient of decreasing moisture exists westward from this Mesquital through the Chihuahuan desert and into the Sonoran desert. By contrast with the Monte, the Sonoran desert seems much more exposed to invasion by taxa which have adapted toward warm-xeric conditions in other contiguous areas.

Past Regional History

The present character of the two desert faunas obviously relates to the past histories of the two regions. For how long has there been selection for a xeric-adapted fauna in each area? What have been the effects on these faunas of secular climatic changes in the Tertiary and Pleistocene? These questions are difficult to answer with any great precision.

According to Axelrod (1948, p. 138, and other papers), "the present desert vegetation of the western United States, as typified by the floras of the Great Basin, Mohave and Sonoran deserts" is no older than middle Pliocene. Prior to the Oligocene, a Neotropical-Tertiary geoflora extended from southeastern Alaska and possibly Nova Scotia south into Patagonia (Axelrod, 1960) and began shrinking poleward as the continent became cooler and drier from the Oligocene onward. With respect to the Monte, Kusnezov (1951), as quoted by Morello (1958), believed that the Monte has existed without major change since "Eocene-Oligocene" times.

Arguments have been presented that there was a Gondwanaland

dry flora prior to the breakup of that land mass in the Cretaceous, which is represented today by xeric relicts in southern deserts (Axelrod, 1970). It seems then that selection for xeric adaptation has been going on in the southern continent and, from paleobotanical evidence, in North America as well (Axelrod, 1970, p. 310) for more than 100 million years. However, major climatic changes have occurred in the geographic areas now known as the Monte and the Sonoran desert. The present desert floras of these two areas are combinations of the old relicts and of types that have evolved as the continents dried and warmed from the Oligocene onward (Axelrod, 1970).

One of the unanswered questions is where the desert-adapted biotas were at times of full glaciations in the Pleistocene. Martin and Mehringer (1965, p. 439) have addressed this question with respect to North American deserts and have concluded that "Sonoran desert plants may have been hard pressed." The question is yet unanswered. The desert plants presumably retreated southward, but the degree of compression of their ranges is unknown. Doubt also exists whether the Monte biota could have remained where it now is at peaks of glaciation in the Southern Hemisphere (Simpson Vuilleumier, 1971).

The Anurans

The number of species of frogs is not greatly different for the two deserts, and, as might be expected, both faunas are relatively small. As we define the two faunas on the basis of present knowledge, the Sonoran desert fauna includes eleven species representing four families and four genera, while that of the Monte includes fourteen species representing three families and seven genera (table 8-1). (Definition of the Monte fauna is less certain and more arbitrary than that of the Sonoran because of scarcity of data. The listings of Monte and Chacoan species used here are based largely on data from Freiberg [1942], Cei [1955a, 1955b, 1959b, 1962], Reig and Cei [1963], and Barrio [1964a, 1964b, 1965a, 1965b, 1968] and on my own observations. Species recorded from Patquia in the province of La Rioja and from Alto Pencoso on the San Luis–Mendoza border [Cei, 1955a, 1955b] are included in the Monte fauna as here considered.)

Table 8-1. *Anuran Faunas: Monte of Argentina and Sonoran Desert of North America*

Sonoran	Monte
Pelobatidae	Ceratophrynidae
Scaphiopus couchi	*Ceratophrys ornata*
S. hammondi	*C. pierotti*
	Lepidobatrachus llanensis
	L. asper
Bufonidae	Bufonidae
Bufo woodhousei	*Bufo arenarum*
B. cognatus	
B. mazatlanensis	
B. retiformis	
B. punctatus	
B. alvarius	
B. microscaphus	
Hylidae	
Pternohyla fodiens	
Ranidae	Leptodactylidae
Rana sp.	*Odontophrynus occidentalis*
(*pipiens* gp.)	*O. americanus*
	Leptodactylus ocellatus
	L. bufonius
	L. prognathus
	L. mystaceus
	Pleurodema cinerea
	P. nebulosa
	Physalaemus biligonigerus

The composition of the two faunas is phylogenetically quite dissimilar. The Sonoran is dominated by members of the genus *Bufo* with seven species. The Monte fauna is dominated by leptodactylids with nine species distributed among four genera of that family.

Ecological similarities are evident between the two pelobatids (*Scaphiopus couchi* and *S. hammondi*) of the Sonoran fauna and the four ceratophrynids (*Ceratophrys ornata*, *C. pierotti*, *Lepidobatrachus asper*, and *L. llanensis*) of the Monte. The Sonoran has a single fossorial hylid (*Pternohyla fodiens*); I have found no evidence of a Monte hylid. However, a remarkably xeric-adapted hylid, *Phyllomedusa sauvagei*, extends at least into the dry Chaco (Shoemaker, Balding, and Ruibal, 1972); and, because of these adaptations, it would not be surprising to find it in the Monte. The canyons of the desert mountains of the Sonoran and Monte have a single species of *Hyla* of roughly the same size and similar habits. In Argentina it is *H. pulchella*; in the United States it is *H. arenicolor*. These are not included in our faunal listing for the two areas. The Sonoran has a ranid (*Rana* sp. [*pipiens* gp.]); the family has penetrated only the northern half of South America (with a single species) from old-world origins and via North America, so has had no opportunity to contribute to the Monte fauna.

The origins of the Monte anuran fauna seem relatively simple. This fauna is principally a depauperate Chacoan fauna (table 8-2). At least thirty-seven species of anurans are included in the Chacoan fauna. Every species in the Monte fauna also occurs in the Chaco. Nine of the fourteen Monte species have ranges that lie mostly within the combined Chaco-Monte. The Monte fauna thus represents that component of a biota which has had a long history of adaptation to xeric or subxeric conditions and is able to occupy the western, xeric end of a moisture gradient that extends from the Atlantic coast west to the base of the Andes. Two of the Monte species (*Leptodactylus mystaceus* and *L. ocellatus*) are wide ranging tropical species that reach both the Monte and the Chaco from the north or east. We are treating *Odontophrynus occidentalis* as a sub-Andean species (Barrio, 1964*a*), but the genus has the Chaco-Monte distribution; and since this species reaches the Atlantic coast in Buenos Aires province, there is no certainty that it evolved in the Monte. *Pleurodema nebulosa* of the Monte is listed by Cei (1955*b*, p. 293) as "a characteristic cordilleran form"; and, as mapped by Barrio (1964*b*), its range barely enters the Chaco, although other members of the same species group occur in the dry Chaco. *Pleurodema cinerea* is treated

Table 8-2. *Comparison of Chaco and Monte Anuran Faunas*

Monte	Chaco
	Hypopachus mulleri
Ceratophrys ornata	*Ceratophrys ornata*
C. pierotti	*C. pierotti*
Lepidobatrachus llanensis	*Lepidobatrachus llanensis*
	L. laevis
L. asper	*L. asper*
Pleurodema nebulosa	*Pleurodema nebulosa*
	P. quayapae
	P. tucumana
P. cinerea	*P. cinerea*
Physalaemus biligonigerus	*Physalaemus biligonigerus*
	P. albonotatus
Leptodactylus ocellatus	*Leptodactylus ocellatus*
	L. chaquensis
L. bufonius	*L. bufonius*
L. prognathus	*L. prognathus*
L. mystaceus	*L. mystaceus*
	L. sibilator
	L. gracilis
	L. mystacinus
Odontophrynus occidentalis	*Odontophrynus occidentalis*
O. americanus	*O. americanus*
Bufo arenarum	*Bufo arenarum*
	B. paracnemis
	B. major
	B. fernandezae
	B. pygmaeus
	Melanophryniscus stelzneri
	Pseudis paradoxus
	Lysapsus limellus

Monte	Chaco
	Phyllomedusa sauvagei
	P. hypochondrialis
	Hyla pulchella
	H. trachythorax
	H. venulosa
	H. phrynoderma
	H. nasica

Note: All species listed for Monte occur also in Chaco.

by Gallardo (1966) as a member of his fauna "Subandina." The genus ranges north to Venezuela.

The Sonoran anurans seemingly have somewhat more diverse geographical origins than those of the Monte, and they have been more thoroughly studied. Most of the ranges can be interpreted as ones that have undergone varying degrees of expansion northward following full glacial displacement into Mexico (Blair, 1958, 1965). Several of these (*Scaphiopus couchi*, *S. hammondi*, and *Bufo punctatus*) have a main part of their range in the Chihuahuan desert (table 8-3). *Bufo cognatus* ranges far northward through the central grasslands to Canada. Three species extend into the Sonoran from the lowlands of western Mexico. One of these is the fossorial hylid *Pternohyla fodiens*. Another, *B. retiformis*, is one of a three-member species group that ranges from the Tamaulipan Mesquital westward through the Chihuahuan desert into the Sonoran. The third, *B. ma zatlanensis*, is a member of a species group that is absent from the Chihuahuan desert but is represented in the Tamaulipan thorn scrub. *Bufo woodhousei* has an almost transcontinental range. *Rana* sp. is an undescribed member of the *pipiens* group.

Two desert-endemic species occur in the Sonoran. One is *Bufo alvarius*, which appears to be an old relict species without any close living relative. *B. microscaphus* occurs in disjunct populations in the Chihuahuan, Sonoran, and southern Great Basin deserts. These populations are clearly relicts from a Pleistocene moist phase extension of the eastern mesic-adapted *B. americanus* westward into the present desert areas (A. P. Blair, 1955; W. F. Blair, 1957).

Table 8-3. *Comparison of Anuran Faunas of Sonoran Desert with Those of Chihuahuan Desert and Tamaulipan Mesquital*

Sonoran Desert	Chihuahuan-Tamaulipan
	Rhinophrynus dorsalis
	Hypopachus cuneus
	Gastrophryne olivacea
Scaphiopus hammondi	*Scaphiopus hammondi*
	S. bombifrons
S. couchi	*S. couchi*
	S. holbrooki
	Leptodactylus labialis
	Hylactophryne augusti
	Syrrhopus marnocki
	S. campi
	Bufo speciosus
Bufo cognatus	*B. cognatus*
B. punctatus	*B. punctatus*
	B. debilis
	B. valliceps
B. woodhousei	*B. woodhousei*
B. retiformis	
B. mazatlanensis	
B. alvarius	
B. microscaphus	
	Hyla cinerea
	H. baudini
	Pseudacris clarki
	P. streckeri
	Acris crepitans
Pternohyla fodiens	
Rana sp. (*pipiens* gp.)	*Rana* sp. (*pipiens* gp.)
	R. catesbeiana

Desert Adaptedness

If taxonomic diversity is taken as a criterion, the Monte fauna presents an impressive picture of desert adaptation. The genera *Odontophrynus* and *Lepidobatrachus* are both xeric adapted and are endemic to the xeric and subxeric region encompassed in this discussion. Three of the four leptodactylid genera which occur in the Monte (*Leptodactylus*, *Pleurodema*, and *Physalaemus*) are characterized by the laying of eggs in foam nests, either on the surface of the water or in excavations on land. This specialization may have a number of advantages, but one of the important ones would be protection from desiccation (Heyer, 1969).

In North America the only genus that can be considered a desert-adapted genus is *Scaphiopus*. This genus has two distinct subgeneric lines which, based on the fossil record, apparently diverged in the Oligocene (Kluge, 1966). Each subgenus is represented by a species in the Sonoran desert. Origin of the genus through adaptation of forest-living ancestors to grassland in the early Tertiary has been suggested by Zweifel (1956). *Pternohyla* is a fossorial hylid that apparently evolved in the Pacific lowlands of Mexico "in response to the increased aridity during the Pleistocene" (Trueb, 1970, p. 698). The diversity of *Bufo* species (*B. mazatlanensis*, *B. cognatus*, *B. punctatus*, and *B. retiformis*) that represent subxeric- and xeric-adapted species groups and the old relict *B. alvarius* implies a long history of *Bufo* evolution in arid and semiarid southwestern North America. Nevertheless, the total anuran diversity of xeric-adapted taxa compares poorly with that in South America.

The greater taxonomic diversity of desert-adapted South American anurans may be attributed to the Gondwanaland origin (Reig, 1960; Casamiquela, 1961; Blair, 1973) of the anurans and the long history of anuran radiation on the southern continent. The taxonomic diversity of anurans in South America vastly exceeds that in North America, which has an attenuated anuran fauna that is a mix of old-world emigrants (Ranidae, possibly Microhylidae) and invaders from South America (Bufonidae, Hylidae, and Leptodactylidae). The drastic effects of Pleistocene glaciations on North American environments may also account for the relatively thin anuran fauna of this continent.

Mechanisms of Desert Adaptation

Limited availability of water to maintain tissue water in adults and unpredictability of rains to permit reproduction and completion of the larval stage are paramount problems of desert anurans. Enough is
known about the ecology, behavior, and physiology of the anurans of
the two deserts to indicate the principal kinds of mechanisms that
have evolved in the two areas.

With respect to the first of these two problems, two major and quite
different solutions are evident in both desert faunas. One is to avoid
the major issue by becoming restricted to the vicinity of permanent
water in the desert environment. The other is to become highly fossorial, to evolve mechanisms of extracting water from the soil, and to
become capable of long periods of inactivity underground. In the
Sonoran desert three of the eleven species fit the first category. The
Rana species is largely restricted to the vicinity of water throughout its
range to the east and is a member of the *R. pipiens* complex, which
is essentially a littoral-adapted group. Ruibal (1962*b*) studied a desert
population of these frogs in California and regards their winter breeding as an adaptation to avoid the desert's summer heat. The relict
endemic *Bufo alvarius* is smooth skinned and semiaquatic (Stebbins, 1951; my data). The relict populations of *B. microscaphus* occur where there is permanent water as drainage from the mountains
or as a result of irrigation. Man's activities in impounding water for irrigation must have been of major assistance to these species in invading a desert region without having to cope with the major water problems of desert life. *Bufo microscaphus*, for example, exists in areas
that have been irrigated for thousands of years by prehistoric cultures
and more recently by European man (Blair, 1955). One species in the
Monte fauna is there by this same adaptive strategy. *Leptodactylus
ocellatus* offers a striking parallel to the *Rana* species. Its existence in
the provinces of Mendoza and San Juan is attributed to extensive agricultural irrigation (Cei, 1955*a*). That a second species, *B. arenarum*,
fits this category is suggested by Ruibal's (1962*a*, p. 134) statement
that "this toad is found near permanent water and is very common
around human habitations throughout Argentina." However, Cei
(1959*a*) has shown experimentally that *B. arenarum* from the Monte

(Mendoza) survives desiccation more successfully than *B. arenarum* from the Chaco (Córdoba), which implies exposure and adaptation to more rigorously desertic conditions for the former.

Most of the anurans of both desert faunas utilize the strategy of sub-terranean life to avoid the moisture-sapping environment of the desert surface. In the Sonoran fauna the two species of *Scaphiopus* have received considerable study. One of these, *S. couchi*, appears to have the greatest capacity for desert existence. Mayhew (1962, p. 158) found this species in southern California at a place where as many as three years might pass without sufficient summer rainfall to "stimulate them to emerge, much less successfully reproduce."

Mayhew (1965) listed a series of presumed adaptations of this species to desert environment:

1. Selection of burial sites beneath dense vegetation where reduced insolation reaching the soil means lower soil temperatures and reduced evaporation from the soil
2. Retention by buried individuals of a cover of dried, dead skin, thus reducing water loss through the skin
3. Rapid development of larvae—ten days from fertilization through metamorphosis (reported also by Wasserman, 1957)

Physiological adaptations of *S. couchi* (McClanahan, 1964, 1967, 1972) include:

1. Storage of urea in body fluids to the extent that plasma osmotic concentration may double during hibernation
2. Muscles showing high tolerance to hypertonic urea solutions
3. Rate of production of urea a function of soil water potential
4. Fat utilization during hibernation
5. Ability to tolerate water loss of 40–50 percent of standard weight
6. Ability to store up to 30 percent of standard body weight as dilute urine to replace water lost from body fluids

The larvae of *S. couchi* are more tolerant of high temperatures than anurans from less-desertic environments, and tadpoles have been observed in nature at water temperatures of 39° to 40°C (Brown, 1969).

Scaphiopus hammondi, as studied by Ruibal, Tevis, and Roig (1969) in southeastern Arizona, shows a pattern of desert adaptation generally comparable to that of *S. couchi* but with some difference in details. These spadefoots burrow underground in September to depths of up to 91 cm and remain there until summer rains come some nine months later. The burrows are in open areas, not beneath dense vegetation as reported for *S. couchi* by Mayhew (1965). *S. hammondi* can effectively absorb soil water through the skin and has greater ability to absorb soil moisture "than that demonstrated for any other amphibian" (Ruibal et al., 1969, p. 571). During the rainy season of July–August, the *S. hammondi* burrows to depths of about 4 cm.

Larval adaptations of *S. hammondi* include rapid development and tolerance of high temperatures (Brown, 1967*a*, 1967*b*), paralleling the adaptations of *S. couchi*.

The adaptations of *Bufo* for life in the Sonoran desert are less well known than those of *Scaphiopus*. Four of the nonsemiaquatic species escape the rigors of the desert surface by going underground. *Bufo cognatus* and *B. woodhousei* have enlarged metatarsal tubercles or digging spades, as in *Scaphiopus*. In southeastern Arizona, *B. cognatus* was found buried at the same sites as *S. hammondi* but in lesser numbers (Ruibal et al., 1969). McClanahan (1964) found the muscles of *B. cognatus* comparable to those of *S. couchi* in tolerance to hypertonic urea solutions, a condition which he regarded as a fossorial-desert adaptation. *Bufo punctatus* has a flattened body and takes refuge under rocks. It has been reported from mammal (*Cynomys*) burrows (Stebbins, 1951). *Bufo punctatus* has the ability to take up water rapidly from slightly moist surfaces through specialization of the skin in the ventral pelvic region ("sitting spot"), which makes up about 10 percent of the surface area of the toad (McClanahan and Baldwin, 1969). *Bufo retiformis* belongs to the arid-adapted *debilis* group of small but very thick-skinned toads (Blair, 1970).

The Sonoran desert species of *Bufo* have not evolved the accelerated larval development that is characteristic of *Scaphiopus*. Zweifel (1968) determined developmental rates for three species of *Scaphiopus*, three species of *Bufo, Hyla arenicolor,* and *Rana* sp. (*pipiens* gp.) in southeastern Arizona. The eight species fell into three groups: most rapid, *Scaphiopus*; intermediate, *Bufo* and *Hyla*; slowest, *Rana*. In my laboratory (table 8-4) *B. punctatus* from central Arizona showed no acceleration of development over the same species from the extreme eastern part of the range in central Texas. *Bufo cognatus* closely paralleled *B. punctatus* in duration of the lar-

Table 8-4. *Duration of Larval Stage of Four of the Sonoran Desert Species of* Bufo

Species	Locality of Origin	Days from Fertilization to Metamorphosis		Lab Stock No.
		First	50%	
B. punctatus	Wimberley, Texas	27	32	B64–173
B. punctatus	Mesa, Arizona	27	36	B64–325
B. cognatus	Douglas, Arizona	28	35	B64–234
B. mazatlanensis	Mazatlan, Sinaloa × Ixtlan, Nayarit	20	26	B63–87
B. alvarius	Tucson × Mesa, Arizona	36	53	B65–271
B. alvarius	Mesa, Arizona	29	33	B64–361

Note: Observations in a laboratory maintained at 24°–27° C.

val stage; *B. mazatlanensis* had a somewhat shorter larval life than these others; and *B. alvarius* spent a slightly longer period as tadpoles, but this could be accounted for by the fact that these are much larger toads. Overall, the impression is that these *Bufo* species have not shortened the larval stage as a desert adaptation. Tevis (1966) found that *B. punctatus* that were spawned in spring in Deep Canyon, California, required approximately two months for metamorphosis.

Developing eggs of *B. punctatus* and *B. cognatus* from Mesa, Arizona, were tested for temperature tolerances by Ballinger and Mc-Kinney (1966). Both of these desert species were limited by lower maxima than was *B. valliceps*, a nondesert toad, from Austin, Texas.

The fossorial anurans of the Monte are much less well known than those of the Sonoran desert. The ceratophrynids appear to be rather similar to *Scaphiopus* in their desert adaptations. Both species of *Lepidobatrachus* are reported to live buried (*viven enterrados*) and emerge after rains (Reig and Cei, 1963). *Lepidobatrachus llanensis* forms a cocoon made of many compacted dead cells of the stratum corneum when exposed to dry conditions (McClanahan, Shoemaker, and Ruibal, 1973). These anurans apparently live an aquatic exist-ence as long as the temporary rain pools exist, in which respect they differ from *Scaphiopus* species, which typically breed quickly and leave the water. The skin of *L. asper* is described (Reig and Cei, 1963) as thin in summer (when they are aquatic) and thicker and more granular in periods of drought. *Ceratophrys* reportedly uses the bur-rows of the viscacha (*Lagidium*), a large rodent (Cei, 1955*b*). How-ever, *C. ornata* does bury itself in the soil, and one was known to stay underground between four and five months and shed its skin after emerging (Marcos Freiberg, 1973, personal communication). *Ceratophrys pierotti* remains near the temporary pools in which it breeds for a considerable time after breeding (my observations). *Odontophrynus* at Buenos Aires makes shallow depressions and may sit in these with only the head showing (Marcos Freiberg, 1973, personal communication). *Leptodactylus bufonius* lives in dens or natural cavities or in viscacha burrows (Cei, 1949, 1955*b*). *Pleuro-dema nebulosa* is a fossorial species with metatarsal spade that spends a major portion of its lifetime living on land in burrows (Rui-bal, 1962*a*; Gallardo, 1965). *Bufo arenarum* "winters buried up to a meter in depth" (Gallardo, 1965, p. 67).

Phyllomedusa sauvagei of the dry Chaco, and possibly the Monte, has achieved a high level of xeric adaptation by excreting uric acid and by controlling water loss through the skin (Shoemaker et al., 1972). Rates of water loss in this arboreal, nonfossorial hylid are com-parable to those of desert lizards rather than to those of other anurans (Shoemaker et al., 1972).

Ruibal (1962*a*) studied the osmoregulation of six of the Chaco-Monte species and found that *P. nebulosa* is capable of producing urine that is hypotonic to the lymph and to the external medium, thus enabling it to store bladder water as a reserve against dehydration. The others, including *P. cinerea*, *L. asper*, and *B. arenarum* of what we are calling the Monte fauna, produced urine that was essentially isotonic to the lymph and the external medium.

Reproductive Adaptations

One of the major hazards of desert existence for an anuran population is the unpredictability of rainfall to provide breeding pools. Two alternative routes are available. One is to be an opportunistic breeder, spending long periods of time underground but responding quickly when suitable rainfall occurs. The alternative is to breed only in permanent water, with the time of breeding presumably set by such cues as temperature or possibly photoperiod. Both strategies are found among the Sonoran desert anurans.

The two *Scaphiopus* species are the epitome of the first of these adaptive routes. *Bufo cognatus*, *B. retiformis*, and *Pternohyla fodiens* are also opportunistic breeders (Lowe, 1964; my data). Two species, *B. punctatus* and *B. woodhousei*, are opportunistic breeders or not, depending on the population. Both are opportunistic in Texas. In the Great Basin desert of southwestern Utah, these two species along with all other local anurans (*B. microscaphus*, *S. intermontanus*, *Hyla arenicolor*, and *Rana* sp. [*pipiens* gp.]) breed without rainfall (Blair, 1955; my data). Peak breeding choruses of *B. punctatus* and *B. alvarius* were found in a stock pond near Scottsdale, Arizona, in the absence of any recent rain (Blair and Pettus, 1954).

The Monte anurans, with the presumed exception of *Leptodactylus ocellatus*, appear to be opportunistic breeders (Cei, 1955a, 1955b; Reig and Cei, 1963; Gallardo, 1965; Barrio, 1964b, 1965a, 1965b). The apparent lesser development of the strategy of permanent water breeders could result from lesser knowledge of the behavior of the Monte anurans. However, the available evidence points to a real difference between the Monte and Sonoran desert faunas in degree of adoption of the habit of breeding in permanent water. *Leptodactylus ocellatus* of the Monte is ecologically equivalent to *R.* sp. (*pipiens* gp.) of the Sonoran desert; both are littoral adapted over a wide geographic range and have been able to penetrate their respective deserts by virtue of this adaptation where permanent water exists. There is no evidence that permanent water breeders are evolving from opportunistic breeders as in *B. punctatus*, *B. woodhousei*, and other North American desert species.

Foam Nests

One mechanism for desert adaptation, the foam nest, has been available for the evolution of the Monte fauna but not for the Sonoran desert fauna. Evolution of the foam-nesting habit has been discussed by various authors, especially Lutz (1947, 1948), Heyer (1969), and Martin (1967, 1970). The presumably more primitive pattern of floating the foam nest on the surface of the water is found among the Monte anurans in the genera *Physalaemus* and *Pleurodema* and in *Leptodactylus ocellatus*. The three other species of *Leptodactylus* in the Monte fauna lay their eggs in foam nests in burrows near water. These have aquatic larvae which are typically flooded out of the nests when pool levels rise with later rainfall. Heyer (1969) discussed advantages of the burrow nests over floating foam nests, among which the most important as adaptations to desert conditions are greater freedom from desiccation, and getting a head start on other breeders in the pool and thus being able to metamorphose earlier than others. Shoemaker and McClanahan (1973) investigated nitrogen excretion in the larvae of *L. bufonius* and found these larvae highly urotelic as an apparent adaptation to confinement in the foam-filled burrow versus the usual ammonotelism of anuran larvae.

Leptodactylids do reach the North American Mesquital (table 8-3), and one burrow-nesting species (*L. labialis*) reaches the southern tip of Texas. The other two genera both have direct, terrestrial development and hence would be unlikely candidates for desert adaptation. *Leptodactylus labialis* with a nesting pattern similar to that of *L. bufonius* would seem to be potential material for desert adaptation.

Cannibalism

An intriguing similarity between the two desert faunas is seen in the occurrence of cannibalism in both areas and in groups (ceratophrynids in South America, *Scaphiopus* in North America) that in other respects show rather similar patterns of desert adaptation.

In *S. bombifrons* and the closely related *S. hammondi*, some larvae have a beaked upper jaw and a corresponding notch in the lower as an apparent adaptation for carnivory (Bragg, 1946, 1950, 1956, 1961, 1964; Turner, 1952; Orton, 1954; Bragg and Bragg, 1959). The larvae of this type have been observed to be cannibalistic in *S. bombifrons* and suspected of being so in *S. hammondi* (Bragg, 1964). Cannibalism could be an important mechanism for concentrating food resources in a part of the population where these are limited and where there is a constant race against drying up of the breeding pool in the desert environment.

The ceratophrynids are much more cannibalistic than *Scaphiopus*. Both larvae and adults are carnivorous and cannibalistic (Cei, 1955*b*; Reig and Cei, 1963; my data). The head of the adult ceratophrynid is relatively large, with wide gape and with enlarged grabbing and holding teeth. Adult *Ceratophrys pierotti* are extremely voracious cannibals; one of these can quickly ingest another individual of its own body size.

Summary

The Monte of Argentina and the Sonoran desert of North America are compared with respect to their anuran faunas. Both deserts are roughly of similar size, but in North America there is a much greater extent of arid lands than in South America, with the Sonoran desert only a part of this expansion. Both deserts are at the dry end of moisture gradients that extend from thorn forest in the east to desert on the west.

Paleobotanical evidence suggests that xeric adaptation may have been occurring in South America prior to the breakup of Gondwanaland in the Cretaceous, while the North American deserts seem no older than middle Pliocene. Both desert systems must have been pressured and shifted during Pleistocene glacial maxima.

The anuran faunas of the two areas are similar in size, eleven species in the Sonoran desert, fourteen in the Monte. All anurans of the Monte occur also in the Chaco, and the fauna of the Monte is simply a depauperate Chacoan fauna. The origins of the Sonoran desert fauna are more diverse than this.

The Monte has the greatest taxonomic diversity, with seven genera versus four for the Sonoran desert. Two of the Monte genera (*Odontophrynus* and *Lepidobatrachus*) are truly desert and subxeric genera, but only one North American genus (*Scaphiopus*) fits this category. The presence of seven species of *Bufo* in the Sonoran desert implies a long history of desert adaptation by this genus in North America.

Mechanisms of desert adaptation are similar in the two areas. In each a littoral-adapted type (*Leptodactylus ocellatus* in the south, *Rana* sp. [*pipiens* gp.] in the north) has invaded the desert area by staying with permanent water. Additionally, the relict North American *B. alvarius* and *B. microscaphus* have followed the same strategy. Several of the North American species have abandoned opportunistic breeding in favor of breeding in permanent water, but no comparable trend is evident for the South American frogs. The most desert-adapted species in the North American desert is *Scaphiopus couchi*, which follows a pattern of highly fossorial life, opportunistic breeding with accelerated larval development, and physiological adaptations of adults to minimal water.

The ceratophrynids of the South American desert show parallel adaptations to those of *Scaphiopus*. In addition to other similarities, both groups employ some degree of cannibalism as an apparent adaptation to desert life.

References

Axelrod, D. I. 1948. Climate and evolution in western North America during middle Pliocene time. *Evolution* 2:127–144.

———. 1960. The evolution of flowering plants. In *Evolution after Darwin: Vol. 1 The evolution of life*, ed. S. Tax, pp. 227–305. Chicago: Univ. of Chicago Press.

———. 1970. Mesozoic paleogeography and early angiosperm history. *Bot. Rev.* 36:277–319.

Ballinger, R. E., and McKinney, C. O. 1966. Developmental temperature tolerance of certain anuran species. *J. exp. Zool.* 161:21–28.

Barbour, M. G., and Díaz, D. V. 1972. *Larrea* plant communities on bajada and moisture gradients in the United States and Argentina. *U.S./Intern. biol. Progn.: Origin and Structure of Ecosystems Tech. Rep.* 72–6:1–27.

Barrio, A. 1964*a*. Caracteres eto-ecológicos diferenciales entre *Odontophrynus americanus* (Dumeril et Bibron) y *O. occidentalis* (Berg) (Anura, Leptodactylidae). *Physis, B. Aires* 24:385–390.

———. 1964*b*. Especies crípticas del género *Pleurodema* que conviven en una misma área, identificados por el canto nupcial (Anura, Leptodactylidae). *Physis, B. Aires* 24:471–489.

———. 1965*a*. El género *Physalaemus* (Anura, Leptodactylidae) en la Argentina. *Physis, B. Aires* 25:421–448.

———. 1965*b*. Afinidades del canto nupcial de las especies cavicolas de género *Leptodactylus* (Anura, Leptodactylidae). *Physis, B. Aires* 25:401–410.

———. 1968. Revisión del género *Lepidobatrachus* Budgett (Anura, Ceratophrynidae). *Physis, B. Aires* 28:95–106.

Blair, A. P. 1955. Distribution, variation, and hybridization in a relict toad (*Bufo microscaphus*) in southwestern Utah. *Am. Mus. Novit.* 1722:1–38.

Blair, W. F. 1957. Structure of the call and relationships of *Bufo microscaphus* Cope. *Coepia* 1957:208–212.

———. 1958. Distributional patterns of vertebrates in the southern United States in relation to past and present environments. In *Zoogeography*, ed. C. L. Hubbs. *Publs Am. Ass. Advmt Sci.* 51:433–468.

———. 1965. Amphibian speciation. In *The Quaternary of the United States*, ed. H. E. Wright, Jr., and D. G. Frey, pp. 543–556. Princeton. Princeton Univ. Press.

———. 1970. Nichos ecológicos y la evolución paralela y convergente de los anfibios del Chaco y del Mesquital Norteamericano. *Acta zool. lilloana* 27:261–267.

———. 1973. Major problems in anuran evolution. In *Evolutionary biology of the anurans: Contemporary research on major problems*, ed. J. L. Vial, pp. 1–8. Columbia: Univ. of Mo. Press.

Blair, W. F., and Pettus, D. 1954. The mating call and its significance in the Colorado River toad (*Bufo alvarius* Girard). *Tex. J. Sci.* 6:72–77.

Bragg, A. N. 1946. Aggregation with cannibalism in tadpoles of

Scaphiopus bombifrons with some general remarks on the probable evolutionary significance of such phenomena. *Herpetologica*
3:89–98.

————. 1950. Observations on *Scaphiopus*, 1949 (Salientia: Scaph-
iopodidae). *Wasmann J. Biol.* 8:221–228.

————. 1956. Dimorphism and cannibalism in tadpoles of *Scaphiopus bombifrons* (Amphibia, Salientia). *SWest. Nat.* 1:105–108.

————. 1961. A theory of the origin of spade-footed toads deduced
principally by a study of their habits. *Anim. Behav.* 9:178–186.

————. 1964. Further study of predation and cannibalism in spadefoot tadpoles. *Herpetologica* 20:17–24.

Bragg, A. N., and Bragg, W. N. 1959. Variations in the mouth parts
in tadpoles of *Scaphiopus* (Spea) *bombifrons* Cope (Amphibia:
Salientia). *SWest. Nat.* 3:55–69.

Brown, H. A. 1967a. High temperature tolerance of the eggs of a desert anuran, *Scaphiopus hammondi. Copeia* 1967:365–370.

————. 1967b. Embryonic temperature adaptations and genetic compatibility in two allopatric populations of the spadefoot toad,
Scaphiopus hammondi. Evolution 21:742–761.

————. 1969. The heat resistance of some anuran tadpoles (Hylidae
and Pelobatidae). *Copeia* 1969:138–147.

Cabrera, A. L. 1953. Esquema fitogeográfico de la República Argentina. *Revta Mus. La Plata (Nueva Serie), Bot.* 8:87–168.

Casamiquela, R. M. 1961. Un pipoideo fósil de Patagonia. *Revta Mus.
La Plata Sec. Paleont. (Nueva Serie)* 4:71–123.

Cei, J. M. 1949. Costumbres nupciales y reproducción de un batracio
caracteristico chaqueño (*Leptodactylus bufonius*). *Acta zool.
lilloana* 8:105–110.

————. 1955a. Notas batracológicas y biogeográficas Argentinas,
I–IV. *An. Dep. Invest. cient., Univ. nac. Cuyo.* 2(2):1–11.

————. 1955b. Chacoan batrachians in central Argentina. *Copeia*
1955:291–293.

————. 1959a. Ecological and physiological observations on poly-
morphic populations of the toad *Bufo arenarum* Hensel, from Argentina. *Evolution* 13:532–536.

————. 1959b. Hallazgos hepetológicos y ampliación de la distribución geográfica de las especies Argentinas. *Actas Trab. Primer
Congr. Sudamericano Zool.* 1:209–210.

————. 1962. Mapa preliminar de la distribución continental de las "sibling species" del grupo *ocellatus* (género *Leptodactylus*). *Revta Soc. argent. Biol.* 38:258–265.

Freiberg, M. A. 1942. Enumeración sistemática y distribución geográfica de los batracios Argentinos. *Physis, B. Aires* 19:219–240.

Gallardo, J. M. 1965. Consideraciones zoogeográficas y ecológicas sobre los anfibios de la provincia de La Pampa Argentina. *Revta Mus. argent. Cienc. nat. Bernardino Rivadavia Inst. nac. Invest. Cienc. nat. Ecol.* 1:56–78.

———. 1966. Zoogeografía de los anfibios chaqueños. *Physis, B. Aires* 26:67–81.

Heyer, W. R. 1969. The adaptive ecology of the species groups of the genus *Leptodactylus* (Amphibia, Leptodactylidae). *Evolution* 23:421–428.

Kluge, A. G. 1966. A new pelobatine frog from the lower Miocene of South Dakota with a discussion of the evolution of the *Scaphiopus-Spea* complex. *Contr. Sci.* 113:1–26.

Kusnezov, N. 1951. *La edad geológica del régimen árido en la Argentina ségun los datos biológicos*. Geográfica una et varia, Publnes esp. Inst. Estud. geogr., Tucumán 2:133–146.

Lowe, C. H., ed. 1964. *The vertebrates of Arizona*. Tucson: Univ. of Ariz. Press.

Lutz, B. 1947. Trends toward non-aquatic and direct development in frogs. *Copeia* 1947:242–252.

———. 1948. Ontogenetic evolution in frogs. *Evolution* 2:29–39.

McClanahan, L. J. 1964. Osmotic tolerance of the muscles of two desert-inhabiting toads, *Bufo cognatus* and *Scaphiopus couchi*. *Comp. Biochem. Physiol.* 12:501–508.

———. 1967. Adaptations of the spadefoot toad, *Scaphiopus couchi*, to desert environments. *Comp. Biochem. Physiol.* 20:73 (?).

———. 1972. Changes in body fluids of burrowed spadefoot toads as a function of soil potential. *Copeia* 1972:209–216.

McClanahan, L. J., and Baldwin, R. 1969. Rate of water uptake through the integument of the desert toad, *Bufo punctatus*. *Comp. Biochem. Physiol.* 29:381–389.

McClanahan, L. J.; Shoemaker, V. H.; and Ruibal, R. 1973. Evaporative water loss in a cocoon-forming South American anuran. Abstract of paper given at 53d Annual Meeting of American Society of Ichthyologists and Herpetologists, at San José, Costa Rica.

Martin, A. A. 1967. Australian anuran life histories: Some evolutionary and ecological aspects. In *Australian inland waters and their fauna*, ed. A. H. Weatherley, pp. 175–191. Canberra: Aust. Nat. Univ. Press.

———. 1970. Parallel evolution in the adaptive ecology of Lepto-dactylid frogs in South America and Australia. *Evolution* 24:643–644.

Martin, P. S., and Mehringer, P. J., Jr. 1965. Pleistocene pollen analysis and biogeography of the southwest. In *The Quaternary of the United States*, ed. H. W. Wright, Jr., and D. G. Frey, pp. 433–451. Princeton: Princeton Univ. Press.

Mayhew, W. W. 1962. *Scaphiopus couchi* in California's Colorado Desert. *Herpetologica* 18:153–161.

———. 1965. Adaptations of the amphibian, *Scaphiopus couchi*, to desert conditions. *Am. Midl. Nat.* 74:95–109.

Morello, J. 1958. La provincia fitogeográfica del Monte. *Op. lilloana* 2:1–155.

Orton, G. L. 1954. Dimorphism in larval mouthparts in spadefoot toads of the *Scaphiopus hammondi* group. *Copeia* 1954:97–100.

Raven, P. H. 1963. Amphitropical relationships in the floras of North and South America. *Q. Rev. Biol.* 38:141–177.

Reig, O. A. 1960. Lineamentos generales de la historia zoogeográfica de los anuros. *Actas Trab. Primer Congr. Sudamericano Zool.* 1:271–278.

Reig, O. A., and Cei, J. M. 1963. Elucidación morfológico-estadística de las entidades del género *Lepidobatrachus* Budgett (Anura, Ceratophrynidae) con consideraciones sobre la extensión del distrito chaqueño del dominio zoogeográfico subtropical. *Physis, B. Aires* 24:181–204.

Ruibal, R. 1962a. Osmoregulation in amphibians from heterosaline habitats. *Physiol. Zoöl.* 35:133–147.

———. 1962b. The ecology and genetics of a desert population of *Rana pipiens*. *Copeia* 1962:189–195.

Ruibal, R.; Tevis, L., Jr.; and Roig, V. 1969. The terrestrial ecology of the spadefoot toad *Scaphiopus hammondi*. *Copeia* 1969:571–584.

Shelford, V. E. 1963. *The ecology of North America*. Urbana: Univ. of Ill. Press.

Shoemaker, V. H., and McClanahan, L. J. 1973. Nitrogen excretion in the larvae of the land-nesting frog (*Leptodactylus bufonius*). *Comp. Biochem. Physiol.* 44A:1149–1156.

Shoemaker, V. H.; Balding, D.; and Ruibal, R. 1972. Uricotelism and low evaporative water loss in a South American frog. *Science, N.Y.* 175:1018–1020.

Sick, W. D. 1969. Geographical substance. In *Biogeography and ecology in South America*, ed. E. J. Fittkau, J. Illies, H. Klinge, G. H. Schwabe, and H. Sioli, 2: 449–474. The Hague: Dr. W. Junk.

Simpson Vuilleumier, B. 1971. Pleistocene changes in the fauna and flora of South America. *Science, N.Y.* 173:771–780.

Solbrig, O. T. 1975. The origin and floristic affinities of the South American temperate desert and semidesert regions. In *Evolution of desert biota*, ed. D. W. Goodall. Austin: Univ. of Texas Press.

Stebbins, R. C. 1951. *Amphibians of western North America*. Berkeley and Los Angeles: Univ. of Calif. Press.

Tevis, L., Jr. 1966. Unsuccessful breeding by desert toads (*Bufo punctatus*) at the limit of their ecological tolerance. *Ecology* 47: 766–775.

Trueb, L. 1970. Evolutionary relationships of casque-headed tree frogs with coossified skulls (family Hylidae). *Univ. Kans. Publs Mus. nat. Hist.* 18:547–716.

Turner, F. B. 1952. The mouth parts of tadpoles of the spadefoot toad, *Scaphiopus hammondi*. *Copeia* 1952:172–175.

Veloso, H. P. 1966. *Atlas florestal do Brasil*. Rio de Janeiro—Guanabara: Ministerio da Agricultura.

Wasserman, A. O. 1957. Factors affecting interbreeding in sympatric species of spadefoots (*Scaphiopus*). *Evolution* 11:320–338.

Zweifel, R. G. 1956. Two pelobatid frogs from the Tertiary of North America and their relationships to fossil and recent forms. *Am. Mus. Novit.* 1762:1–45.

———. 1968. Reproductive biology of anurans of the arid southwest, with emphasis on adaptation of embryos to temperature. *Bull. Am. Mus. nat. Hist.* 140:1–64.

Notes on the Contributors

John S. Beard was born in England and educated at Oxford University. After graduation he spent nine years in the Colonial Forest Service in the Caribbean and during this period studied for the degree of D.Phil. at Oxford. Soon after the end of the Second World War, he went to South Africa, where he was engaged in research on crop improvement in the wattle industry. In 1961 he was appointed director of King's Park in Perth, Western Australia, where, among other things, he was responsible for establishing a botanical garden. Ten years later he became director of the National Herbarium of New South Wales, a post from which he has recently retired. He is now devoting much of his time to the preparation of a series of detailed maps of Australian vegetation.

Dr. Beard has published books and papers on the vegetation of tropical America and has wide interests in plant ecology, biogeography, and systematics.

W. Frank Blair is professor of zoology at the University of Texas at Austin. He was born in Dayton, Texas. His first degree was taken at the University of Tulsa; he was awarded the M.S. at the University of Florida, and the Ph.D. at the University of Michigan. After eight years as a research associate there, he moved to a faculty position at the University of Texas, where he has been ever since.

At the inception of the International Biological Program, Dr. Blair became director of the Origin and Structure of Ecosystems section and, soon afterward, national chairman for the whole program. He also served as vice-president of the IBP on the international scale.

He has been involved in a wide range of personal research on vertebrate ecology and has worked extensively in Latin America as well as the United States. He is senior author of *Vertebrates of the United States* and has edited *Evolution in the Genus "Bufo"*—a subject on which much of his most recent research has concentrated.

David W. Goodall is Senior Principal Research Scientist at CSIRO Division of Land Resources Management, Canberra, Australia. Born and brought up in England, he studied at the University of London where he was awarded the Ph.D. degree, and, after a period of research in what is now Ghana, took up residence in Australia, of which country he is a citizen. He was awarded the D.Sc. degree of Melbourne University in 1953. He came to the United States in 1967 and the following year was invited to become director of the Desert Biome section of the International Biological Program then getting under way. This position he continued to hold until the end of 1973. During most of this period, and until the end of 1974, he held a position as professor of systems ecology at Utah State University.

His main research interests were initially in plant physiology, particularly in its application to agriculture and horticulture; but later he shifted his interest to plant ecology, especially statistical aspects of the subject, and to systems ecology.

Bobbi S. Low is associate professor of resource ecology at the University of Michigan. She was born in Kentucky and took her first degree at the University of Louisville and her doctorate at the University of Texas at Austin. After postdoctoral work at the University of British Columbia, she spent three years as a Research Fellow at Alice Springs, Australia, and returned to the United States in 1972. Her main research interests have been in evolutionary ecology and in ecology of vertebrates in arid areas, both in the United States and in Australia.

James A. MacMahon is professor of biology at Utah State University and assistant director of the Desert Biome section of the International Biological Program. He was born in Dayton, Ohio, and took his first degree at Michigan State University and his doctorate at Notre Dame University, Indiana, in 1963. He then was appointed to a professorial position at the University of Dayton, and in 1971 he moved to Utah.

Though much of his research has been devoted to reptiles and Amphibia, he has also been concerned with plants, mammals, and invertebrates. In all these groups of organisms, he has mainly been interested in their ecology, particularly at community level, in relation to the arid-land environment.

A. R. Main is professor of zoology at the University of Western Australia. He was born in Perth; after military service during the Second World War, he returned there to take a first degree and then a doctorate at the University of Western Australia. He is a Fellow of the Australian Academy of Science.

He has done extensive research on the ecology of mammals and Amphibia in the Australian deserts, and he and his students have published numerous papers on the subject.

Guillermo Sarmiento is associate professor in the Faculty of Science, Universidad de Los Andes, Mérida, Venezuela. He was born in Mendoza, Argentina, and was educated at the University of Buenos Aires, where he was awarded a doctorate in 1965. He was appointed assistant professor and moved to Venezuela two years later.

His main research interests have been in tropical plant ecology, particularly as applied to savannah and to the vegetation of arid lands.

Otto T. Solbrig was born in Buenos Aires, Argentina. He took his first degree at the Universidad de La Plata and his Ph.D. at the University of California, Berkeley, in 1959. He worked at the Gray Herbarium, Harvard University, for seven years (during which period he became a U.S. citizen); after a period as professor of botany at the University of Michigan, he returned to Harvard University in 1969 as professor of biology and chairman of the Sub-Department of Organismic and Evolutionary Biology. Within the U.S. contribution to the International Biological Program, he served as director of the Desert Scrub subprogram of the Origin and Structure of Ecosystems section.

He has wide field experience in various parts of Latin America as well as in the United States. His main research interests have been in plant biosystematics, biogeography, and population biology.

Index

Evolution of Desert Biota

Deposition site:
- 1 = Temporary ponds
- 2a = Permanent ponds
- 2b = Permanent streams
- 3a = Burrows, requiring rain to hatch
- 3b = Burrows, not requiring rain to hatch
- 4a = Terrestrial (seeps, etc.)
- 4b = On leaves above water
- 4c = On submerged leaves
- 5 = With parent: brood pouch, on back, etc.

[c]Deposition site:
1 = Temporary ponds
2a = Permanent ponds
2b = Permanent streams
3a = Burrows, requiring rain to hatch
3b = Burrows, not requiring rain to hatch
4a = Terrestrial (seeps, etc.)
4b = On leaves above water
4c = On submerged leaves
5 = With parent: brood pouch, on back, etc.

[a]Habitat:
- 1 = North America
- 2 = Central America
- 3 = South America
- 4 = Europe
- 5 = Asia
- 6 = Africa
- 7 = Australia
- 8 = New Zealand

[c]Deposition site:
- 1 = Temporary ponds
- 2a = Permanent ponds
- 2b = Permanent streams
- 3a = Burrows, requiring rain to hatch
- 3b = Burrows, not requiring rain to hatch
- 4a = Terrestrial (seeps, etc.)
- 4b = On leaves above water
- 4c = On submerged leaves
- 5 = With parent: brood pouch, on back, etc.